VOLUME ONE HUNDRED AND THIRTY EIGHT

INTERNATIONAL REVIEW OF NEUROBIOLOGY

Neurobiology of the Placebo Effect Part I

INTERNATIONAL REVIEW OF NEUROBIOLOGY

VOLUME 138

SERIES EDITOR

PATRICIA JANAK
Janak Lab, Dunning Hall
Baltimore, MD, USA

PETER JENNER
Division of Pharmacology and Therapeutics
GKT School of Biomedical Sciences
King's College, London, UK

VOLUME ONE HUNDRED AND THIRTY EIGHT

INTERNATIONAL REVIEW OF NEUROBIOLOGY

Neurobiology of the Placebo Effect Part I

Edited by

LUANA COLLOCA

Department of Pain Translational Symptom Science, School of Nursing; Departments of Anesthesiology and Psychiatry, School of Medicine;
Center to Advance Chronic Pain Research, University of Maryland, Baltimore, MD, United States

ELSEVIER

ACADEMIC PRESS
An imprint of Elsevier

Academic Press is an imprint of Elsevier
50 Hampshire Street, 5th Floor, Cambridge, MA 02139, United States
525 B Street, Suite 1800, San Diego, CA 92101-4495, United States
The Boulevard, Langford Lane, Kidlington, Oxford OX5 1GB, United Kingdom
125 London Wall, London, EC2Y 5AS, United Kingdom

First edition 2018

ISBN: 978-0-12-814325-4
ISSN: 0074-7742

For information on all Academic Press publications
visit our website at https://www.elsevier.com/books-and-journals

Publisher: Zoe Kruze
Acquisition Editor: Sam Mahfoudh
Editorial Project Manager: Andrea Gallego Ortiz
Production Project Manager: Abdulla Sait
Cover Designer: Mark Rogers
Cover Image Courtesy: Giancarlo Colloca

Typeset by SPi Global, India

CONTENTS

CONTRIBUTORS

Titilola Akintola
Program in Toxicology, University of Maryland School of Medicine; University of Maryland School of Nursing, Baltimore, MD, United States

Olivia Aspiras
University of Toledo, Toledo, OH, United States

Lauren Y. Atlas
Section on Affective Neuroscience and Pain, National Center for Complementary and Integrative Health; National Institute on Drug Abuse, NIH, Bethesda, MD, United States

Diletta Barbiani
University of Turin Medical School, Turin, Italy; Plateau Rosà Laboratories, Plateau Rosà, Italy/Switzerland

Fabrizio Benedetti
University of Turin Medical School, Turin, Italy; Plateau Rosà Laboratories, Plateau Rosà, Italy/Switzerland

Sven Benson
Institute of Medical Psychology and Behavioral Immunobiology, University Hospital Essen, University of Duisburg–Essen, Essen, Germany

Ulrike Bingel
University Hospital Essen, University of Duisburg–Essen, Essen, Germany

Charlotte Blease
School of Psychology, University College Dublin, Dublin, Ireland; Program in Placebo Studies and the Therapeutic Encounter, Harvard Medical School, Boston, MA, United States

Pablo Briñol
Universidad Autónoma de Madrid, Madrid, Spain

Eleonora Camerone
Institute of Psychiatry, Psychology & Neuroscience, King's College, London, United Kingdom

Fawn C. Caplandies
University of Toledo, Toledo, OH, United States

Elisa Carlino
University of Turin Medical School, Turin, Italy

Ben Colagiuri
School of Psychology, The University of Sydney, Sydney, NSW, Australia

Luana Colloca
Center to Advance Chronic Pain Research; School of Nursing; School of Medicine, University of Maryland, Baltimore, MD, United States

Alia J. Crum
Stanford University, Stanford, CA, United States

Jan De Houwer
Ghent University, Ghent, Belgium

Bettina K. Doering
Catholic University Eichstätt-Ingolstadt, Eichstätt, Germany

Sigrid Elsenbruch
Institute of Medical Psychology and Behavioral Immunobiology, University Hospital Essen, University of Duisburg-Essen, Essen, Germany

Andrea W.M. Evers
Health, Medical and Neuropsychology Unit, Faculty of Social and Behavioural Sciences, Institute of Psychology, Leiden University; Leiden University Medical Center, Leiden; Radboud University Medical Center, Nijmegen, The Netherlands

Jens Gaab
Clinical Psychology and Psychotherapy, University of Basel, Basel, Switzerland

Andrew L. Geers
University of Toledo, Toledo, OH, United States

Julia A. Glombiewski
University of Koblenz–Landau, Landau, Germany

Martin Hadamitzky
Institute of Medical Psychology and Behavioral Immunobiology, University Hospital Essen, University of Duisburg–Essen, Essen, Germany

Ad R.M.M. Hermus
Radboud University Medical Center, Nijmegen, The Netherlands

Jeremy Howick
Faculty of Philosophy, University of Oxford, Oxford, United Kingdom

Karin B. Jensen
Karolinska Institute, Stockholm, Sweden

Asaf Keller
Program in Neuroscience, University of Maryland School of Medicine; Center to Advance Chronic Pain Research, University of Maryland, Baltimore, MD, United States

Irving Kirsch
Harvard Medical School, Boston, MA, United States

Julian Kleine-Borgmann
University Hospital Essen, University of Duisburg–Essen, Essen, Germany

Franziska Labrenz
Institute of Medical Psychology and Behavioral Immunobiology, University Hospital Essen, University of Duisburg-Essen, Essen, Germany

Cosima Locher
Clinical Psychology and Psychotherapy, University of Basel, Basel, Switzerland

Elizabeth A. Necka
Section on Affective Neuroscience and Pain, National Center for Complementary and Integrative Health, NIH, Bethesda, MD, United States

Marta Pecina
University of Pittsburgh, Pittsburgh, PA, United States

Richard E. Petty
Ohio State University, Columbus, OH, United States

Veronica F. Quinn
School of Psychology, The University of Sydney, Sydney, NSW, Australia

Winfried Rief
Philipps-University Marburg, Marburg, Germany

Sean H.P.P. Roerink
Radboud University Medical Center, Nijmegen, The Netherlands

Manfred Schedlowski
Institute of Medical Psychology and Behavioral Immunobiology, University Hospital Essen, University of Duisburg–Essen, Essen, Germany; Karolinska Institute, Stockholm, Sweden

Wiebke Sondermann
West German Cancer Center, University Hospital Essen, University of Duisburg–Essen, Essen, Germany

Fred C.G.J. Sweep
Radboud University Medical Center, Nijmegen, The Netherlands

Judith Tekampe
Health, Medical and Neuropsychology Unit, Faculty of Social and Behavioural Sciences, Institute of Psychology, Leiden University, Leiden; Radboud University Medical Center, Nijmegen, The Netherlands

Lene Vase
School of Business and Social Sciences, Aarhus University, Aarhus, Denmark

Erin A. Vogel
University of California, San Francisco, San Francisco, CA, United States

Sean R. Zion
Stanford University, Stanford, CA, United States

Jon-Kar Zubieta
University Neuropsychiatric Institute, University of Utah Health Sciences Center, Salt Lake City, UT, United States

Henriët van Middendorp
Health, Medical and Neuropsychology Unit, Faculty of Social and Behavioural Sciences, Institute of Psychology, Leiden University, Leiden; Radboud University Medical Center, Nijmegen, The Netherlands

PREFACE
The Fascinating Mechanisms and Implications of the Placebo Effect

It is becoming clear that any treatment is significantly modulated by placebo effects in clinical settings. Placebo effects are positive outcomes that are attributable to the psychosocial context and individual treatment expectations rather than the action of the medication or intervention (Colloca & Benedetti, 2005). Placebo effects also occur when placebos are given following the administration of active and effective medications (e.g., painkillers) creating drug-like effects (e.g., dose-extending placebos) (Colloca, Enck, & DeGrazia, 2016). Pharmacological studies indicate that placebos mimic the action of active treatments and promote the endogenous release of opioids and nonopioids in both humans and animals. Expectations of benefit facilitate the activation of pain and nonpain control systems leading to release of endogenous substances crucially involved in placebo-induced benefits. Indeed, neurobiological studies have identified dopaminergic, opioidergic, vasopressinergic, and endocannabinoidergic pathways as promising systems contributing to the modulation of pain experience and other symptoms. This book presents recent behavioral and neurobiological advances on placebo effects and expands on well established proposed conceptual frameworks and theories. Since placebo effects act as "boosters" of treatment expectancy and clinical outcomes, gaining deeper understanding of the top-down mechanisms of symptoms modulation and well-being has enormous implications for personalizing and optimizing pain management and other symptoms.

One area that reserves further investigation to understand the mechanisms of placebo effects is research in animals that would allow the creation of much needed molecular models to determine the underlying brain and peripheral mechanisms. Despite the opportunities and excitement of animal research, there are aspects that need to be addressed in order to advance placebo research. Gaps and controversies have been acknowledged in pain placebo research (Keller, Akintola, & Colloca, 2018), but solid models and findings are provided for placebo effects in the immune system. Well-established learning paradigms have been developed in the context of the immune system in both animals and humans (Hadamitzky, Sondermann,

Benson, & Schedlowski, 2018), where a novel taste is used as conditioned stimulus (CS) along with the administration of immunosuppressive drugs (e.g., cyclosporine A) that act as the unconditioned stimulus (US). Subsequently, the presentation of the CS alone elicits, after a certain number of associations, the conditioned response (CR) including hormonal and body responses (Ader & Cohen, 1975; Tekampe et al., 2018).

Importantly, when a person expects and experiences a placebo analgesic effect, cognitive and emotional circuitries are activated with experience of pain reduction and improvements in other symptoms. Molecular neuroimaging studies using positron emission tomography and the selective μ-opioid receptor tracer [^{11}C]carfentanil have greatly contributed to current understanding of the neurobiology of the placebo effect. The self-healing capacity to activate endogenous opioid and nonopioid networks associated with the administration of a placebo (Pecina & Zubieta, 2018), or other surgical and pharmacological interventions, points to a sort of inner pharmacy with survival and evolutionary meanings. Partially determined by genetic factors (Colagiuri, Schenk, Kessler, Dorsey, & Colloca, 2015), maintained through learning mechanisms, and sustained by the cognitive dynamic integration of expectations surrounding the therapeutic environment, patient–clinician relationship, and the act of administering an intervention, placebo effects promote symptomatic improvements.

The delineation of these mechanistic advances has been guided by theories, models, and frameworks posited to understand components of the placebo effect (e.g., motivation) as well as grab its complexity in real-world settings. For example, basic tenets of the response expectancy theory, and its distinction between response expectancies and stimulus expectancies, help uncover the individual experience of placebo-induced benefits (Kirsch, 2018). On the other hand, conditioning and learning principles are relevant for placebo research because they clarify the change occurring at the behavioral and brain levels when someone is taking a pill that can contain either an active pharmacological substance or merely a drop of saline solution (De Houwer, 2018).

Placebo effects and positive outcomes resulting from expectations about a treatment outcome should be considered as powerful components of modern medicine. Persuasion that involves changes in beliefs or attitudes as a result of providing critical information can be embedded in daily clinical communication. Therefore, it is clear that the practitioner's attitudes, his confidence and competence, can at least in part favor the formation and magnitude of placebo effects (Geers et al., 2018). Similarly, mindsets may represent a critical mental construct that illustrates the role of the treatment contexts in shaping placebo effects (Zion & Crum, 2018). Moreover, the patient–provider therapeutic alliance contributes to placebo effects and health outcomes. Overall, the patient's *perception* of the clinician, his psychosocial orientation (e.g., loneliness, poor patient–clinician relationship), and perceptions of their interpersonal relationships may broadly influence placebo effects (Necka & Atlas, 2018).

Recently, compelling research is attempting to unravel how placebo effects are elicited in critical contexts with an understanding of what is minimally required to observe placebo effects from intellectual disability, to high altitudes to open-label placebos that challenge the common sense that placebo effects rely on deceptively administering placebos to patients. However, the effect of the placebo effect does not reside in the sham treatment itself; rather, it relies on expectancies that surround the patient and the intervention. Therefore, recent data suggest that patients with impaired cognitive functions may respond to placebos by virtue of implicit cognitive processes that go beyond desire, suggestions, or verbal communication (e.g., conditioning). There are minimum requirements for eliciting placebo effects, both from the view of conscious awareness and from the perspective of brain functionality (Jensen, 2018).

When the brain preserves the primary ability of integrating incoming sensory information with the inner world that reflects the individual's prior experiences, placebo effects are generated primarily to promote adaptation to old and new environments and minimize trials and errors. Placebo effects reflect the ability to merge prior experience and ideas about treatment outcomes with sensory perception reconciling mismatches between *what is expected* and *what is experienced*. Expectations and sensory inputs are both finely modulated to provide the variety of nuances used in the interpretation of facts and experiences.

With this in mind, it is not surprising that critical life functions, like ventilation, oxygenation, circulation, and perfusion, can be shaped by placebo effects also in extreme contexts such as at an altitude as high as 3500, 4500, and 5500 m, where oxygen pressures are 64%, 57%, and 50%, respectively, compared to the sea level. Oxygen-related body responses can be conditioned at high altitudes (Benedetti, Barbiani, & Camerone, 2018).

Placebo effects can also occur when patients know they have been given placebos. Open-label placebos can be interpleading in paradigms with therapeutic treatments so that learning and nondeceptive approaches are harnessed to minimize drug intake while reducing side effects and costs. Although there are distinct differences between open-label and dose-extending placebos, use of such placebos can be preauthorized (e.g., patients agree to receive, at some point, placebos along with medication) in accordance with professional norms governing disclosure and informed consent, and be used in combination with a standard use of treatment (Colloca & Howick, 2018).

Without invoking the use of any placebos, expectations can be shaped to make a treatment and/or an outcome most effective via psychotherapy. Indeed, there is a compelling notion that the placebo effect and psychotherapy represent two psychological interventions that share much more than their first letter (Gaab, Locher, & Blease, 2018). Through psychotherapy, communication, and framing styles, expectations can be positively manipulated as indicated by a recent randomized-controlled trial (RCT). A presurgery expectation optimization program, applied to patients scheduled for coronary artery bypass grafting, resulted in lower disability scores at 6-month post-surgery follow-ups (Rief et al., 2017). Expectations can be changed by a series of procedures with the scope to optimize outcome in medical settings (Doering, Glombiew, & Rief, 2018).

However, expectancy, patient–clinician relationships, and prior therapeutic histories can be negative, thus compromising clinical outcomes. The negative counterpart of the placebo effect is named "nocebo" effect (Colloca, 2017a, 2017b; Klinger, Blasini, Schmitz, & Colloca, 2017), and despite its relevance for modulation of mechanisms and clinical implications, the nocebo phenomenon has received less attention over the past decades than the placebo effect. However, laboratory and translational research is unraveling some of the behavioral and biological mechanisms as well as the clinical implications. Nocebo effects have been investigated in the field of pain (e.g., experimental acute and visceral pain) (Elsenbruch & Labrenz, 2018; Kleine-Borgmann & Bingel, 2018), nausea (Quinn & Colagiuri, 2018), and other symptoms. Particular emphasis has been given to visceral pain, a model used for behavioral and neuroimaging studies that has helped foster the understanding of nocebo as compared to placebo mechanisms. The recent findings facilitate the transition of experimental research from bench to bedside and ways in which the environment can be manipulated to prevent, reduce, or eradicate nocebo effects (Elsenbruch & Labrenz, 2018). Preexposure to placebo stimulation (latent inhibition) prevents nocebo effects in experimental conditioned nausea (Quinn & Colagiuri, 2018).

Learning strategies, including overshadowing, latent inhibition, extinction, and contingency degradation, represent future research avenues that, if transitioned from bench to bedside, might promote strategies to reduce unintended nocebo effects to be used by clinicians and researchers while designing trials to test new interventions (Quinn & Colagiuri, 2018). Indeed, randomized controlled trials have shown an increased trend to fail, which may be due to both nocebo and especially placebo responses. The neurobiology of placebo effects highlights that patients' expectations interfere with the response to medications and therefore, new clinical trial approaches should be used to improve drug development (Vase & Carlino, 2018).

This series of 18 articles found their inspiration during the first meeting in 2017 of the Society for Interdisciplinary Placebo Studies (SIPS, https://www.placebosociety.org/) that is a newly created international association of scholars who share the goal of understanding the placebo effect in medical treatment, psychotherapy, and complementary and integrative medicine by promoting communication and cooperation between research centers and scholars. By using multidisciplinary approaches including neuroscience, psychology, anthropology, and philosophy, it may be possible to expand knowledge on the neurobiological mechanisms and brain functions as well as fully appreciate the implications of this research to develop ethically acceptable ways to harness placebo effects with the scope to optimize clinical trial designs, treatment outcomes, and therapeutic strategies.

Luana Colloca
Department of Pain Translational Symptom Science, School of Nursing, University of Maryland, Baltimore, MD, United States
Departments of Anesthesiology and Psychiatry, School of Medicine, University of Maryland, Baltimore, MD, United States
Center to Advance Chronic Pain Research, University of Maryland, Baltimore, MD, United States

REFERENCES

Ader, R., & Cohen, N. (1975). Behaviorally conditioned immunosuppression. *Psychosomatic Medicine*, *37*(4), 333–340.

Benedetti, F., Barbiani, D., & Camerone, E. (2018). Critical life functions: Can placebo replace oxygen? In L. Colloca (Ed.), *The neurobiology of the placebo effect: Vol. 138*. Oxford: Elsevier.

Colagiuri, B., Schenk, L. A., Kessler, M. D., Dorsey, S. G., & Colloca, L. (2015). The placebo effect: From concepts to genes. *Neuroscience*, *307*, 171–190. https://doi.org/10.1016/j.neuroscience.2015.08.017.

Colloca, L. (2017a). Nocebo effects can make you feel pain. *Science*, *358*(6359), 44. https://doi.org/10.1126/science.aap8488.

Colloca, L. (2017b). Tell me the truth and I will not be harmed: Informed consents and nocebo effects. *The American Journal of Bioethics*, *17*(6), 46–48. https://doi.org/10.1080/15265161.2017.1314057.

Colloca, L., & Benedetti, F. (2005). Placebos and painkillers: Is mind as real as matter? *Nature Reviews. Neuroscience*, *6*(7), 545–552. https://doi.org/10.1038/nrn1705.

Colloca, L., Enck, P., & DeGrazia, D. (2016). Relieving pain using dose-extending placebos: A scoping review. *Pain*, *157*(8), 1590–1598. https://doi.org/10.1097/j.pain.0000000000000566.

Colloca, L., & Howick, J. (2018). Placebos without deception: Outcomes, mechanisms, and ethics. In L. Colloca (Ed.), *The neurobiology of the placebo effect: Vol. 138*. Oxford: Elsevier.
Doering, B. K., Glombiew, J. A., & Rief, W. (2018). Expectation focused psychotherapy to improve clinical outcomes. In L. Colloca (Ed.), *The neruobiology of the placebo effect: Vol. 138*. Oxford: Elsevier.
Elsenbruch, S., & Labrenz, F. (2018). Nocebo effects in visceral pain. In L. Colloca (Ed.), *The neurobiology of the placebo effect: Vol. 138*. Oxford: Elsevier.
Gaab, J., Locher, C., & Blease, C. (2018). Placebo and psychotherapy: Differences, similarities and implications. In L. Colloca (Ed.), *The neurobiology of the placebo effect: Vol. 138* (p. 13). Oxford: Elsevier.
Geers, A. L., Briñol, P., Vogel, E., Aspiras, O. G., Caplandies, F., & Petty, R. (2018). The application of persuasion theory to placebo effect. In L. Colloca (Ed.), *The neurobiology of the placebo effect: Vol. 138*. Oxford: Elsevier.
Hadamitzky, M., Sondermann, W., Benson, S., & Schedlowski, M. (2018). Placebo effects in the immune system. In L. Colloca (Ed.), *The neurobiology of the placebo effect: Vol. 138* (p. 322). Oxford: Elsevier.
De Houwer, J. (2018). Conditioning as a higher-order cognitive phenomenon: Implications for placebo research. In L. Colloca (Ed.), *The neurobiology of the placebo effect: Vol. 138*. Oxford: Elsevier.
Jensen, K. (2018). What is minimally required to elicit placebo effects? In L. Colloca (Ed.), *The neurobiology of the placebo effect: Vol. 138*. Oxford: Elsevier.
Keller, A., Akintola, T., & Colloca, L. (2018). Placebo analgesia in rodents: Current and future research. In L. Colloca (Ed.), *The Neurobiology of the placebo effect: Vol. 138* (p. 322). Oxford: Elsevier.
Kirsch, I. (2018). Response expectancy and the placebo effect. In L. Colloca (Ed.), *The neurobiology of the placebo effect: Vol. 138*. Oxford: Elsevier.
Kleine-Borgmann, J., & Bingel, U. (2018). Nocebo effects: Neurobiological mechanisms and strategies for prevention and optimizing treatment. In L. Colloca (Ed.), *The neurobiology of the placebo effect: Vol. 138*. Oxford: Elsevier.
Klinger, R., Blasini, M., Schmitz, J., & Colloca, L. (2017). Nocebo effects in clinical studies: Hints for pain therapy. *Pain Reports, 2*(2), e586. https://doi.org/10.1097/PR9.0000000000000586.
Necka, E., & Atlas, L. Y. (2018). The role of social and interpersonal factors in placebo analgesia. In L. Colloca (Ed.), *The neurobiology of the placebo effect: Vol. 138*. Oxford: Elsevier.
Pecina, M., & Zubieta, J. K. (2018). Expectancy modulation of opioid neurotransmission. In L. Colloca (Ed.), *The neurobiology of the placebo effect: Vol. 138*. Oxford: Elsevier.
Quinn, V. F., & Colagiuri, B. (2018). Using learning mechanisms to inhibit the nocebo effect. In L. Colloca (Ed.), *The neurobiology of the placebo effect: Vol. 138*. Oxford: Elsevier.
Rief, W., Shedden-Mora, M. C., Laferton, J. A., Auer, C., Petrie, K. J., Salzmann, S., et al. (2017). Preoperative optimization of patient expectations improves long-term outcome in heart surgery patients: Results of the randomized controlled PSY-HEART trial. *BMC Medicine, 15*(1), 4. https://doi.org/10.1186/s12916-016-0767-3.
Tekampe, J., van Middendorp, H., Sweep, C. G. J., Roerink, S. H. P. P., Hermus, A. R. M. M., & Evers, A. R. M. M. (2018). Human pharmacological conditioning of the immune and endocrine system: Challenges and opportunities. In L. Colloca (Ed.), *The neurobiology of the placebo effect: Vol. 138*. Oxford: Elsevier.
Vase, L., & Carlino, E. (2018). Can knowledge of placebo and nocebo mechanisms help improve randomized clinical trials? In L. Colloca (Ed.), *The neurobiology of the placebo effect: Vol. 138*. Oxford: Elsevier.
Zion, S. R., & Crum, A. (2018). Mindsets matter: A new framework for harnessing the placebo effect in modern medicine. In L. Colloca (Ed.), *The neurobiology of the placebo effect: Vol. 138*. Oxford: Elsevier.

CHAPTER ONE

Placebo Analgesia in Rodents: Current and Future Research

Asaf Keller*,†,1, **Titilola Akintola**‡,§, **Luana Colloca**†,§

*Program in Neuroscience, University of Maryland School of Medicine, Baltimore, MD, United States
†Center to Advance Chronic Pain Research, University of Maryland, Baltimore, MD, United States
‡Program in Toxicology, University of Maryland School of Medicine, Baltimore, MD, United States
§University of Maryland School of Nursing, Baltimore, MD, United States
[1]Corresponding author: e-mail address: akeller@som.umaryland.edu

Contents

Abstract

The investigation of placebo effects in animal pain models has received less attention than human research. This may be related to a number of difficulties, including the fact that animals lack the ability to use language and establish expectancies verbally, that animals cannot report and rate the extent to which they experience pain, and the inadequacy of current models of pain. Here, we describe the relatively small number of studies that have been published, communicating the opportunities and excitement of this research. We critically discuss pitfalls and limitations with the hope that this will advance future animal placebo-related research.

International Review of Neurobiology, Volume 138
ISSN 0074-7742
https://doi.org/10.1016/bs.irn.2018.02.001

1. INTRODUCTION

Unlike many other fields of biomedical sciences, the field of placebo analgesia has been led primarily by advances in human research (Colloca & Benedetti, 2005; Colloca, Enck, & DeGrazia, 2016; Wager & Atlas, 2015). Animal studies—that provide critical synergy to human studies—have lagged in this field. Indeed, the first systematic studies of placebo analgesia in rodents have appeared only in the current decade. This may simply reflect the historical origins of placebo research that emerged from attempts to better control clinical trials. However, it is also likely that prejudices against the capacity of nonhuman animals to display placebo analgesia have contributed to the paucity of research in this field (McMillan, 1999).

This deficiency is unfortunate because of the distinct advantages of animal research for revealing the underlying mechanisms of clinical phenomena, in particular those related to pain (Burma, Leduc-Pessah, Fan, & Trang, 2017; Mogil, 2009; Mogil, Davis, & Derbyshire, 2010; Munro, Jansen-Olesen, & Olesen, 2017). These advantages include the ability to produce reproducible models of pain, the use of transgenic animals to probe-specific mechanisms and brain circuits, and the ability to perform experimental manipulations that cannot be performed in humans for technical or ethical reasons.

Nevertheless, recent progress suggests that animal models of placebo analgesia with face validity can be developed and utilized to understand the mechanisms underlying this clinically important phenomenon. In this review, we illustrate the most relevant contributions on pain modulation induced by pharmacological conditioning in which the unconditioned stimulus (UCS) (e.g., the drug) is replaced by a placebo to elicit a conditioned analgesic response in both acute and chronic pain. We critically review pitfalls and strengths of published studies with a careful description of the methodology and discussion of the findings. We trust that this critical review of animal studies may help advance future methodology and approaches to explore placebo effects in animals with sophisticated molecular, behavioral, and brain imaging techniques.

2. PLACEBO ANALGESIA AND ACUTE PAIN IN RODENTS: PIONEERING ATTEMPTS

One of the earliest attempts to induce placebo analgesia in rodents achieved the opposite effect. Krank, Hinson, and Siegel (1981) reported that pairing morphine administration with an environmental cue resulted in significantly *more* reactivity to an aversive heat stimulus.

In contrast, Bardo and collaborators have shown that when a gustatory cue is used as a conditioned stimulus and paired with morphine as an UCS, rats exhibit placebo analgesia (Bardo & Valone, 1994; Miller, Kelly, Neisewander, McCoy, & Bardo, 1990). Bardo and colleagues tested rats responses on a hot plate test by administering morphine (UCS) paired with an aversive taste cue (conditioned stimulus, CS). First, they showed a decrease in CS intake across repeated testing, as expected. After three pairings, subsequent presentations of the aversive taste cue increased paw-lick latencies. They confirmed that the placebo response was not due to stress-induced analgesia from the injections because similar injections using lithium as the CS produced conditioned aversion, but not placebo analgesia.

Bardo et al. subsequently demonstrated that repeated applications of an odor (every other day for 4 days) can be used as the CS with morphine to produce placebo heat analgesia (Valone, Randall, Kraemer, & Bardo, 1998).

Surprisingly, Bryant et al. reported that *a single* association of a novel context with fentanyl resulted in a placebo effect (Bryant et al., 2009). The authors tested the prediction that mice conditioned with an opioid in a novel context would exhibit multiple, conditioned opioid-like behaviors, such as hyperlocomotion, Straub tail, and analgesia. Naïve C57 male mice in the conditioned group were administered fentanyl in a novel context (an unheated hot plate) and saline in the home cage, while control animals received saline in both locations. A third, "unconditioned" group of animals received fentanyl in the home cage and saline in the novel context. On test day, after receiving a vehicle injection, the conditioned mice showed an increased latency to lick, compared to controls, when placed on the heated hot plate. This result is somewhat difficult to assess because the contexts used for the conditioned and control groups were different, and because sample sizes used are not apparent.

Moreover, it has been demonstrated that repeating associations between conditioning stimuli and unconditioned stimuli is critical for the emergence of learned placebo responses (Colloca, 2014; Pacheco-Lopez, Engler, Niemi, & Schedlowski, 2006).

3. PAVLOVIAN CONDITIONING AND ACUTE PAIN

Some of the seminal rodent work on placebo analgesia in acute pain was performed by Guo's group at the Chinese Academy of Sciences. The authors paired tactile and visual cues with morphine administration for 4 days in female mice placed on a hot plate, while measuring the animal's latency to respond to the noxious heat (Guo, Wang, & Luo, 2010). Animals were exposed to noxious heat (55°C) on a hot plate 30 min after receiving an injection of either morphine or saline. During this 30 min window, the mice were placed in the conditioning chamber that included a grid floor (tactile) and light (visual) cue only after they had received 10 mg/kg of morphine. On day 5, the mice were administered only saline and then placed in the conditioning chamber. As expected, morphine conditioning increased the latency of the nocifensive responses (paw lick or jump). Response latencies after saline injection on the 5th day were longer than those recorded before the contextual pairing in the morphine-conditioned group, suggesting that these mice developed placebo analgesia. Importantly, this effect was prevented by the opioid receptor inverse agonist, naloxone, when it was administered in place of saline on the 5th day.

Similarly, Guo et al. (2010) reported that placebo analgesia was evoked by an exposure to the conditioned cues previously paired with aspirin (400 mg/kg), but that this effect was *not* blocked by naloxone. This study replicated human findings on naloxone that blocked placebo analgesia only when conditioning with opioids was performed (Amanzio & Benedetti, 1999). However, despite the fact that aspirin reduces pain, it is somewhat surprising that aspirin produced heat analgesia in naïve, inflammation-free, and unsensitized animals, a necessary condition for evoking the placebo reported in this chapter. Indeed, higher doses of NSAIDs are reported to be ineffective in modulating responses to heat (Zelcer, Kolesnikov, Kovalyshyn, Pasternak, & Pasternak, 2005). There is also a concern that the use of a multiple comparison statistical test—in analyzing both the morphine and aspirin data—that violates the assumption of independence of observations, and that is prone to Type I errors, may have complicated the interpretation of the results (Guo et al., 2010).

In a follow-up study, Guo et al. (2011) used the same paradigm to associate cues with administration of morphine, with the aim to examine the effects of placebo analgesia on depressive behaviors. As before, the mice developed robust placebo heat analgesia after pairing morphine with contextual cues (based on their published data, the effect size is large; Cohen's $d = 1.2$). Mice trained to associate the contextual cue with an increased paw-lick latency were subsequently tested on the tail suspension test and forced swim test. Interestingly, this contextual paradigm was reported to also produce antidepressant effects on test day, evidenced by a modest reduction in the duration of immobility measured with both tests. Indeed, the opioid system involved in the production of placebo analgesia is also implicated in mood disorders (Ide et al., 2010). Thus, this finding is consistent with their hypothesis that the placebo effect could be transferable from one domain to another, that is, from pain to emotion. These placebo effects were similar in magnitude to those obtained by administering to mice the serotonin–norepinephrine reuptake inhibitor, clomipramine, which has both antidepressant and direct antinociceptive actions (Guo et al., 2011). These behavioral findings were also accompanied by reductions in levels of the stress-related hormone, adrenocorticotropic hormone during the placebo testing, providing additional support for the interpretation that placebo analgesia reduced stress levels.

In their third study, Guo's group asked whether placebo analgesia could be expressed also in male rats (Zhang, Zhang, Wang, & Guo, 2013). Consistent with their results in mice (Guo et al., 2010, 2011), they demonstrated placebo analgesia after 4 days of preconditioning with morphine. On day 5, naloxone, delivered systemically at relatively high dosage (5 mg/kg), produced a small attenuation of the placebo effect. Injection of naloxone in the rostral anterior cingulate cortex (rACC) was reported to produce a more robust, and dose-dependent reversal of the placebo effect. They also attempted to determine which opioid receptors were involved in the placebo analgesic response with microinjections of μ-, ∂-, or κ-opioid receptor antagonists into the rACC. Curiously, Zhang et al. (2013) do not appear to have included in the statistical analyses data from the important saline injection group. These saline-control data were also not included in their analyses of the effects of antagonists of μ-, ∂-, or κ-opioid receptor antagonists. Nevertheless, they conclude from these experiments that the placebo effect is mediated exclusively via the μ-opioid receptor.

In an attempt to avoid the potential distress associated with injections of analgesics during the conditioning phase of inducing placebo analgesia, Lee et al. (2015) developed an alternative conditioning approach, employing a conditioned place preference paradigm—a form of Pavlovian conditioning that allows assessing the motivational effects associated with experiences. The authors conditioned rats to develop preference for a chamber that was associated with a subsequent exposure to low heat (45°C), and to develop aversion to a chamber associated with a subsequent exposure to noxious heat (50°C). We note that the change in preferences reported, about 10%, was modest, and their interpretation complicated by the authors' inexplicable decision to exclude from analysis periods in which the rats were in neither chamber, and by their use of a data normalization approach that may inappropriately amplify observed differences. Injections of haloperidol (a dopamine receptor antagonist), but not of naloxone (an opioid receptor antagonist), after the conditioning phase and immediately before testing abolished the expression of the learned preferences. Sample sizes (three animals per group) were rather small.

Using a similar analysis, the authors found that this drug-free conditioning process produced a reduction in nociceptive responses to heat in animals that learned to associate their conditioned chamber with lower heat exposure. This effect was small, about 5% in magnitude, and was unaffected by prior treatment with either naloxone or haloperidol.

Replication, and more rigorous analyses, is necessary before this novel approach can be evaluated as a method for producing rodent placebo analgesia.

4. AN OPERANT CONDITIONING MODEL

An important advance in developing a rodent model of placebo analgesia was the demonstration that placebo can be conditioned with an operant learning paradigm with pain. Neubert and collaborators took advantage of an operant conflict paradigm they have previously adapted to assess orofacial pain without experimenter initiated stimuli, to reduce handling related stress, and to produce cortically dependent behavior that is indicative of pain intensity (Neubert et al., 2005; Nolan, Price, Caudle, Murphy, & Neubert, 2012). The paradigm establishes a behavioral outcome whereby an animal decides between receiving a food reward and escaping a mildly aversive thermal stimulus. Thus, unlike prior preclinical models of placebo analgesia that relied on purely reflexive responses to relatively intense noxious stimuli as pain assays, Nolan et al. (2012) were able to address the emotional aspects of placebo-induced analgesia and executive cortical control placed over them.

Specifically, Nolan et al. (2012) demonstrated that associating, in hairless male rats, a simple placebo manipulation—handling and a subcutaneous injection—with the analgesic effect of morphine (1 mg/kg) for only two sessions produced a significant placebo effect that was reversed by naloxone. As the authors readily admit, their study included a relatively small number of subjects, and, as they correctly state, this "may have impacted the reliability of the statistical analyses."

Nolan and colleagues report also a large variability in response size among their rats and suggested that some rats may have a higher propensity to behave as placebo responders (Nolan et al., 2012). This is consistent with human studies demonstrating large variations in placebo effects among individuals, opening up new research avenues such as the investigation of critical phenotypes that characterize placebo responders and nonresponders (Belcher, Ferre, Martinez, & Colloca, 2017; Colloca & Benedetti, 2005; Colloca & Miller, 2011).

5. PLACEBO-CONDITIONED RESPONSES IN CHRONIC PAIN MODELS

To our knowledge, the first published study on an attempt to develop a model of placebo analgesia in chronic pain was by McNabb, White, Harris, and Fuchs (2014). They used a well-established and characterized model of persistent peripheral neuropathic pain—unilateral spinal nerve ligation in female rats (Kim & Chung, 1992). Pain was defined, in this study, as significant reduction in thresholds to withdrawal from tactile stimuli applied to the hind paw affected by the nerve ligation. The authors explain that they chose this model because the resulting hypersensitivity to mechanical stimulation is long lasting, stable over time, and responsive to treatment with analgesics (gabapentin, loperamide, or morphine) that were used in this study as unconditioned stimuli. The conditioning stimuli were contextual cues, including environmental (a novel testing room), olfactory, visual, tactile (restraint and drug injections), and temporal (fixed time of day and injection-test latency). In each experimental group of rats, these cues were paired with one of the analgesics for 4 days. On the 5th day, the cues were paired with saline administration to determine if placebo analgesia had developed. It did not. No significant differences were found on test day, when saline was administered, between control and placebo groups, indicating a lack of a conditioned placebo analgesic response.

The authors carefully concluded that placebo analgesia of chronic pain is "not particularly robust." They also postulated that the failure to evoke placebo analgesia may reflect: (1) the severity of the spinal nerve injury model, suggesting that intense, chronic pain may not be subject to placebo analgesia; (2) the relatively slow time course of the analgesics used, resulting in inadequate temporal alignment of the conditioning cues; and (3) the reliance on only a reflexive measure of tactile sensitivity, suggesting that addressing these issues might reveal placebo analgesia in animal models of chronic pain (McNabb et al., 2014).

6. FUTURE DIRECTIONS

The demonstration of placebo analgesia in the studies described earlier is valuable and exciting. However, these studies present at least two constraints. First, many of them rely on analyses of nociceptive tests, such as the tail-flick test. These tests reflect, at least in part, spinal reflexes, as they persist after section or cold block of upper parts of the spinal cord (Irwin, Houde, Bennett, Hendershot, & Seevers, 1951; Sinclair, Main, & Lo, 1988). Although spinal mechanisms are involved in placebo analgesia (Matre, Casey, & Knardahl, 2006), clearly supraspinal—and most likely cortical—structures orchestrate these effects, indicating that studies of placebo analgesia need to incorporate nonreflexive nociceptive measures as well. The study by Nolan et al. (2012), described earlier, is of particular importance in this regard.

Equally important, studies of placebo analgesia in animals have focused primarily on models of acute pain. Although clearly informative, these studies may not have immediate relevance to chronic pain conditions that present a far more intractable, intolerable, and complex clinical problem. Chronic pain affects over 100 million Americans—more than affected by heart disease, cancer, and diabetes combined and is the most common complaint of patients in outpatient clinics (Upshur, Luckmann, & Savageau, 2006). Chronic pain is also an important contributor to the opioid epidemic ravishing many communities, especially in the United States (Skolnick & Volkow, 2016; Volkow & McLellan, 2016).

It is, therefore, desirable to develop models of placebo analgesia in chronic neuropathic pain to determine the possibility that a placebo given after pharmacological conditioning elicits relief of pain intensity and pain unpleasantness.

Whereas McNabb et al. (2014) used 4 days of conditioning, it would be important to increase the number of conditioning days, because the strength of the conditioned placebo response may increase with the number of cue associations (Ader & Cohen, 1982; Niemi et al., 2007).

Finally, it would be of interest to include a behavioral assay of neuropathic pain as well as quantitative measurements of facial expressions, as measurements of the evolutionarily conserved ability to express emotions related to pain (Crook, Dickson, Hanlon, & Walters, 2014; Darwin, 1872; Williams, 2002). To allow reliable and reproducible quantification of pain-related expressions, Mogil and collaborators developed facial grimace scales for mice (Langford et al., 2010) and for rats (Sotocinal et al., 2011) and demonstrated that these objective metrics have high accuracy and reliability for detecting ongoing pain of relatively short duration (minutes to hours). This approach has been shown to be a reliable and sensitive metric for the assessment of ongoing pain in a rodent model of chronic, trigeminal neuropathic pain in both rats and mice (Akintola et al., 2017). With these experimental components in hand, it would be possible to set out to test the hypothesis that placebo analgesia can be conditioned in animals suffering from chronic pain.

6.1 Conditioned Compensatory Responses

Another relevant aspect to be considered is represented by learning paradigms that are the applicable ways to induce placebo effects in animals. When learning paradigms are adopted (e.g., classical conditioning), conditioned compensatory responses can occur. These responses are opposite to those induced by the pharmacological drug administered during the acquisition phase (exposure period). Compensatory mechanisms also underpin tolerance, defined as a decreased response to a drug within the course of administration (Siegel, Baptista, Kim, McDonald, & Weise-Kelly, 2000). The literature on pharmacological conditioning indicates that opposite behavioral responses may happen. For example, dogs treated with epinephrine every few days developed tolerance and *tachycardic* responses that tended to decrease over time. When placebo (e.g., an inert solution) replaced the epinephrine, an opposite *bradycardic* response was noted (Subkov & Zilov, 1937). As mentioned earlier, Krank and colleagues observed that rats that received morphine in the presence of the environmental cue showed an increased sensitivity to the heat stimulus, suggesting conditioned hyperalgesia (Krank et al., 1981). They performed a study in an experimental model of acute pain—response to thermal heat sensitivity on a hot plate test. Rats were administered saline (placebo) and tested on the hot plate test in the presence of an environmental stimulus which had either been paired or unpaired with either morphine or saline. Rats were divided into four groups: receiving morphine injection with and without the cue or saline injections with and without the cue. Rats conditioned with morphine and the cue showed hypersensitivity, whereas rats conditioned with morphine without the cue were not significantly different from both control groups (Krank et al., 1981). These paradoxical responses in relationship with pharmacological exposure indicate the initiation of adaptive responses that aim at compensating body responses for the primary action of drug.

6.2 Evolutionary Perspective

Are placebo effects regulated by natural selection? Why do some animals (as well as humans) respond to placebo manipulations, while others do not respond? Placebo effects elicited in animals predominantly rely on learning principles and indicate that the capability to release endogenous substances may have promoted survival (Colloca & Miller, 2011). Exploring the evolution of placebo effects is relevant to understanding the foundation of body pharmacological memories and, perhaps, biological functions and factors that have favored survival in the face of threats to life and ameliorated symptoms that are modulated by self-healing mechanisms (Miller, Colloca, & Kaptchuk, 2009). The studies discussed in this review suggest that there exists an innate capacity to enhance adaptation and mimic drug responses. The modulation of endogenous systems to promote healing by learning would, presumably, enhance survival. Human research suggests that the genetic makeup may shape the vulnerability to environmental features and situational factors. Despite the high complexity of the placebo phenomenon, recent evidence supports the idea that placebo responses may be associated with genetics (Colagiuri, Schenk, Kessler, Dorsey, & Colloca, 2015). Further animal research is needed to understand whether genetic polymorphisms—that have been linked to the neurobiology of the placebo effects—may also account for variations among animals in placebo studies.

The propensity to be conditioned to placebo manipulations to modify behaviors and disease may also be part of the biological heritage across species. An organism as simple as *Aplysia californica* responds to classical and instrumental conditioning (Carew, Hawkins, & Kandel, 1983). However, associative learning cannot explain the full range of placebo effects. The ability to experience placebo effects as a by-product of the interaction among peers, in group solidarity behaviors and prolonged nurturance adds, from an evolutionary perspective, a social dimension to the placebo effect (Humphrey, 2002). Internal healing mechanisms may counteract the otherwise biologically useful defense mechanisms, such as emotions associated with pain and anxiety reactions. Placebo effects are based on an innate ability to learn from direct and vicarious experience with a potential survival value that, in humans, is mingled with cultural features.

7. CONCLUSIONS

Recent research efforts in humans are targeting distinct pain disorders, including irritable bowel syndrome (Kaptchuk et al., 2008; Vase, Robinson, Verne, & Price, 2003, 2005), idiopathic and neuropathic pain (Petersen et al., 2012, 2014; Vase, Petersen, & Lund, 2014), low back pain (Carvalho et al., 2016; Hashmi et al., 2012), migraine (Kam-Hansen et al., 2014), and knee osteoarthritis (Gollub et al., 2018; Tetreault et al., 2016). Studying placebo analgesia in models of chronic pain is important because the pathogenesis of chronic pain involves brain structures and mechanisms that differ substantially from those related to acute pain (Baliki & Apkarian, 2015; Tracey, 2016). This suggests that mechanisms underlying placebo analgesia in chronic pain may differ substantially from those related to placebo analgesia in acute pain, emphasizing a need for further research in animal models of chronic pain.

This chapter reviews current published studies in rodents with the scope to provide critical thoughts for future animal research that offers the unique possibility to investigate new molecular mechanisms and CNS pain modulation systems involved in pharmacological (conditioned) placebo effects. In particular, the ability to create pharmacological memories with behavioral and neurobiological drug-like effects throughout conditioning points to a potential innovative approach to clinical pain management. Combining drugs and placebos in conditioning paradigms (e.g., a dose-extending placebo regimen) is clinically relevant because it may provide us with strategies to optimize clinical outcomes, while dependence, cost, and other unwanted side effects of opioid and nonopioid therapeutics are minimized (Colloca et al., 2016).

The current widespread epidemic of opioid addiction and annual deaths driven by the misuse of opioids adds clinical translational value to this research approach (Skolnick & Volkow, 2016; Volkow & McLellan, 2016). The development of novel animal placebo models for acute and chronic pain is relevant for addressing issues ranging from the optimization of placebo-conditioned paradigms (e.g., number of conditioning trials; occurrence of extinction) to the development of biological markers for responders and nonresponders.

ACKNOWLEDGMENTS

Work reported here was supported by grants from the National Institute of Neurological Disorders and Stroke (R01NS099245 to A.K.), and the National Institute of Dental and Craniofacial Research (R01DE025946 to L.C.) of the National Institutes of Health. This work was supported also by the MPowering the State Grant from the State of Maryland (to A.K. and L.C.). The content is solely the responsibility of the authors and does not necessarily represent the official views of the funding agencies. The authors report no conflict of interest. The funding sources had no role in study design; the collection, analysis, and interpretation of data; the writing of the report; or in the decision to submit the chapter for publication.

REFERENCES

Ader, R., & Cohen, N. (1982). Behaviorally conditioned immunosuppression and murine systemic lupus erythematosus. *Science*, *215*(4539), 1534–1536.

Akintola, T., Raver, C., Studlack, P., Uddin, O., Masri, R., & Keller, A. (2017). The grimace scale reliably assesses chronic pain in a rodent model of trigeminal neuropathic pain. *Neurobiology of Pain*, *2*, 13–17.

Amanzio, M., & Benedetti, F. (1999). Neuropharmacological dissection of placebo analgesia: Expectation-activated opioid systems versus conditioning-activated specific subsystems. *The Journal of Neuroscience*, *19*(1), 484–494.

Baliki, M. N., & Apkarian, A. V. (2015). Nociception, pain, negative moods, and behavior selection. *Neuron*, *87*(3), 474–491. https://doi.org/10.1016/j.neuron.2015.06.005.

Bardo, M. T., & Valone, J. M. (1994). Morphine-conditioned analgesia using a taste cue: Dissociation of taste aversion and analgesia. *Psychopharmacology*, *114*(2), 269–274.

Belcher, A. M., Ferre, S., Martinez, P. E., & Colloca, L. (2017). Role of placebo effects in pain and neuropsychiatric disorders. *Progress in Neuro-Psychopharmacology & Biological Psychiatry*, in press. https://doi.org/10.1016/j.pnpbp.2017.06.003.

Bryant, C. D., Roberts, K. W., Culbertson, C. S., Le, A., Evans, C. J., & Fanselow, M. S. (2009). Pavlovian conditioning of multiple opioid-like responses in mice. *Drug and Alcohol Dependence*, *103*(1–2), 74–83. https://doi.org/10.1016/j.drugalcdep.2009.03.016.

Burma, N. E., Leduc-Pessah, H., Fan, C. Y., & Trang, T. (2017). Animal models of chronic pain: Advances and challenges for clinical translation. *Journal of Neuroscience Research*, *95*(6), 1242–1256. https://doi.org/10.1002/jnr.23768.

Carew, T. J., Hawkins, R. D., & Kandel, E. R. (1983). Differential classical conditioning of a defensive withdrawal reflex in Aplysia californica. *Science*, *219*(4583), 397–400.

Carvalho, C., Caetano, J. M., Cunha, L., Rebouta, P., Kaptchuk, T. J., & Kirsch, I. (2016). Open-label placebo treatment in chronic low back pain: A randomized controlled trial. *Pain*, *157*(12), 2766–2772. https://doi.org/10.1097/j.pain.0000000000000700.

Colagiuri, B., Schenk, L. A., Kessler, M. D., Dorsey, S. G., & Colloca, L. (2015). The placebo effect: From concepts to genes. *Neuroscience*, *307*, 171–190. https://doi.org/10.1016/j.neuroscience.2015.08.017.

Colloca, L. (2014). Placebo, nocebo, and learning mechanisms. *Handbook of Experimental Pharmacology*, *225*, 17–35. https://doi.org/10.1007/978-3-662-44519-8_2.

Colloca, L., & Benedetti, F. (2005). Placebos and painkillers: Is mind as real as matter? *Nature Reviews. Neuroscience*, *6*(7), 545–552. https://doi.org/10.1038/nrn1705.

Colloca, L., Enck, P., & DeGrazia, D. (2016). Relieving pain using dose-extending placebos: A scoping review. *Pain*, *157*(8), 1590–1598. https://doi.org/10.1097/j.pain.0000000000000566.

Colloca, L., & Miller, F. G. (2011). How placebo responses are formed: A learning perspective. *Philosophical Transactions of the Royal Society of London. Series B, Biological Sciences*, *366*(1572), 1859–1869. https://doi.org/10.1098/rstb.2010.0398.

Crook, R. J., Dickson, K., Hanlon, R. T., & Walters, E. T. (2014). Nociceptive sensitization reduces predation risk. *Current Biology*, *24*(10), 1121–1125. https://doi.org/10.1016/j.cub.2014.03.043.

Darwin, C. (1872). *The expression of the emotions in man and animals*. London: J. Murray.

Gollub, R. L., Kirsch, I., Maleki, N., Wasan, A. D., Edwards, R. R., Tu, Y., et al. (2018). A functional neuroimaging study of expectancy effects on pain response in patients with knee osteoarthritis. *The Journal of Pain*, in press. https://doi.org/10.1016/j.jpain.2017.12.260.

Guo, J. Y., Wang, J. Y., & Luo, F. (2010). Dissection of placebo analgesia in mice: The conditions for activation of opioid and non-opioid systems. *Journal of Psychopharmacology, 24*(10), 1561–1567. https://doi.org/10.1177/0269881109104848.

Guo, J. Y., Yuan, X. Y., Sui, F., Zhang, W. C., Wang, J. Y., Luo, F., et al. (2011). Placebo analgesia affects the behavioral despair tests and hormonal secretions in mice. *Psychopharmacology, 217*(1), 83–90. https://doi.org/10.1007/s00213-011-2259-7.

Hashmi, J. A., Baria, A. T., Baliki, M. N., Huang, L., Schnitzer, T. J., & Apkarian, A. V. (2012). Brain networks predicting placebo analgesia in a clinical trial for chronic back pain. *Pain, 153*(12), 2393–2402. https://doi.org/10.1016/j.pain.2012.08.008.

Humphrey, N. (2002). Great expectations: The evolutionary psychology of faith healing and the placebo effect. In N Humphrey (Ed.), *The mind made flesh: Frontiers of psychology and evolution* (pp. 255–285). Oxford: Oxford University Press. ISBN-13: 978-0192802279.

Ide, S., Fujiwara, S., Fujiwara, M., Sora, I., Ikeda, K., Minami, M., et al. (2010). Antidepressant-like effect of venlafaxine is abolished in mu-opioid receptor-knockout mice. *Journal of Pharmacological Sciences, 114*(1), 107–110.

Irwin, S., Houde, R. W., Bennett, D. R., Hendershot, L. C., & Seevers, M. H. (1951). The effects of morphine methadone and meperidine on some reflex responses of spinal animals to nociceptive stimulation. *The Journal of Pharmacology and Experimental Therapeutics, 101*(2), 132–143.

Kam-Hansen, S., Jakubowski, M., Kelley, J. M., Kirsch, I., Hoaglin, D. C., Kaptchuk, T. J., et al. (2014). Altered placebo and drug labeling changes the outcome of episodic migraine attacks. *Science Translational Medicine, 6*(218), 218ra215. https://doi.org/10.1126/scitranslmed.3006175.

Kaptchuk, T. J., Kelley, J. M., Conboy, L. A., Davis, R. B., Kerr, C. E., Jacobson, E. E., et al. (2008). Components of placebo effect: Randomised controlled trial in patients with irritable bowel syndrome. *BMJ, 336*(7651), 999–1003. https://doi.org/10.1136/bmj.39524.439618.25.

Kim, S. H., & Chung, J. M. (1992). An experimental model for peripheral neuropathy produced by segmental spinal nerve ligation in the rat. *Pain, 50*(3), 355–363.

Krank, M. D., Hinson, R. E., & Siegel, S. (1981). Conditional hyperalgesia is elicited by environmental signals of morphine. *Behavioral and Neural Biology, 32*(2), 148–157.

Langford, D. J., Bailey, A. L., Chanda, M. L., Clarke, S. E., Drummond, T. E., Echols, S., et al. (2010). Coding of facial expressions of pain in the laboratory mouse. *Nature Methods, 7*(6), 447–449. https://doi.org/10.1038/nmeth.1455.

Lee, I. S., Lee, B., Park, H. J., Olausson, H., Enck, P., & Chae, Y. (2015). A new animal model of placebo analgesia: Involvement of the dopaminergic system in reward learning. *Scientific Reports, 5*, 17140. https://doi.org/10.1038/srep17140.

Matre, D., Casey, K. L., & Knardahl, S. (2006). Placebo-induced changes in spinal cord pain processing. *The Journal of Neuroscience, 26*(2), 559–563. https://doi.org/10.1523/JNEUROSCI.4218-05.2006.

McMillan, F. D. (1999). The placebo effect in animals. *Journal of the American Veterinary Medical Association, 215*(7), 992–999.

McNabb, C. T., White, M. M., Harris, A. L., & Fuchs, P. N. (2014). The elusive rat model of conditioned placebo analgesia. *Pain, 155*(10), 2022–2032. https://doi.org/10.1016/j.pain.2014.07.004.

Miller, F. G., Colloca, L., & Kaptchuk, T. J. (2009). The placebo effect: Illness and interpersonal healing. *Perspectives in Biology and Medicine, 52*(4), 518–539. https://doi.org/10.1353/pbm.0.0115.

Miller, J. S., Kelly, K. S., Neisewander, J. L., McCoy, D. F., & Bardo, M. T. (1990). Conditioning of morphine-induced taste aversion and analgesia. *Psychopharmacology, 101*(4), 472–480.

Mogil, J. S. (2009). Animal models of pain: Progress and challenges. *Nature Reviews. Neuroscience, 10*(4), 283–294. https://doi.org/10.1038/nrn2606.

Mogil, J. S., Davis, K. D., & Derbyshire, S. W. (2010). The necessity of animal models in pain research. *Pain, 151*(1), 12–17. https://doi.org/10.1016/j.pain.2010.07.015.

Munro, G., Jansen-Olesen, I., & Olesen, J. (2017). Animal models of pain and migraine in drug discovery. *Drug Discovery Today, 22*(7), 1103–1111. https://doi.org/10.1016/j.drudis.2017.04.016.

Neubert, J. K., Widmer, C. G., Malphurs, W., Rossi, H. L., Vierck, C. J., Jr., & Caudle, R. M. (2005). Use of a novel thermal operant behavioral assay for characterization of orofacial pain sensitivity. *Pain, 116*(3), 386–395. https://doi.org/10.1016/j.pain.2005.05.011.

Niemi, M. B., Harting, M., Kou, W., Del Rey, A., Besedovsky, H. O., Schedlowski, M., et al. (2007). Taste-immunosuppression engram: reinforcement and extinction. *Journal of Neuroimmunology*, 188(1–2), 74–79. https://doi.org/10.1016/j.jneuroim.2007.05.016.

Nolan, T. A., Price, D. D., Caudle, R. M., Murphy, N. P., & Neubert, J. K. (2012). Placebo-induced analgesia in an operant pain model in rats. *Pain, 153*(10), 2009–2016. https://doi.org/10.1016/j.pain.2012.04.026.

Pacheco-Lopez, G., Engler, H., Niemi, M. B., & Schedlowski, M. (2006). Expectations and associations that heal: Immunomodulatory placebo effects and its neurobiology. *Brain, Behavior, and Immunity, 20*(5), 430–446. https://doi.org/10.1016/j.bbi.2006.05.003.

Petersen, G. L., Finnerup, N. B., Grosen, K., Pilegaard, H. K., Tracey, I., Benedetti, F., et al. (2014). Expectations and positive emotional feelings accompany reductions in ongoing and evoked neuropathic pain following placebo interventions. *Pain, 155*(12), 2687–2698. https://doi.org/10.1016/j.pain.2014.09.036.

Petersen, G. L., Finnerup, N. B., Norskov, K. N., Grosen, K., Pilegaard, H. K., Benedetti, F., et al. (2012). Placebo manipulations reduce hyperalgesia in neuropathic pain. *Pain, 153*(6), 1292–1300. https://doi.org/10.1016/j.pain.2012.03.011.

Siegel, S., Baptista, M. A., Kim, J. A., McDonald, R. V., & Weise-Kelly, L. (2000). Pavlovian psychopharmacology: The associative basis of tolerance. *Experimental and Clinical Psychopharmacology, 8*(3), 276–293.

Sinclair, J. G., Main, C. D., & Lo, G. F. (1988). Spinal vs. supraspinal actions of morphine on the rat tail-flick reflex. *Pain, 33*(3), 357–362.

Skolnick, P., & Volkow, N. D. (2016). Re-energizing the development of pain therapeutics in light of the opioid epidemic. *Neuron, 92*(2), 294–297. https://doi.org/10.1016/j.neuron.2016.09.051.

Sotocinal, S. G., Sorge, R. E., Zaloum, A., Tuttle, A. H., Martin, L. J., Wieskopf, J. S., et al. (2011). The rat grimace scale: A partially automated method for quantifying pain in the laboratory rat via facial expressions. *Molecular Pain*, 7, 55. https://doi.org/10.1186/1744-8069-7-55.

Subkov, A. A., & Zilov, G. N. (1937). The role of conditioned reflex adaptation in the origin of hyperergic reactions. *Bulletin of Experimental Biology and Medicine, 4*, 294–296.

Tetreault, P., Mansour, A., Vachon-Presseau, E., Schnitzer, T. J., Apkarian, A. V., et al. (2016). Brain connectivity predicts placebo response across chronic pain clinical trials. *PLoS Biology, 14*(10), e1002570. https://doi.org/10.1371/journal.pbio.1002570.

Tracey, I. (2016). A vulnerability to chronic pain and its interrelationship with resistance to analgesia. *Brain, 139*(Pt 7), 1869–1872. https://doi.org/10.1093/brain/aww147.

Upshur, C. C., Luckmann, R. S., & Savageau, J. A. (2006). Primary care provider concerns about management of chronic pain in community clinic populations. *Journal of General Internal Medicine, 21*(6), 652–655. https://doi.org/10.1111/j.1525-1497.2006.00412.x.

Valone, J. M., Randall, C. K., Kraemer, P. J., & Bardo, M. T. (1998). Olfactory cues and morphine-induced conditioned analgesia in rats. *Pharmacology, Biochemistry, and Behavior, 60*(1), 115–118.

Vase, L., Petersen, G. L., & Lund, K. (2014). Placebo effects in idiopathic and neuropathic pain conditions. *Handbook of Experimental Pharmacology, 225*, 121–136. https://doi.org/10.1007/978-3-662-44519-8_7.

Vase, L., Robinson, M. E., Verne, G. N., & Price, D. D. (2003). The contributions of suggestion, desire, and expectation to placebo effects in irritable bowel syndrome patients. An empirical investigation. *Pain, 105*(1–2), 17–25.

Vase, L., Robinson, M. E., Verne, G. N., & Price, D. D. (2005). Increased placebo analgesia over time in irritable bowel syndrome (IBS) patients is associated with desire and expectation but not endogenous opioid mechanisms. *Pain, 115*(3), 338–347. https://doi.org/10.1016/j.pain.2005.03.014.

Volkow, N. D., & McLellan, A. T. (2016). Opioid abuse in chronic pain—Misconceptions and mitigation strategies. *The New England Journal of Medicine, 374*(13), 1253–1263. https://doi.org/10.1056/NEJMra1507771.

Wager, T. D., & Atlas, L. Y. (2015). The neuroscience of placebo effects: Connecting context, learning and health. *Nature Reviews. Neuroscience, 16*(7), 403–418. https://doi.org/10.1038/nrn3976.

Williams, A. C. (2002). Facial expression of pain: An evolutionary account. *The Behavioral and Brain Sciences, 25*(4), 439–455. discussion 455-488.

Zelcer, S., Kolesnikov, Y., Kovalyshyn, I., Pasternak, D. A., & Pasternak, G. W. (2005). Selective potentiation of opioid analgesia by nonsteroidal anti-inflammatory drugs. *Brain Research, 1040*(1–2), 151–156. https://doi.org/10.1016/j.brainres.2005.01.070.

Zhang, R. R., Zhang, W. C., Wang, J. Y., & Guo, J. Y. (2013). The opioid placebo analgesia is mediated exclusively through mu-opioid receptor in rat. *The International Journal of Neuropsychopharmacology, 16*(4), 849–856. https://doi.org/10.1017/S1461145712000673.

FURTHER READING

Colloca, L. (2017). Sham opioids relieve multidimensional aspects of chronic back pain. *Pain, 158*(10), 1849–1850. https://doi.org/10.1097/j.pain.0000000000000996.

Nolan, T. A., Caudle, R. M., & Neubert, J. K. (2011). Effect of caloric and non-caloric sweet reward solutions on thermal facial operant conditioning. *Behavioural Brain Research, 216*(2), 723–725. https://doi.org/10.1016/j.bbr.2010.08.023.

CHAPTER TWO

Expectancy Modulation of Opioid Neurotransmission

Marta Pecina*,[1], Jon-Kar Zubieta†
*University of Pittsburgh, Pittsburgh, PA, United States
†University Neuropsychiatric Institute, University of Utah Health Sciences Center, Salt Lake City, UT, United States
[1]Corresponding author: e-mail address: pecinam@upmc.edu

Contents

Abstract

Expectancies are powerful modulators of cognitive and emotional experiences, as well as the neurobiological responses linked to these processes. In medicine, placebo effects are a clear example of how expectancies activate opioid neurotransmission in a treatment context, leading to the experience of analgesia and the improvement of emotional states, among other symptoms. Molecular neuroimaging techniques using positron emission tomography (PET) and the selective μ-opioid receptor tracer [^{11}C] carfentanil have significantly contributed to our understanding of the neurobiological systems involved in the formation of placebo effects. This line of research has described neural and neurotransmitter networks implicated in placebo effects and provided the technical tools to examine inter-individual differences in the function of placebo responsive mechanisms. As a consequence, the capacity to activate endogenous opioid networks during the administration of placebos has been linked to the concept of resiliency mechanisms, partially determined by genetic factors, and uncovered by the cognitive emotional integration of the expectations created by the therapeutic environment and its maintenance through learning mechanisms. This evidence has contributed to the understanding of the biological bases of the cognitive and psychological mechanisms implicated in the response to treatments, and opened up new

International Review of Neurobiology, Volume 138
ISSN 0074-7742
https://doi.org/10.1016/bs.irn.2018.02.003

opportunities for drug development and the enhancement of treatment responses. Further, delineation of these processes within and across diseases is critical to understand neural systems that could be enhanced to promote symptomatic improvement and modify disease progression.

1. EXPECTATION MODULATION OF OPIOID NEUROTRANSMISSION

Positive expectations of improvement—the so-called placebo effect—are powerful modulators of clinical outcomes. This effect has been particularly well described during the treatment of clinical and experimental pain and depression, although it extends to a large number of other conditions, such as Parkinson disease and functional bowel disorders (Benedetti, 2008; Weimer, Colloca, & Enck, 2015). Work in this area has demonstrated that responses to positive expectations are associated with the activation of specific brain regions and neurotransmitter systems (Scott et al., 2008; Wager et al., 2004; Wager, Scott, & Zubieta, 2007; Zubieta et al., 2005), which seem to overlap with those affected by the pathology and treatments under study (de la Fuente-Fernandez et al., 2001; de la Fuente-Fernandez et al., 2002; Kennedy et al., 2006; Mayberg et al., 2002; Scott et al., 2008; Volkow et al., 2003).

Over three decades of research have linked placebo analgesia to the activation of endogenous opioid system. The first demonstration of the effects of positive expectations of improvement on changes in neurotransmitter systems showed that placebo analgesic effects could be blocked after the administration of the μ-opioid receptor (MOPR) antagonist naloxone (Levine, Gordon, & Fields, 1978). Placebo effects were then understood as a mechanism of pain transmission inhibition through the descending pain modulating system that originates in the cerebral cortex. Further research demonstrated the activation of several cortical areas in response to placebos, including the anterior cingulate cortex (ACC) and the dorsolateral prefrontal cortex (DLPFC) (Petrovic et al., 2002; Wager et al., 2004). This activation then extends into the descending pain modulating system, involving the hypothalamus, the periaqueductal gray (PAG), and the rostroventromedial medulla (Eippert et al., 2009a), and reaches down to the spinal cord, where inhibition of dorsal horn neurons are likely to occur (Eippert et al., 2009b). Neuropharmacological studies further confirmed the role of the opioid system in placebo analgesia using opioid antagonists blockade (Amanzio &

Benedetti, 1999; Benedetti, 1996; Eippert et al., 2009a, 2009b; Levine et al., 1978), and in vivo receptor binding of MOPRs (Wager, Scott, & Zubieta, 2007; Zubieta et al., 2005).

In addition, substantial evidence has supported the activation of reward-related networks in the context of placebo administration, both in (Scott et al., 2007) and outside the pain field (e.g., Parkinson's disease) (de la Fuente-Fernandez et al., 2001). It has been argued that the placebo effects represent a form of reward expectation which activates the same circuitry that is involved in the expectation of rewards in general (de la Fuente-Fernandez et al., 2001). This hypothesis essentially states that a person's expectations about their subsequent response to a placebo are central to the placebo effect (Kirsch, 1997). More recently, and along similar lines, it has been shown that individual positive or negative beliefs regarding the efficacy of a particular treatment modality influenced the formation of placebo effects (Watkinson, Chapman, & Horne, 2017).

This dual neural representation of placebo analgesia in pain and reward-related regions is consistent with the notion that positive expectations might have a twofold effect: the reduction of negative outcomes (e.g., pain, depression) and/or the increase likelihood of positive outcomes (e.g., therapeutic benefit) (Benedetti, 2014). Not surprisingly, placebo-induced increased opioid neurotransmission has been demonstrated in both pain- and reward-related networks as well as across disorders (e.g., pain, depression), consistently with the role of the opioid system in the regulation of stress and reward responses, as it will be pointed out later.

This review examines some of the findings we just summarized, in the context of a broader understanding of the role of the opioid receptor system and its function in regulating stress and reward responses, as a way to understand the neurobiology of placebo effects.

2. OPIOID NEUROTRANSMISSION AND FUNCTION

The opioid system consists of a large number of opioid peptides (β-endorphin, the endomorphins, enkephalins, and dynorphins) and their opioid receptor sites (μ: β-endorphin, the endomorphins, and enkephalins, δ: enkephalins; κ: dynorphins). In particular, the MOPRs are broadly distributed and are critically involved in the induction of endogenous and exogenous analgesia, reward and stress responsiveness (Kreek & Koob, 1998; Vaccarino & Kastin, 2000), as well as the regulation of emotion (Zubieta et al., 2003), hedonic responses to natural rewards (Pecina & Berridge, 2005), and social interactions (Herman & Panksepp, 1978; Hsu et al., 2013). MOPRs are widely distributed in the brain and attain their highest concentrations in the thalamus and PAG, where they regulate pain and stress responses, and in the amygdala, nucleus accumbens, and the cingulate cortex, where these receptors modulate reward, emotion, and in the case of the amygdala, also sensory processing (Oroszi & Goldman, 2004).

While opioids are best recognized for their ability to modulate pain processing, this effect might be, in fact, an expression of a broader function to counteract stress responses. Pain, as any other stressor, signals physical threat and elicits many of the same responses as nonnoxious stressors, including increased arousal, changes in autonomic activity, avoidance behaviors, and negative affect (Ribeiro et al., 2005). This "antistress" activity of endogenous opioids seems to be specifically mediated by MOPRs, whereas a stress-like aversion has been associated with the dynorphin–κ-opioid receptor (KOPR) system (Chavkin, 2013). These studies suggest that opioids mediate responses to stress broadly, providing biological mechanisms of susceptibility or resilience to physical or emotional stressors.

Emerging evidence from the human and animal literatures also points to the possibility that opioids contribute to the processing of social information. According to the brain opioid theory of social attachment (Panksepp et al., 1978), endogenous opioids are released by experiences of social bonding and mediate the pleasant feelings stemming from social bonding and affiliation (Ikemoto & Panksepp, 1992; Panksepp, 2003; Panksepp & Beatty, 1980). Furthermore, social isolation decreases the expression of proenkephalin in the nucleus accumbens, caudate, and putamen, regions known to be involved in reward and motivational processing (Angulo et al., 1991), and patients with depression show reduced endogenous opioid release in response to social rejection in brain regions implicated in stress, mood, and motivation (Hsu et al., 2015).

There is also a large body of evidence demonstrating the key role of the endogenous opioid system regulating reward and addictive behaviors. In this context, it is generally accepted that systemic and local region-specific administration of MOPR agonists, and to a lesser extent δ-opioid receptor (DOPR) agonists, stimulate positive reinforcement and euphoria, whereas KOPR agonists inhibit positive reinforcement and induce aversion and dysphoria. In fact, MOPR and DOPR antagonists have direct aversive–anxiogenic effects and can also suppress the positive reinforcing properties of natural rewards (Colasanti et al., 2011), whereas KOPR antagonists have been shown to facilitate these effects (Van Ree et al., 2000).

This ubiquitous role of the endogenous opioid system and opioid receptors modulating the adaptation of the organism to physical or emotional challenges which threaten its internal homeostasis, as well as the ability to regulate reward responses, place this system at the core of how positive expectations in the context of a "psychosocial" therapeutic encounter under stressful conditions (e.g., illness), might result in a cascade of neurobiological responses that promote the recovery of an organism undergoing allostatic challenges.

3. OPIOID PHARMACOLOGICAL AND NEUROIMAGING STUDIES OF PLACEBO EFFECTS

As briefly described earlier, pharmacological (Amanzio & Benedetti, 1999; Gracely et al., 1983; Levine et al., 1978) and neuroimaging studies (Petrovic et al., 2002; Scott et al., 2008; Zubieta et al., 2005) have extensively demonstrated the role of the opioid system in placebo analgesia. The first study that confirmed this relationship showed that placebo analgesic effects could be blocked by the administration of the MOPR antagonist naloxone (Levine et al., 1978). In a subsequent study Amanzio and Benedetti (1999) extended these findings investigating the role of opioid and nonopioid mechanisms in the formation of placebo analgesic effects. The contribution of the opioid system to the formation of placebo analgesic effects was further studied using blood flow measures and a pharmacological challenge (Petrovic et al., 2002). This study compared the effects of the short-acting MOPR agonist remifentanil on regional cerebral blood flow (rCBF) as measured with positron emission tomography (PET), with the effects of a placebo under expectations of analgesia. The results of this study demonstrated overlapping brain activity in the rostral ACC under the placebo and remifentanil conditions. Placebo administration also increased the correlation between the activity of this region and that of the midbrain PAG, a region known to exert modulatory effects on pain transmission. Individuals with high placebo analgesic responses showed greater rCBF responses to remifentanil, suggesting that individual differences in placebo analgesia may involve differences in the concentration or functionality of MOPRs (Petrovic et al., 2002). Soon after, Wager et al. (2004) and Eippert, Bingel, Schoell, Yacubian, Büchel (2008) confirmed the role of the rACC in the formation of placebo responses. Furthermore, on the basis of previous experiments on the blockade of placebo analgesia by the opioid antagonist naloxone (Amanzio & Benedetti, 1999), Eippert et al. (2009a) conducted a study to investigate the location of naloxone action in the brain. Using a pharmaco-fMRI intervention, they found that naloxone reduced behavioral placebo effects as well as placebo-induced responses in pain-modulatory cortical structures, such as the DLPFC and the rostral ACC. They also found a similar naloxone modulation of placebo-induced responses in key structures of the descending pain control system, including the hypothalamus, the PAG, and the rostral ventromedial medulla. Most importantly, naloxone abolished the increase in coupling between the rostral ACC and the PAG that was induced by the placebo. Finally, the same group demonstrated the first direct evidence that psychological factors can influence nociceptive processing at the earliest stage of the central nervous system, namely the dorsal horn of the spinal cord and that one mechanism of placebo analgesia is inhibition of spinal cord nociceptive processing.

4. PLACEBO EFFECTS AND μ-OPIOID NEUROTRANSMISSION

Among the different approaches to investigate the role of the opioid system in the neurobiology of placebo effects in humans, the use of selective radioligands and PET has resulted in major contributions to the field. Cerebral MOPRs availability and endogenous opioid release can be quantified in vivo using PET and the MOPR selective radiotracer [^{11}C]carfentanil (Fig. 1), or the nonselective opioid receptor radioligands, such as [^{11}C]buprenorphine,

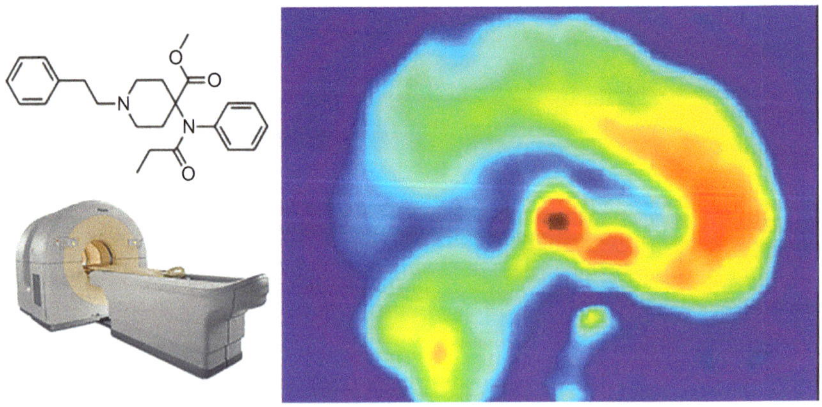

Fig. 1 Brain regions with high density of μ-opioid receptors and high affinity for the radioligand [^{11}C]carfentanil during positron emission tomography. *Credit: Data from Dr. Jon-Kar Zubieta's Lab.*

$[^{11}C]$-/$[^{18}F]$diprenorphine (DPN), or the μ- and K-antagonist ^{11}C-cyclofoxy. In these types of functional molecular assays, acute reductions in the in vivo receptor availability (binding potential) from a pain to a pain with placebo condition reflect placebo-induced activation of endogenous opioid neurotransmission; however, the sensitivity of these radiotracers to displacement by the endogenous ligands is quite variable, and greater for the agonist $[^{11}C]$ carfentanil among the available agents.

The first direct evidence of the administration of a placebo with expectations of analgesia was associated with the activation of the endogenous opioid system and MOPRs in vivo was published in 2005 (Zubieta et al., 2005). Here, the administration of the placebo was associated with increased endogenous opioid release in the rostral and subgenual ACC, the DLPFC, anterior insular cortex, and the nucleus accumbens. This activation was also associated with quantifiable reductions in physical and emotional elements of the pain experience. The regions implicated in this phenomenon included some involved in cognitive and emotional integration (DLPFC, rostral ACC); the representation and modulation of pain and emotionally salient stimuli (insula); and reward and saliency assessments (nucleus accumbens). The DLPFC was not found to be related to changes in the psychophysical properties of the pain challenge, but instead was negatively associated with the expected analgesic effect of the placebo, suggesting that the reduction in the inhibitory effect of MOPRs in this cognitive and antinociceptive region was allowing the top down engagement of subcortical pain regulatory regions through changes in the activation of MORs. These findings were replicated in a follow-up study (Scott et al., 2008), where the administration of the placebo was also associated with significant endogenous opioid activation in the pre- and subgenual ACC, orbitofrontal cortex, anterior and posterior insula, medial thalamus, nucleus accumbens, amygdala, and PAG. There was a notable lack of involvement of the DLPFC in this particular study. Regional magnitudes of activation correlated with the expected level of subjects expected analgesia (nucleus accumbens, PAG), the update of these verbally induced expectations by the subjectively perceived efficacy of the placebo (the ratio between observed and expected efficacy) (nucleus accumbens, amygdala), as well as with placebo induced changes in pain intensity (rostral ACC, nucleus accumbens, orbitofrontal cortex) and positive affect (nucleus accumbens).

Further studies examined the contribution of expectations and of learning processes, the latter defined as the differential between initial expectations and self-assessed effectiveness (prediction error) on the activity of the endogenous opioid system during placebo administration. These analyses incorporated 48 healthy subjects (20 of them were previously studied; Scott et al., 2008) using PET and the MOPR selective radiotracer [^{11}C] carfentanil (Pecina, Stohler, & Zubieta, 2014), during the same challenge. In order to create a measure of expectations and expectations/outcomes comparisons, subjects were assigned to a Low (≤50) or High (>50) expectations or effectiveness groups based on their expected analgesic effects (0–100 VAS) before the experiment and their perceived effectiveness of the placebo (0–100 VAS) after the experiment. When the two variables were combined, these resulted in four groups: a High Expectation/Low Effectiveness group, Low Expectations/Low Effectiveness group, High Expectations/High Effectiveness group, and Low Expectations/High Effectiveness group. This study showed a lack of significant relationships between the subjects expected analgesic effects and placebo-associated reductions in pain ratings. Instead, individuals with high expectations showed greater μ-opioid system activation in the DLPFC which were not associated with placebo analgesic effects. Conversely, a learning mechanism defined by the discrepancy between expected analgesia and subjectively perceived effectiveness, a prediction error signal was associated with placebo analgesic responses, and with the activation of regional μ-opioid neurotransmission in a substantial number of regions implicated in opioid-mediated antinociception (ACC, orbitofrontal cortex, amygdala, thalamus, and insula; Fig. 2A). The largest placebo responses were observed in those with low expectations and high

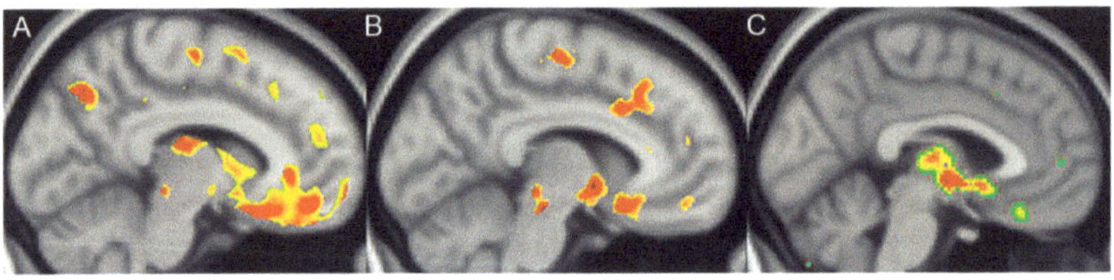

Fig. 2 Placebo-induced changes in μ-opioid neurotransmission in the ventromedial prefrontal cortex in response to different placebo manipulations. (A) Activation of regional μ-opioid neurotransmission during a pain challenge in patients with low expectancies of improvement and high effectiveness. (B) Activation of regional μ-opioid neurotransmission during a pain challenge in patients high levels of Ego-Resiliency, NEO Altruism, NEO Straightforwardness, and low levels of NEO Angry Hostility. (C) Activation of regional μ-opioid neurotransmission in patients with major depression after the administration of intravenous and oral placebos with expectations of fast-acting antidepressant effects. *Credit: Data from Dr. Jon-Kar Zubieta's Lab.*

subjective effectiveness (positive prediction error signal), whereas "nocebo" hyperalgesic responses were observed in those reporting high expectations and low reported effectiveness (negative prediction error signal). The magnitude of μ-opioid system activation in the dorsal ACC mediated the effect of prediction error on placebo analgesia.

In an attempt to investigate the role of the opioid system in the maintenance of placebo analgesia, a subsequent study examined whether the magnitude of regional opioid activation would also be associated with the strength of the recall of the placebo-induced analgesic responses (e.g., the persistence of a memory of placebo-induced analgesia) (Pecina, Stohler, & Zubieta, 2013). In this study, participants were asked to recall their pain experience by completing the McGill Pain Questionnaire (MPQ) in a phone interview 24 h after the study completion using the same scanning protocol used in previous studies (Scott et al., 2008). Subjects were further assigned to a positive placebo effect recall or a negative placebo effect recall based on their responses to the MPQ 24 h after each scanning procedure.

This data showed that in addition to its immediate placebo analgesic effects, the MOPR system is involved in the subsequent recall of the analgesic experience. This is consistent with animal models showing an effect of the enkephalinergic system and MOPRs in learning and memory when activated at the time of conditioning (Rigter, 1978). Specifically, the accurate or enhanced recall of analgesic effects 24 h after the studies (the recall of the placebo effect) was associated with μ-opioid system activation during placebo administration in the ventral tegmental area and the Papez circuit, implicated in reward-motivated learning and memory processing, respectively (Adcock et al., 2006). This report highlighted a novel role of this system in the formation of memories and potentially the sustainability of placebo analgesic effects through reinforcement learning (Au Yeung et al., 2014; Colloca & Miller, 2011; Lui et al., 2010).

5. OPIOID-MEDIATED PERSONALITY PREDICTORS OF PLACEBO EFFECTS

Neurobiological systems implicated in cognitive and psychological responses are likely to be influenced by more stable layers of neurobiology (e.g., personality traits), which ultimately affect those same cognitive and psychological responses. These endophenotypes can therefore be used to predict these behavioral responses.

To address this question, we examined the contribution of personality traits to the variability in placebo analgesic responses in a sample of 50 healthy volunteers using the sustained pain paradigm described earlier (Pecina et al., 2013b). Our primary hypothesis was that placebo analgesia would be associated with stress resiliency personality-related traits, an effect likely mediated, at least in part, by the endogenous opioid system, given the role of the MOPR-mediated neurotransmission in the maintenance of homeostasis during various forms of stress, including sustained pain (Ribeiro et al., 2005). We examined the predictive value of a variety of scales assessing emotional, psychological, and social well-being, dispositional optimism, satisfaction with life, and ego-resiliency. We also evaluated overall personality traits (NEO Personality Inventory Revised; Costa & McRae, 1992) and traits specifically related to the trait anxiety and reward processing.

Using the change in average VAS score of pain intensity as the dependent variable, the most predictive traits of placebo analgesia in a univariate model were Ego-Resiliency, NEO-Agreeableness, and NEO-Neuroticism, which, respectively, explained 16%, 14%, and 12% of the variance in placebo response. The former two were positive predictors, while the latter was a negative predictor. A multivariate model, which decomposed the Agreeableness and Neuroticism NEO-domains into their respective facets, showed that a composite measure of Ego-Resiliency, and the NEO facets Altruism, Straightforwardness (positive predictors), and Angry Hostility (negative predictor) accounted for 25% of the variation in placebo analgesic responses and had a predictive ability of 18%. Subjects scoring above the median in the composite of those trait measures also presented greater placebo-induced activation of μ-opioid neurotransmission in the subgenual and dorsal ACC, orbitofrontal cortex, insula, nucleus accumbens, amygdala, and PAG (Fig. 2B). Additionally, we found significant reductions in cortisol plasma levels during placebo administration, which were correlated with reductions in subjective pain reports and μ-opioid system activation in the dorsal ACC and PAG. These results and others (for a review, see Jaksic, Aukst-Margetic, & Jakovljevic, 2013) suggest that stable personality traits related to stress resiliency and interpersonal relationships have a substantial impact on the capacity to develop placebo effects and could be employed to stratify analysis and reduce variability in treatment trials where placebo effects can be particularly prominent, obscuring the effects of potentially active treatments.

6. GENETIC MODULATION OF PLACEBO EFFECTS: *A118G* ASN40ASP AND *FAAH* PRO129THR

Given the involvement of the endogenous opioid system and MOPRs in the formation of placebo effects, it would be expected that genetic variation impacting this neurotransmitter system would influence placebo-activated processes. A candidate genetic variant examined by our group was the *OPRM1* A118G variant (Asn40Asp, rs1799971; (Bergen et al., 1997), whereby the 118G variant (Asp40) is thought to be expressed at lower levels (Zhang et al., 2005). Consistent with findings in animal models, we found that *OPRM1* G carriers, compared with AA homozygotes, showed an overall reduction of baseline MOPR availability in regions implicated in pain and affective regulation (Fig. 3). In response to a sustained painful stimulus, we found no effect of A118G on pain-induced endogenous opioid release. G carriers showed lower placebo-induced μ-opioid system activation in the anterior insula, the amygdala, the nucleus accumbens, the thalamus, and the brainstem. At a trait level, G carriers reported higher NEO-Neuroticism scores, a personality trait previously associated with increased pain and lower placebo responses, which were negatively correlated with baseline MOPR availability in the anterior insula and subgenual ACC. These results demonstrated that the A118G *OPRM1* polymorphism contributes to interindividual variations in the function of neurotransmitters responsive to pain, as well as their regulation through cognitive–emotional influences in the context of placebo effects.

In addition, genetic variation within neural systems, known to colocalize and interact with the endogenous opioid system, has the potential to discover targets that would allow for the modulation (e.g., enhancement in clinical practice, reduction in clinical trials) of placebo activated processes. One such system is the endocannabinoid, comprised of cannabinoid CB1 and CB2 receptors and their endogenous ligands, including *N*-arachidonoylethanolamine (anandamide, AEA) and 2-arachidonoylglycerol (2-AG) (Kogan & Mechoulam, 2006). This system is thought to be involved in analgesia (Hohmann, 2002) and reward/reinforcement (Gardner & Vorel, 1998) mechanisms, both of which are thought to be engaged during the development of placebo effects (Colloca, Sigaudo, & Benedetti, 2008). Previous work had shown that in the context of a conditioning paradigm, the cannabinoid receptor 1 (CBR1) antagonist SR 141716A (Rimonabant) blocked nonopioid, ketorolac-conditioned placebo analgesia, but not opioid placebo

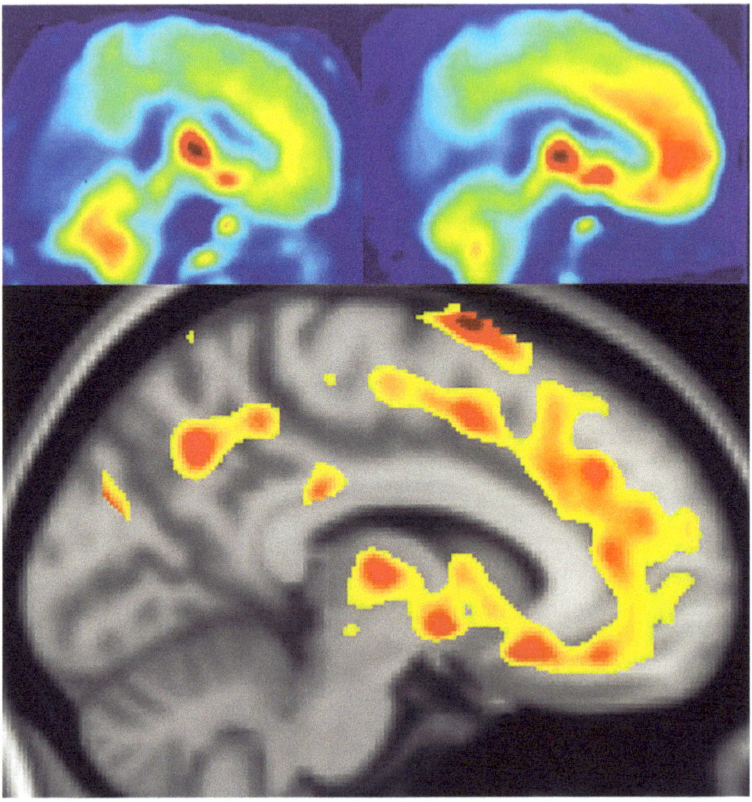

Fig. 3 Effects of the mu opioid receptor polymorphism (*OPRM1* A118G) on μ-opioid receptor binding potential. *Top left*: μ-opioid receptor binding potential in G carriers. *Top right*: μ-opioid receptor binding potential in AA homozygotes. *Bottom*: AA homozygotes, compared to G carriers, had greater μ-opioid receptor binding at baseline in the anterior cingulate cortex, nucleus accumbens, and thalamus. *Credit: Data from Dr. Jon-Kar Zubieta's Lab.*

responses after morphine conditioning (Benedetti et al., 2011). However, our study showed that genetic variation in the functional missense variant Pro129Thr of the gene coding *fatty acid amide hydrolase* (*FAAH*), the major degrading enzyme of endocannabinoids, modulated opioid neurotransmission measures during a placebo analgesia experiment (Pecina et al., 2014b), independent of other aspects of pain. Surprisingly, Pro129/Pro129 homozygotes, which have increased activity of the *FAAH* and therefore lower cannabinoid levels (Chiang et al., 2004), showed significantly greater psychophysical placebo responses, a more positive internal affective state during the placebo condition and greater recall of the placebo experience 24 h after the pain challenge compared to Thr129 carriers. The neuroimaging data showed that during placebo administration *FAAH* Pro129/Pro129 homozygotes had greater placebo-induced endogenous opioid system activation, but not DA, in areas of the prefrontal cortex, including the DLFPC; the dorsal and ventromedial PFC; the lateral and medial orbitofrontal cortex; the inferior frontal gyrus; the dorsal, rostral, and subgenual ACC; the anterior and posterior insula; and the hippocampus and parahippocampal gyrus. Subcortically, regions where *FAAH* Pro129/Pro129 homozygotes had greater endogenous opioid release included the nucleus accumbens/mammillary region, the dorsal and ventral putamen, and the anterior and posterior thalamus (Fig. 2). The effects of *FAAH* on placebo-induced regional activation of μ-opioid neurotransmission were significantly correlated with psychophysical responses to placebo and with an enhancement of the recall of placebo effects 24 h after the pain challenge.

These results suggested that functional *FAAH* genotype variation selectively influenced psychophysical placebo effects and placebo-induced activation of MOPR-mediated neurotransmission in a network of regions previously involved in placebo-induced analgesia (Scott et al., 2008; Zubieta et al., 2005). *FAAH* Thr129 carriers, despite their chronic greater tonic eCB concentrations, showed lower psychophysical placebo responses and regional μ-opioid activation during placebo administration, compared with Pro129/Pro129 homozygotes, which suggest a downregulation of CBR1 sites as potentially mediating these reduced placebo responses. These results demonstrated an interaction between eCB and μ-opioid neurotransmission in the formation of placebo responses in the absence of previous conditioning and provided new insights into the neurobiology of placebo effects in conditions where these interactions play a critical role, such as substance use disorders (Belcher et al., 2017; Fattore et al., 2005, 2007; Navarro et al., 2001).

7. OPIOID-MEDIATED PREDICTORS OF PLACEBO EFFECTS IN MAJOR DEPRESSION

As briefly described in Section 1, the opioid system has a significant role in the neurobiology of emotion, stress responses, and memory. This system has been involved in the physiological regulation of affective experiences in healthy humans as well as in cognitive mechanisms of treatment response in patients diagnosed with major depression.

In a study that examined the effects of a sadness-induction challenge on the changes in μ-opioid neurotransmission using [^{11}C]carfentanil in 14 healthy female volunteers (Zubieta et al., 2003), the sustained sadness condition was associated with reduced endogenous opioid activity in the rostral ACC, ventral pallidum, amygdala, and inferior temporal cortex. Deactivation of μ-opioid neurotransmission in the rACC, ventral pallidum, and amygdala was correlated with increases in negative affect ratings and reductions in positive affect ratings during the sustained sadness state, confirming the role of the MOPR system in the physiological regulation of affective experiences in humans. In a follow-up study in 14 patients with major depressive disorder (MDD) (Kennedy et al., 2006), patients with depression showed significantly greater opioid release in the left inferior temporal cortex, which correlated with increased negative affect ratings experienced during the condition. MDD patients, compared to controls, showed significantly lower baseline MOPR binding in the posterior thalamus, which also associated with poorer treatment responses. Healthy controls instead showed larger sadness-induced opioid release in the ACC, temporal cortex, ventral basal ganglia, hypothalamus, amygdala, and periamygdalar cortex. These two pieces of literature represented the first human evidence implicating the μ-opioid system in the response to an emotional challenge in the absence of a painful stimuli both in healthy controls and patients with depression, and suggested a dysregulation of emotional stress regulatory mechanisms. In a separate sample, and using a peer rejection and acceptance paradigm, patients with major depression, compared to healthy controls, demonstrated lower levels of endogenous opioid release and μ-receptor activation in response to both acceptance and rejection, which was associated with lower levels of stress tolerance, as measures with standardized scales (Hsu et al., 2013, 2015).

A recent novel study directly tested the contribution of the opioid system to the formation of placebo effects in unmedicated patients with moderate–severe major depression (Pecina et al., 2015). The rationale of this study followed the well-described role of opioid neurotransmission in placebo analgesic effects (Pecina & Zubieta, 2015; Zubieta et al., 2005), as well as the role of this system in the biology of depression, as summarized earlier. The study design incorporated a commonly used placebo lead-in phase with the administration of two identical placebos: one described as potentially having fast-acting antidepressant effects (active) and the other described as being a placebo with no antidepressant effects (inactive). In addition to evaluating the effects of sustained placebo pills, an intravenous (IV) placebo administration followed the 1-week active placebo to investigate the effects of acute placebo administration on μ-opioid neurotransmission. Following each placebo intervention, patients underwent a 10-week open-label trial with a common SSRI antidepressant.

In this study, higher baseline MOPR binding in the nucleus accumbens was associated with better response to 10 weeks of open-label antidepressant treatment. Reductions in depressive symptoms after 1 week of placebo treatment were associated with increased placebo-induced opioid release in the subgenual ACC, nucleus accumbens, midline thalamus, and amygdala (Fig. 2C). Importantly, placebo-induced activation of the opioid system in these stress and emotion regulatory regions predicted 43% of the variance in the response to antidepressant treatment after 10 weeks. Similarly, subjective clinical placebo effects predicted 46% of the response to antidepressant treatment, while the combination of both predicted 57% of the total antidepressant response. Still, by weeks 8 and 10, depression severity scores were roughly twice as high among placebo nonresponders compared with placebo responders. Furthermore, achievement of remission was also significantly higher among placebo responders compared with nonresponders, an observation that potentially challenged a common tenant that eliminating placebo responders in clinical trials with placebo lead-in phases or novel sequential parallel comparison designs would help to more clearly interpret RCT results.

This study demonstrated that opioid neurotransmission is not only relevant to the formation of placebo analgesia responses but also to the formation of placebo responses in other conditions, such as depression. Furthermore, interindividual differences in opioid neurotransmission explained a substantial amount of the variance in response to common antidepressant treatments, by its role in the placebo response inherent in any treatment or by its interaction with common antidepressants.

8. FUTURE DIRECTIONS

Many questions remain to apply this knowledge to therapeutic interventions.

Are there processes that may enhance placebo effects in clinical practice? The available information points to various genetic polymorphisms that could be targeted, directly or indirectly, to promote the engagement of these internal processes of recovery. For example, and as noted earlier, lesser functional variants of the *FAAH* and the *OPRM1* genes were associated with lower endogenous opioid-mediated placebo analgesic responses, while those of the catechol-O-methyltransferase (COMT) enzyme were linked to higher placebo responses in patients diagnosed with irritable bowel syndrome. The latter effect was particularly prominent in a treatment arm enhanced by increased patient–clinician interaction (Hall et al., 2012). While the field of placebo genetics is still in its infancy, this information provides targets that could be modulated in practice to enhance clinical responses (a desirable effect in unblinded clinical treatments). In RCTs, genetic information can be utilized for the opposite effect, the stratification of study participants as more or less likely to respond to the placebo arms, hence aiding in the separation of active and "inactive" effects and in the separation of placebo (or sham) and study drug (or procedure)-associated treatment outcomes.

Does the modulation of placebo-activated mechanisms change outcomes in chronic conditions, such as persistent pain syndromes or major depression, for example? In clinical trials, the placebo responses invariably parallel those of active treatments, suggesting that perhaps through processes related to reward learning (Pecina, Stohler, & Zubieta, 2014a), placebo-activated mechanisms can improve the clinical status of patients with illnesses such as major depression, Parkinson's disease, or various forms of persistent pain, which register high placebo response rates in RCTs. At the present time, however, there is no neurobiological data showing that pathophysiological processes present in, for example, major depression, or persistent pain, can be reversed by continued placebo treatment in placebo responders. For example, while reductions in MOPR concentrations have been observed in fibromyalgia (Harris et al., 2007), and the capacity to activate endogenous opioid neurotransmission has been shown reduced in chronic low back pain (Martikainen et al., 2013)and major depression (Hsu et al., 2015; Pecina et al., 2015), it is presently unknown whether placebo administration, while known to affect those mechanisms, can reverse the observed deficits.

Does placebo responsiveness predict better responses to psychosocial interventions, such as cognitive behavioral or other therapies, as opposed to traditional drug-based approaches? MOPR-mediated neurotransmission is centrally implicated in prosocial, affiliative behaviors and, as noted earlier, is one of the neurotransmitter systems known to mediate placebo responses in humans, an effect clearly shown in the absence of conditioning procedures. Social processes that form part of patient–clinician interactions have been shown to enhance the responsiveness to treatment. A study of patients with irritable bowel syndrome showed that nonspecific effects could produce statistically and clinically significant improvement in outcomes, and that the patient–practitioner relationship was the most important component of those nonspecific effects (Kaptchuk et al., 2008). While it makes logical sense that placebo responsiveness and the expectations created by a positive patient–clinician relationship may be mediated by similar neurobiological processes, it remains to be determined whether the capacity to respond to placebo administration would also be linked to a higher likelihood of response to psychosocial or cognitive (e.g., cognitive behavioral therapy) interventions, hence reducing the need for more invasive or medication-based treatments.

Do interindividual variations in placebo responsiveness at the neurobiological or psychophysical levels interact or promote responses to standard therapeutic approaches? It has been shown that placebo responders also demonstrate a greater response to open antidepressant administration in major depression, an effect that was largely accounted for by the activation of endogenous opioid neurotransmission (Pecina et al., 2015). As noted by Enck et al. (2013) this suggests that neurobiological processes activated by placebo administration could be additive, or even interactive, with those impacted by active drugs or procedures. For example, antidepressants are known to act on neurotransmitter systems such as the serotonergic, noradrenergic, and dopaminergic systems, but also indirectly modulate endogenous opioid neurotransmission. A number of studies have shown that the analgesic effects of antidepressants are at least partly mediated by their modulation of the endogenous opioid system, while in animal models of depression, their antidepressant effects can also be antagonized by opioid receptor blockade. In this manner, the additive or interactive effects of placebo-activated processes with those of various treatments could in fact enhance or potentiate positive outcomes in clinical studies.

How can this information clarify the analysis of outcomes in clinical trials of novel interventions? At the very least, the study of placebo neurobiology points to specific mechanisms of resiliency that should not be disregarded as noise, but as processes that when engaged or enhanced, could impact the recovery from common chronic and highly impactful disorders.

9. CONCLUSIONS

The studies reviewed in this chapter point to the involvement of specific neurotransmitter systems and neural networks in the formation of placebo analgesic and antidepressant responses, with the endogenous opioid system and MOPRs playing a central role in these processes. In the absence of conditioning, interindividual variation in neurobiological and psychophysical placebo effects is associated with expectations (a result of the interaction between the individual and the "treatment" environment and potential preexisting biases) and maintained by learning processes (driven by the relationship between initial expectations and subjectively experienced outcomes (Pecina et al., 2014b)). They are also determined by the individual neurobiology, in particular genetic variations in common polymorphisms that modulate endogenous opioid system function. As would be expected for genetically modulated traits, specific personality characteristics associated with stress tolerance and the capacity to respond favorably (or unfavorably) to changes in the environment, also predicted a substantial proportion of the variance in the development of both, subjective and objectively measured placebo responses.

Largely disregarded as measurement "noise" in clinical trials, placebo neurobiology rather points at mechanisms that can be activated by cognitive–emotional processes that take place in the therapeutic environment to reduce illness burden by the engagement of stress regulatory mechanisms (either physical, such as pain, or emotional, as in the case of depression), where the endogenous opioid system plays a central role.

REFERENCES

Adcock, R. A., et al. (2006). Reward-motivated learning: Mesolimbic activation precedes memory formation. *Neuron*, *50*(3), 507–517.

Amanzio, M., & Benedetti, F. (1999). Neuropharmacological dissection of placebo analgesia: Expectation-activated opioid systems versus conditioning-activated specific subsystems. *The Journal of Neuroscience*, *19*(1), 484–494.

Angulo, J. A., et al. (1991). Isolation stress increases tyrosine hydroxylase mRNA in the locus coeruleus and midbrain and decreases proenkephalin mRNA in the striatum and nucleus accumbens. *Brain Research. Molecular Brain Research*, *11*(3–4), 301–308.

Au Yeung, S. T., et al. (2014). Partial reinforcement, extinction, and placebo analgesia. *Pain*, *155*(6), 1110–1117.

Belcher, A. M., et al. (2017). Role of placebo effects in pain and neuropsychiatric disorders. *Progress in Neuro-Psychopharmacology & Biological Psychiatry*, pii: S0278-5846(17)30164-1. https://doi.org/10.1016/j.pnpbp.2017.06.003. [Jun 8; Epub ahead of print].

Benedetti, F. (1996). The opposite effects of the opiate antagonist naloxone and the cholecystokinin antagonist proglumide on placebo analgesia. *Pain*, *64*(3), 535–543.

Benedetti, F. (2008). Mechanisms of placebo and placebo-related effects across diseases and treatments. *Annual Review of Pharmacology and Toxicology*, *48*, 33–60.

Benedetti, F. (2014). Placebo effects: From the neurobiological paradigm to translational implications. *Neuron*, *84*(3), 623–637.

Benedetti, F., et al. (2011). Nonopioid placebo analgesia is mediated by CB1 cannabinoid receptors. *Nature Medicine*, *17*(10), 1228–1230.

Bergen, A. W., et al. (1997). Mu opioid receptor gene variants: Lack of association with alcohol dependence. *Molecular Psychiatry*, *2*(6), 490–494.

Chavkin, C. (2013). Dynorphin—Still an extraordinarily potent opioid peptide. *Molecular Pharmacology*, *83*(4), 729–736.

Chiang, K. P., et al. (2004). Reduced cellular expression and activity of the P129T mutant of human fatty acid amide hydrolase: Evidence for a link between defects in the endocannabinoid system and problem drug use. *Human Molecular Genetics*, *13*(18), 2113–2119.

Colasanti, A., et al. (2011). Opioids and anxiety. *Journal of Psychopharmacology*, *25*(11), 1415–1433.

Colloca, L., & Miller, F. G. (2011). How placebo responses are formed: A learning perspective. *Philosophical Transactions of the Royal Society of London. Series B, Biological Sciences*, *366*(1572), 1859–1869.

Colloca, L., Sigaudo, M., & Benedetti, F. (2008). The role of learning in nocebo and placebo effects. *Pain*, *136*(1–2), 211–218.

Costa, P., & McRae, R. (1992). Normal personality assessment in clinical practice: The NEO personality inventory. *Psychological Assessment*, *4*(1), 5–13.

de la Fuente-Fernandez, R., et al. (2001). Expectation and dopamine release: Mechanism of the placebo effect in Parkinson's disease. *Science*, *293*(5532), 1164–1166.

de la Fuente-Fernandez, R., et al. (2002). Dopamine release in human ventral striatum and expectation of reward. *Behavioural Brain Research*, *136*(2), 359–363.

Eippert, F., Bingel, U., Schoell, E., Yacubian, J., & Büchel, C. (2008). Blockade of endogenous opioid neurotransmission enhances acquisition of conditioned fear in humans. *The Journal of Neuroscience*, *28*, 5465–5472.

Eippert, F., et al. (2009a). Activation of the opioidergic descending pain control system underlies placebo analgesia. *Neuron*, *63*(4), 533–543.

Eippert, F., et al. (2009b). Direct evidence for spinal cord involvement in placebo analgesia. *Science*, *326*(5951), 404.

Enck, P., et al. (2013). The placebo response in medicine: Minimize, maximize or personalize? *Nature Reviews. Drug Discovery*, *12*(3), 191–204.

Fattore, L., et al. (2005). Endocannabinoid system and opioid addiction: Behavioural aspects. *Pharmacology, Biochemistry, and Behavior*, *81*(2), 343–359.

Fattore, L., et al. (2007). Bidirectional regulation of mu-opioid and CB1-cannabinoid receptor in rats self-administering heroin or WIN 55,212-2. *The European Journal of Neuroscience*, *25*(7), 2191–2200.

Gardner, E. L., & Vorel, S. R. (1998). Cannabinoid transmission and reward-related events. *Neurobiology Disease*, *5*(6 Pt. B), 502–533.

Gracely, R. H., et al. (1983). Placebo and naloxone can alter post-surgical pain by separate mechanisms. *Nature*, *306*(5940), 264–265.

Hall, K. T., et al. (2012). Catechol-O-methyltransferase val158met polymorphism predicts placebo effect in irritable bowel syndrome. *PLoS One*, 7(10), e48135.

Harris, R. E., et al. (2007). Decreased central mu-opioid receptor availability in fibromyalgia. *The Journal of Neuroscience*, *27*(37), 10000–10006.

Herman, B. H., & Panksepp, J. (1978). Effects of morphine and naloxone on separation distress and approach attachment: Evidence for opiate mediation of social affect. *Pharmacology, Biochemistry, and Behavior*, *9*(2), 213–220.

Hohmann, A. G. (2002). Spinal and peripheral mechanisms of cannabinoid antinociception: Behavioral, neurophysiological and neuroanatomical perspectives. *Chemistry and Physics of Lipids*, *121*(1–2), 173–190.

Hsu, D. T., et al. (2013). Response of the mu-opioid system to social rejection and acceptance. *Molecular Psychiatry*, *18*(11), 1211–1217.

Hsu, D. T., et al. (2015). It still hurts: Altered endogenous opioid activity in the brain during social rejection and acceptance in major depressive disorder. *Molecular Psychiatry*, *20*(2), 193–200.

Ikemoto, S., & Panksepp, J. (1992). The effects of early social isolation on the motivation for social play in juvenile rats. *Developmental Psychobiology, 25*(4), 261–274.
Jaksic, N., Aukst-Margetic, B., & Jakovljevic, M. (2013). Does personality play a relevant role in the placebo effect? *Psychiatria Danubina, 25*(1), 17–23.
Kaptchuk, T. J., et al. (2008). Components of placebo effect: Randomised controlled trial in patients with irritable bowel syndrome. *BMJ, 336*(7651), 999–1003.
Kennedy, S. E., et al. (2006). Dysregulation of endogenous opioid emotion regulation circuitry in major depression in women. *Archives of General Psychiatry, 63*(11), 1199–1208.
Kirsch, I. (1997). Response expectancy theory and application: A decennial review. *Applied and Preventive Psychology, 6*(2), 69–79.
Kogan, N. M., & Mechoulam, R. (2006). The chemistry of endocannabinoids. *Journal of Endocrinological Investigation, 29*(3 Suppl), 3–14.
Kreek, M. J., & Koob, G. F. (1998). Drug dependence: Stress and dysregulation of brain reward pathways. *Drug and Alcohol Dependence, 51*(1–2), 23–47.
Levine, J. D., Gordon, N. C., & Fields, H. L. (1978). The mechanism of placebo analgesia. *Lancet, 2*(8091), 654–657.
Lui, F., et al. (2010). Neural bases of conditioned placebo analgesia. *Pain, 151*(3), 816–824.
Martikainen, I. K., et al. (2013). Alterations in endogenous opioid functional measures in chronic back pain. *The Journal of Neuroscience, 33*(37), 14729–14737.
Mayberg, H. S., et al. (2002). The functional neuroanatomy of the placebo effect. *The American Journal of Psychiatry, 159*(5), 728–737.
Navarro, M., et al. (2001). Functional interaction between opioid and cannabinoid receptors in drug self-administration. *The Journal of Neuroscience, 21*(14), 5344–5350.
Oroszi, G., & Goldman, D. (2004). Alcoholism: Genes and mechanisms. *Pharmacogenomics, 5*(8), 1037–1048.
Panksepp, J. (2003). Neuroscience. Feeling the pain of social loss. *Science, 302*(5643), 237–239.
Panksepp, J., & Beatty, W. W. (1980). Social deprivation and play in rats. *Behavioral and Neural Biology, 30*(2), 197–206.
Panksepp, J., et al. (1978). The biology of social attachments: Opiates alleviate separation distress. *Biological Psychiatry, 13*(5), 607–618.
Pecina, S., & Berridge, K. C. (2005). Hedonic hot spot in nucleus accumbens shell: Where do mu-opioids cause increased hedonic impact of sweetness? *The Journal of Neuroscience, 25*(50), 11777–11786.
Pecina, M., Stohler, C. S., & Zubieta, J. K. (2013a). Role of mu-opioid system in the formation of memory of placebo responses. *Molecular Psychiatry, 18*(2), 135–137.
Pecina, M., Stohler, C. S., & Zubieta, J. K. (2014a). Neurobiology of placebo effects: Expectations or learning? *Social Cognitive and Affective Neuroscience, 9*(7), 1013–1021.
Pecina, M., & Zubieta, J. K. (2015). Molecular mechanisms of placebo responses in humans. *Molecular Psychiatry, 20*(4), 416–423.
Pecina, M., et al. (2013b). Personality trait predictors of placebo analgesia and neurobiological correlates. *Neuropsychopharmacology: Official Publication of the American College of Neuropsychopharmacology, 38*(4), 639–646.
Pecina, M., et al. (2014b). FAAH selectively influences placebo effects. *Molecular Psychiatry, 19*(3), 385–391.
Pecina, M., et al. (2015). Association between placebo-activated neural systems and antidepressant responses: Neurochemistry of placebo effects in major depression. *JAMA Psychiatry, 72*(11), 1087–1094.
Petrovic, P., et al. (2002). Placebo and opioid analgesia—Imaging a shared neuronal network. *Science, 295*(5560), 1737–1740.
Ribeiro, S. C., et al. (2005). Interface of physical and emotional stress regulation through the endogenous opioid system and mu-opioid receptors. *Progress in Neuro-Psychopharmacology & Biological Psychiatry, 29*(8), 1264–1280.

Rigter, H. (1978). Attenuation of amnesia in rats by systemically administered enkephalins. *Science, 200*(4337), 83–85.

Scott, D. J., et al. (2007). Individual differences in reward responding explain placebo-induced expectations and effects. *Neuron, 55*(2), 325–336.

Scott, D. J., et al. (2008). Placebo and nocebo effects are defined by opposite opioid and dopaminergic responses. *Archives of General Psychiatry, 65*(2), 220–231.

Vaccarino, A. L., & Kastin, A. J. (2000). Endogenous opiates: 1999. *Peptides, 21*(12), 1975–2034.

Van Ree, J. M., et al. (2000). Endogenous opioids and reward. *European Journal of Pharmacology, 405*(1–3), 89–101.

Volkow, N. D., et al. (2003). Expectation enhances the regional brain metabolic and the reinforcing effects of stimulants in cocaine abusers. *The Journal of Neuroscience, 23*(36), 11461–11468.

Wager, T. D., Scott, D. J., & Zubieta, J. K. (2007). Placebo effects on human mu-opioid activity during pain. *Proceedings of the National Academy of Sciences of the United States of America, 104*(26), 11056–11061.

Wager, T. D., et al. (2004). Placebo-induced changes in FMRI in the anticipation and experience of pain. *Science, 303*(5661), 1162–1167.

Watkinson, A., Chapman, S. C. E., & Horne, R. (2017). Beliefs about pharmaceutical medicines and natural remedies explain individual variation in placebo analgesia. *The Journal of Pain, 18*(8), 908–922.

Weimer, K., Colloca, L., & Enck, P. (2015). Placebo effects in psychiatry: Mediators and moderators. *Lancet Psychiatry, 2*(3), 246–257.

Zhang, Y., et al. (2005). Allelic expression imbalance of human mu opioid receptor (OPRM1) caused by variant A118G. *The Journal of Biological Chemistry, 280*(38), 32618–32624.

Zubieta, J. K., et al. (2003). Regulation of human affective responses by anterior cingulate and limbic mu-opioid neurotransmission. *Archives of General Psychiatry, 60*(11), 1145–1153.

Zubieta, J. K., et al. (2005). Placebo effects mediated by endogenous opioid activity on mu opioid receptors. *The Journal of Neuroscience, 25*(34), 7754–7762.

CHAPTER THREE

Placebo Effects in the Immune System

Martin Hadamitzky*[,1], Wiebke Sondermann[†], Sven Benson*, Manfred Schedlowski*[,‡]

*Institute of Medical Psychology and Behavioral Immunobiology, University Hospital Essen, University of Duisburg–Essen, Essen, Germany
[†]West German Cancer Center, University Hospital Essen, University of Duisburg–Essen, Essen, Germany
[‡]Karolinska Institute, Stockholm, Sweden
[1]Corresponding author: e-mail address: martin.hadamitzky@uk-essen.de

Contents

Abstract

Even though knowledge and systematic application of placebo responses in the immune system are sparse, this topic is of particular importance since it may aim at drug-dose reduction while maintaining therapeutic efficacy of treatment in clinical settings. Placebo responses in the immune system are inducible by associated learning paradigms, such as behaviorally conditioned immunosuppression. One established learning paradigm in both rats and humans is conditioned taste avoidance (CTA), where a novel taste as conditioned stimulus (CS) is paired with the administration of the immunosuppressive drug cyclosporine A (CsA) as unconditioned stimulus. By representing the CS alone at a later time point, the conditioned response is reflected by avoidance behavior toward the taste (CTA). Simultaneously, diminished cytokine production and proliferative capacity of T cells are observed, closely mimicking the pharmacological effects of CsA. This chapter provides an overview on placebo responses in the immune system and delineates actual approaches, translational aspects, and limitations of learning paradigms in clinical settings.

ISSN 0074-7742
https://doi.org/10.1016/bs.irn.2018.01.001

1. INTRODUCTION

Positive disease outcomes observed after treatment with nonspecific treatment ingredients can be defined as placebo responses (Enck, Bingel, Schedlowski, & Rief, 2013). For estimating the extent to which a drug's effect exceeds nonspecific placebo treatment effects, clinical trials compare the effects of a drug to an "inert" placebo under double blind conditions (Price, Finniss, & Benedetti, 2008). However, many approaches discovered that placebo responses are rather complex psychoneurobiological phenomena, modulated and regulated centrally by distinct brain areas and peripherally by different physiological processes (Benedetti, Carlino, & Pollo, 2011; Klosterhalfen & Enck, 2006; Price et al., 2008). In general, placebo responses are primarily driven by the formation of a patient's expectation toward the benefit of a treatment and by associative learning processes, whereas the importance of these two aspects may differ in the different physiological systems and end-organ functions (Colloca & Miller, 2011; Price et al., 2008; Stewart-Williams & Podd, 2004).

Systematic evocation of placebo responses in the immune system, affecting neuroendocrine and immunological functions, can be achieved by learning paradigms such as behavioral conditioning but not via cognitive factors such as mere expectation (Albring et al., 2012; Benedetti et al., 2003). In this context, studies already gathered evidence for the efficacy of placebo-controlled dose reduction (PCDR) in humans by employing associative learning procedures. Certainly, they also revealed the difficulties and limitations that may arise during application (Doering & Rief, 2012). In this chapter, the different paradigms and translational approaches in the field of conditioned immunomodulatory placebo effects will be summarized. In particular, the well-established paradigm of behaviorally conditioned immunosuppression in rodents and humans, using the calcineurin inhibitor cyclosporine A (CsA), will be highlighted with regard to its therapeutic potential and possible application in clinical settings.

2. CONDITIONING OF IMMUNE RESPONSES—APPROACHES AND PARADIGMS ACROSS SPECIES

For long time, the CNS and the immune system were assumed to be independent acting systems. However, within the last decades a growing body of evidence revealed that both systems share a common chemical language and are continuously exchanging information (Dantzer, O'Connor, Freund, Johnson, & Kelley, 2008; Tracey, 2010). Indeed, it has been suggested that the immune system acts as a sensory organ with immune cells being mobile sentinels that inform the brain about the immune status in the periphery (Blalock & Smith, 2007). Behavioral conditioning of immune functions is one fascinating example for this bidirectional communication between the two (Ader & Cohen, 1975; Schedlowski & Pacheco-Lopez, 2010; Wayner, Flannery, & Singer, 1978). Associative learning processes such as immune conditioning have probably evolved over years as adaptive mechanisms to protect an organism from potentially harmful immune responses by avoiding ingestion or contact with immunomodulating substances (Ader, 2003; Bermudez-Rattoni, 2004; Schedlowski & Pacheco-Lopez, 2010). In a less classical sense, immune responses can also be induced experimentally by using artificial substances or drugs in associative learning or Pavlovian (classical) conditioning (Ader, 1976, 2003). For instance, a conditioned immune response was first demonstrated in guinea pigs by coworkers of Ivan Petrovich Pavlov, who reported a conditioned increase in peripheral leucocyte numbers following repeated injections of the plant extract Tapioca together with a slight slit or heating of the skin (Vits et al., 2011). These early observations were rediscovered and experimentally elaborated by Ader and Cohen (1975), who showed that the increase of antibody titers after injection of sheep erythrocytes was diminished in rats conditioned to the immunosuppressive properties of the compound cyclophosphamide.

Since those pioneer studies, a large number of animal research revealed that various immunological parameters can be influenced by conditioning protocols, while a flavor or odor is typically employed as CS along with a pharmacological agent used as unconditioned stimulus (US). This phenomenon, to behaviorally condition immune functions, is known as conditioned taste avoidance (CTA). It most commonly pairs a novel taste as a conditioned stimulus (CS) with the injection of an immunosuppressive drug as US. Following one or several CS/US pairings, representation of the CS alone evokes the conditioned response. The conditioned response is reflected on the behavioral level by avoiding ingestion of the CS (CTA) (Garcia, Kimeldorf, & Koelling, 1955; Garcia, Lasiter, Bermudez-Rattoni, & Deems, 1985). Concomitantly to this response, animals display a conditioned suppression of immune functions, similar to that formerly induced by the immunosuppressive drug administered as US (Ader, 2003). More precisely, the parameters affected comprise humoral immune functions such as antibody responses (Alvarez-Borda, Ramirez-Amaya, Perez-Montfort, & Bermudez-Rattoni, 1995; Cohen, Ader, Green, & Bovbjerg, 1979) as well as cellular immune functions such as mitogen-induced lymphocyte proliferation (Lysle, Cunnick, Fowler, & Rabin, 1988), leucocyte numbers (Klosterhalfen & Klosterhalfen, 1987), the circulation of lymphocyte subpopulations (Husband, King, & Brown, 1987), the activity of natural killer cells (Ghanta, Hiramoto, Solvason, & Spector, 1985), or acute phase reactions (Exton, Bull, King, & Husband, 1995; Janz et al., 1996).

Interestingly, studies demonstrated that it is not only possible to suppress immune functions via conditioning protocols but also to induce a learned immune enhancement. A conditioned increase in natural killer cell activity has been shown in rodents (Solvason, Ghanta, & Hiramoto, 1988) and in healthy humans who received a sherbet sweet as a CS together with epinephrine injections (Buske-Kirschbaum, Kirschbaum, Stierle, Jabaij, & Hellhammer, 1994; Buske-Kirschbaum, Kirschbaum, Stierle, Lehnert, & Hellhammer, 1992). Furthermore, it was demonstrated in immunized rodents that enhanced activity of cytotoxic T lymphocytes could be conditioned by pairing the immunization procedure with camphor odor (Hiramoto et al., 1993). In humans, it has been demonstrated that the immune activation in response to human recombinant interferon-gamma (IFN-γ) could be conditioned in healthy subjects, by using oral propylene glycol as CS (Longo et al., 1999). However, employing the endotoxin lipopolysaccharide as a US in a single-trial taste-immune conditioning paradigm in humans demonstrated that behaviorally conditioning induced smell aversion but lacked on inducing learned alterations in cytokine levels (Grigoleit et al., 2012).

Different lines of research discovered that conditioning protocols affected allergic reactions such as bronchial asthma and histamine release in guinea pigs (Justesen, Braun, Garrison, & Pendleton, 1970; Russell et al., 1984). In human studies, type 1 hypersensitivity reactions have been shown to be modulated by behavioral conditioning, as measured by mast cell tryptase (Gauci, Husband, Saxarra, & King, 1994), nasal airflow, histamine release, and subjective symptoms (Barrett, King, & Pang, 2000) and wheal size (Booth, Petrie, & Brook, 1995). One study that is particularly interesting with regard to its possible clinical applicability showed an antiallergic response in patients with house dust mite allergy who underwent a conditioning protocol with the antihistamine desloratadine as US (Goebel, Meykadeh, Kou, Schedlowski, & Hengge, 2008). As desloratadine was replaced by a placebo pill, wheal sizes and subjective measures were influenced even in the absence of a distinct CS (which points to an expectation-induced placebo response). However, reduced basophil activation was only observed in the group that was reexposed to a distinct CS (a novel-tasting drink). This finding supports the assumption that conditioning can modulate unconscious immunological functions, whereas conscious processes can be modulated also by expectation (Benedetti et al., 2003).

3. BEHAVIORALLY CONDITIONED IMMUNOSUPPRESSION

3.1 The Taste Avoidance (CTA) Paradigm With CsA

As stated, conditioning of immune functions commonly involves the pairing of immunomodulatory compounds such as antigens or immunosuppressive drugs (US) with a neutral (conditioning) stimulus (CS). A robust paradigm in rats was established with the calcineurin (CaN) inhibitor and potent immunosuppressive drug CsA as US and a saccharine taste as a CS.

CsA is a drug frequently used in the clinic when immunosuppressive therapy is required (Kapturczak, Meier-Kriesche, & Kaplan, 2004), since it specifically inhibits the activity of the calcium/calmodulin-activated serine-threonine phosphatase CaN in T cells. By dephosphorylating the transcription factor nuclear factor of activated T cells (NF-AT), CaN promotes transcription and production of the cytokines interleukin-2 (IL-2), IFN-γ, and related genes (Steinbach, Reedy, Cramer, Perfect, & Heitman, 2007). CsA inhibits the CaN activity and thus prevents NF-AT from dephosphorylation. Thus, gene and protein expression of the aforementioned cytokines are diminished, leading to suppression of T-lymphocyte proliferation (Batiuk & Halloran, 1997; Halloran, Helms, Kung, & Noujaim, 1999; McCaffrey et al., 1993). However, when the two stimuli CsA and the saccharin solution are repeatedly paired during the acquisition (*learning*) phase of the conditioning paradigm, two things occur during the retrieval (*memory*) phase: (1) on the behavioral level, the rats avoid drinking the saccharine solution. But, more importantly, (2) reexposure to the taste is inducing a learned immunosuppression (Fig. 1A) This paradigm has been repeatedly applied in rodents showing a conditioned reduction in spleen and thymus weight, reduced proliferation rates of spleen lymphocytes, as well as decreased IL-2 and IFN-γ cytokine concentrations (Fig. 1B) (Exton, von Horsten, Schult, et al., 1998; Exton, Von Horsten, Voge, et al., 1998; von Horsten et al., 1998).

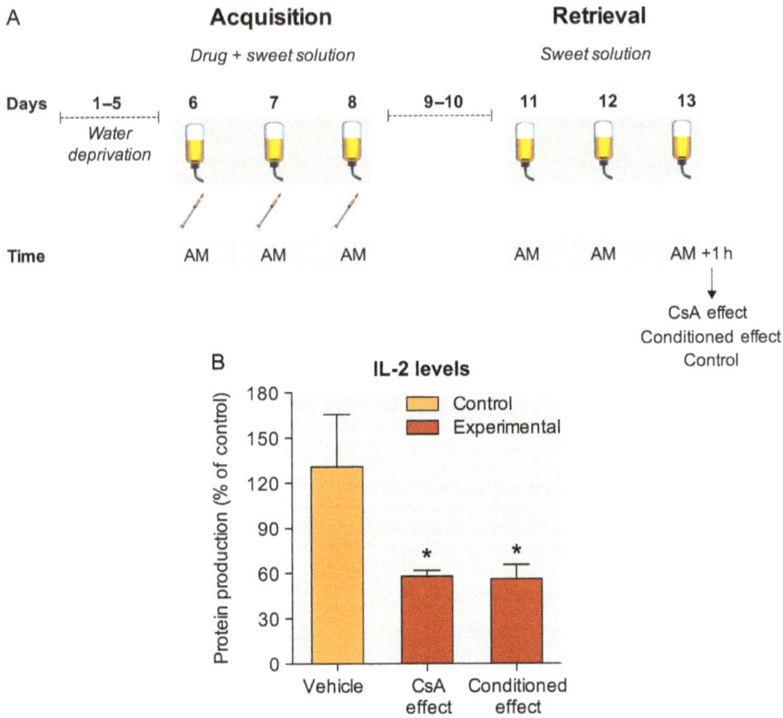

Fig. 1 Principle of behaviorally conditioned immunosuppression with CsA in the rat. (A) Schematic representation of the commonly used experimental protocol. Animals are subjected to a water-deprivation regime for 5 days, allowing them to drink for 15 min in the morning/afternoon each day. During the morning session on the fifth day, the *acquisition* phase starts and a saccharin solution as conditioned stimulus (CS) is paired with an injection of 20 mg/kg CsA as the unconditioned stimulus (US). This protocol is repeated three times with CS–US learning trials separated by 72 h. During *retrieval*, conditioned rats receive the saccharin solution in the morning and water in the afternoon. Before and after each drinking session, the drinking bottles are weighed to measure fluid consumption. One hour after the third CS reexposure animals are sacrificed for immunological analyses. (B) After three CS reexposures, conditioned animals display a marked conditioned suppression of IL-2 cytokine production measured ex vivo in anti-CD3(+)-stimulated splenocytes. *Adapted and modified after Hadamitzky, M., Bosche, K., Wirth, T., Buck, B., Beetz, O., Christians, U., et al. (2016). Memory-updating abrogates extinction of learned immunosuppression.* Brain, Behavior, and Immunity, 52, *40–48.*

3.2 Brain Regions and Structural Pathways

How does an immunopharmacological compound become an US? Either the drug itself directly acts on the brain or the CNS may "sense" the drug-induced immunological changes via two pathways: On the humoral way the involved messengers such as cytokines or prostaglandins can either cross the blood–brain barrier (Banks, 2005) or reach the brain via the circumventricular organs (Goehler, Erisir, & Gaykema, 2006). The neural pathway on the other hand would require activation of neurons in the periphery that signal to the brain. The vagus nerve is well positioned to detect and inform the brain about changes in the immune status (Tracey, 2002). It has been experimentally demonstrated that many effects in the CNS that are induced by changes in the immune system can be alleviated by prior vagotomy (Konsman, Luheshi, Bluthe, & Dantzer, 2000; Luheshi et al., 2000). Thus, it is proposed that in CsA-induced conditioned immunosuppression, the vagus nerve may act as a major afferent route, as well. A former study analyzed whether CsA is sensed via the chemosensitive vagus nerve or whether CsA directly acts on the brain. These data revealed that a single peripheral administration of CsA increases neuronal activity in the insular cortex and the amygdala as reflected by changes in intracerebral EEG recordings, c-Fos expression, and amygdaloid noradrenaline release. However, this increased neuronal activity was not affected by prior vagal deafferentation but rather seems to be partially induced by direct action of CsA on cortico-amygdaloid structures (Pacheco-Lopez et al., 2013). Together, these data indicate that CsA as an US may directly act on the brain by a still unknown transduction mechanism.

The sympathetic nervous system is one of the pathways via which the CNS is communicating with the peripheral immune system in general. Specifically, this seems to be the major efferent pathway in the taste-immune conditioning paradigm, since surgical denervation of the sympathetic nerve innervating the spleen blocked the conditioned reduction of cytokine production and lymphocyte proliferation (Exton, von Horsten, Schult, et al., 1998). However, a different study suggests that this is not the only efferent pathway for conditioned immunosuppressive effects as denervation of the splenic nerve still allows a conditioned reduction of ear swelling in contact hypersensitivity (Exton et al., 2000). Regarding involved transmitters and receptors, it has been revealed that noradrenaline is the predominant neurotransmitter mediating the learned immunosuppressive effects via beta-adrenoceptors (Exton et al., 2002). Moreover, activation of beta-adrenoceptors inhibits calcineurin activity in CD4(+) T lymphocytes (Riether et al., 2011), demonstrating that both the conditioned response and the pharmacological agent CsA are inducing immunosuppression via inhibition of calcineurin activity, however, using different intracellular pathways.

The insular cortex and the amygdala as well as the ventromedial nucleus of the hypothalamus (VMH) are essential brain structures centrally involved in conditioned immunosuppression (Pacheco-Lopez et al., 2005). Experimentally induced excitotoxic lesions confirmed that the insular cortex is necessary for both acquisition and retrieval of the conditioned response (Bermudez-Rattoni & McGaugh, 1991; Cubero, Thiele, & Bernstein, 1999; Pacheco-Lopez et al., 2005). In contrast, the amygdala only seems to mediate the input of visceral information necessary for acquisition, whereas the VMH appears to be involved only in the output pathway to the immune system.

4. TRANSLATIONAL ASPECTS AND CLINICAL RELEVANCE OF ASSOCIATIVE LEARNING PROTOCOLS

The findings obtained from experimental animal studies together with deeper understanding of the mechanisms responsible for conditioned immune functions provided the basis for expanding this interesting placebo phenomenon to being applied in humans (Vits et al., 2011). In these studies, healthy men received CsA as US together with an unusual flavored and colored drink (CS). After several CS–US pairings, CsA was replaced by placebo capsules, which were then consumed together with the distinct CS (Fig. 2A). A conditioned immunosuppression following repeated reexposure to the CS was demonstrated by reduced IL-2 (Fig. 2B) and IFN-γ cytokine levels and mRNA expression, as well as inhibited T cell proliferation (Goebel et al., 2002). Importantly, it was possible to repeatedly recall this conditioned immunosuppression by exposing the conditioned subjects to the CS again after an 11-day break (Wirth et al., 2011). Moreover, plasma noradrenaline concentration, state anxiety, and baseline levels of IL-2 predicted nearly 60% of the conditioned immunosuppressive response (Ober et al., 2012), indicating the presence of biological and psychological predictors of conditioned placebo responses in general and the conditioned immune responses in particular. Experimental evidence in humans further confirmed previous observations in rodents (Niemi et al., 2007), showing that the induction of a reliable conditioned response on the immune system requires multiple CS–US combinations during acquisition and evocation (Albring et al., 2012; Goebel et al., 2005; Grigoleit et al., 2012) (Fig. 2A). Importantly, by using the CsA-taste conditioning paradigm in humans, another study revealed that a conditioned immunosuppression could not be induced by just manipulating the expectancy of the test subjects (Albring

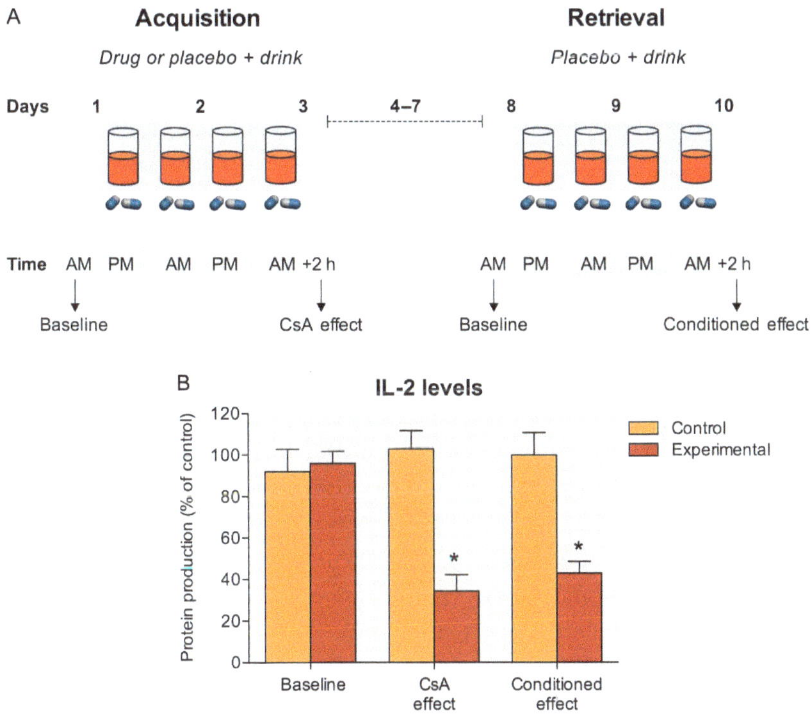

Fig. 2 Principle of behaviorally conditioned immunosuppression with CsA in humans. (A) Schematic representation of the commonly used experimental protocol. During *acquisition* on days 1–3, subjects receive CsA (2.5 mg/kg/12 h) as conditioned stimulus (CS) paired with a colored, novel-tasting drink as conditioned stimulus (CS). *Retrieval* starts 4 days following the *acquisition* phase with participants undergoing four reexposures to the drink only, now paired with placebo capsules on days 8–10. (B) Reduction of IL-2 cytokine production of anti-CD3(+)-stimulated peripheral blood mononuclear cells measured at baseline (day 1), after four pairings of CsA (2.5 mg/kg/12 h) with a novel drink (CS) (CsA effect on day 3), and after four reexposures every 12 h to the novel drink (CS) without CsA (conditioned effect on day 10). *Adapted and modified after Goebel, M. U., Trebst, A. E., Steiner, J., Xie, Y. F., Exton, M. S., Frede, S., et al. (2002). Behavioral conditioning of immunosuppression is possible in humans.* FASEB Journal, 16, *1869–1873; Albring, A., Wendt, L., Benson, S., Witzke, O., Kribben, A., Engler, H., et al. (2012). Placebo effects on the immune response in humans. The role of learning and expectation.* PLoS One, 7, *e49477.*

et al., 2012). The underlying pathways and biological mechanisms steering the learned placebo response on immune functions for this specific model are suggested to be similar to the ones elucidated and described in rodents (Schedlowski & Pacheco-Lopez, 2010).

Behavioral conditioning of immune functions is not only an excellent model to study the bidirectional communication between the CNS and the peripheral immune system, but it also offers a new therapeutic approach that might allow reducing the amount of medication required by supporting the therapeutic benefit through learned placebo effects. Studies in rodents clearly document the therapeutic potential of conditioned immunosuppression, showing that the outcome of various diseases can be positively affected through learning protocols. For instance, the mortality rate and the development of proteinuria in mice with lupus erythematosus were retarded in conditioned animals receiving a saccharine solution (CS) together with the immunosuppressive drug cyclophosphamide (US) (Ader & Cohen, 1982; Klosterhalfen & Klosterhalfen, 1990). Likewise, other studies in rats showed that conditioning procedures reduced inflammatory processes in a adjuvant arthritis model (Klosterhalfen & Klosterhalfen, 1983), prolonged survival of skin allograft (Gorczynski, 1990; Grochowicz et al., 1991), and delayed tumor growth (Ghanta, Hiramoto, Solvason, Soong, & Hiramoto, 1990; Ghanta, Hiramoto, Solvason, & Spector, 1987; Ghanta, Miura, Hiramoto, & Hiramoto, 1988).

In humans, antihistaminergic properties of the H_1-receptor antagonist desloratadine were behaviorally conditioned in patients suffering from allergic house dust mite rhinitis. Here, learned placebo effects significantly reduced subjective symptoms, the allergic response to the skin prick test, and basophile activation (Goebel et al., 2008). This finding was moreover confirmed in a different study where conditioned placebo effects induced by both expectation and learning affected allergic responses (Vits et al., 2013).

5. PERSISTENCE OF CONDITIONED PLACEBO EFFECTS

However, a prerequisite for applying learning protocols in clinical settings not just as temporal but also as continuous intervention is the reproducibility and/or preservation of conditioned immunosuppressive effects over time. Regarding the recall of the learned immunosuppression, one study in healthy participants showed that the conditioning effect is not present after one reexposure to the CS, but after four (Albring et al., 2012). This is in line with an earlier study in rats proposing that the engram is strengthened with repeated reexposures to the CS (Niemi et al., 2007). A different acknowledgeable and still unsolved challenge not only in CTA but also in the context of conditioned immunosuppression is the fact that the frequency or intensity of a conditioned response in general will naturally decrease over time following the withdrawal of reinforcement (Berman & Dudai, 2001;

Pavlov, 1927). Hence, before considering a systematic application of conditioning procedures as supportive therapeutic option in clinical settings, the process of extinction in taste-immune learning with CsA and the associated conditioned immunosuppressive response in particular need to be elucidated in more detail with sophisticated experimental studies (Hadamitzky, Engler, & Schedlowski, 2013). In this regard, it has recently been demonstrated that extinction of a CsA-induced taste avoidance and, more importantly, the concomitantly conditioned suppression of IL-2 and IFN-γ cytokine responses could be abrogated when a subtherapeutic dose of the US, the immunosuppressant CsA, was administered as a reminder cue in parallel with the CS during retrieval (Hadamitzky et al., 2016). Importantly, this updated learned immunosuppressive response and its maintenance is also of clinical relevance, since it significantly prolonged in rats the survival time of heterotopically transplanted, major histocompatibility complex-incompatible hearts (Exton, von Horsten, Schult, et al., 1998; Hadamitzky et al., 2016). Here, neural processes mediating memory updating have apparently been transferred to peripheral immune functions via this treatment protocol (Lückemann, Unteroberdorster, Kirchhof, Schedlowski, & Hadamitzky, 2017). Putting these findings into learning theory, this synergistic effect may be attributable to memory updating or a reconsolidation-like process of the memory trace (immunosuppression), when it is activated via reexposure to the CS (Agren, 2014; Alberini & LeDoux, 2013; de Carvalho Myskiw, Furini, Schmidt, Ferreira, & Izquierdo, 2015; Schiller et al., 2010; Tronson & Taylor, 2007). For memory updating it is suggested that a recall of memories turns them into a transient labile state in which they are susceptible for alterations and thus may be modified, erased, or strengthened within a short time window (Nader & Einarsson, 2010; Nader, Schafe, & Le Doux, 2000). Thus, a consistent "reminder" given by the otherwise ineffective subtherapeutic dose of CsA might have strengthened the engram, counteracting the extinction processes thereby preserving the conditioned response (Lückemann et al., 2017).

However, before participation in an immunosuppressive learning paradigm as a supportive therapy, one needs to overcome another important obstacle. In clinical routine, the majority of patients are already under pharmacological therapy with CaN inhibitors such as CsA, and it is not known whether and to what extent these patients could still develop a conditioned immunosuppression. A distinct experimental approach addressed this question and preexposed animals either to the US or the CS three times before initiating acquisition of a CsA-induced taste avoidance paradigm. On the behavioral level a significantly diminished CTA was observed when animals were preexposed to the US or the CS prior to conditioning. However, in contrast, US or CS preexposure did not affect the conditioned suppression of IL-2 cytokine production (Lückemann et al., 2016). These results support earlier observations, showing that reexposure to the taste, but not a significant taste avoidance, is essential for a conditioned immunosuppression indicated by weak or even absent correlations between the CTA and the learned immunosuppressive response (Ader & Cohen, 1982; Bovbjerg, Ader, & Cohen, 1984; Bovbjerg, Kim, Siskind, & Weksler, 1987; Exton, Schult, Donath, et al., 1998; Schedlowski & Pacheco-Lopez, 2010).

6. FUTURE DIRECTIONS AND LIMITATIONS

Employing associative learning protocols in clinical routine, the goal will be to support pharmacological therapies rather than to replace it. PCDR, in which a certain amount of the usual dose of a drug is replaced by a placebo, is a concept that might maintain the efficacy of a treatment while reducing the dose of medication and unwanted side effects (Doering & Rief, 2012). In one study, the partial replacement (25% or 50%) of corticosteroids with a placebo in patients with psoriasis induced a similar symptom reduction compared to the full-dose treatment (Ader et al., 2010). Using this "partial reinforcement" paradigm (Acosta, Thiel, Sanabria, Browning, & Neisewander, 2008), it was also demonstrated in kids with Attention Deficit Hyperactivity Disorder (Sandler, Glesne, & Bodfish, 2010), and in a study of experimentally induced coughing (Leech, Mazzone, & Farrell, 2012), that drug efficiency can be maintained, while drug dosage was reduced.

While these examples confirm positive short-term effects, the potential for reducing negative consequences of long-term drug applications needs to be analyzed in more detail. More precisely, the physiological systems prone for conditioning need to be explored, and the reinforcement schedules to be used to achieve optimal effects by preventing habituation/extinction of learned immunosuppression must be determined (Lückemann et al., 2017). For some conditions, partial reinforcement with full-dose treatment results show stable effects, while for other conditions presentation of the CS together with subtherapeutic doses of the US may be the better choice. In this context, a therapeutic approach has been tested in multiple sclerosis where patients received cyclophosphamide treatment together with a CS.

Interestingly, they displayed a decrease in peripheral leukocyte counts when reexposed to the CS together with a subtherapeutic dose of the drug (Giang et al., 1996). This finding indicates that the memory trace of the conditioned immunosuppression can also be preserved in humans by triggering memory updating or reconsolidation-like mechanisms during reactivating by reexposure to the CS together with the subtherapeutic reminder cues.

Even though the use of learning paradigms may provide a valuable tool in a supportive therapy together with conventional pharmacological regimens, its clinical application in immune therapy is still controversial and limited. However, caution is needed in generalizing learned immunopharmacological effects. Conditioned compensatory responses that are opposite to those induced by the medication may arise as a result of tolerance, toward the drug (Siegel, Baptista, Kim, McDonald, & Weise-Kelly, 2000). Due to the fact that pharmacological stimulations can initiate adaptive responses that compensate for the primary drug effect, further studies need to clarify whether predrug cues may elicit paradoxical responses (Colloca, Enck, & DeGrazia, 2016). Another major issue is raised by the question whether the partially severe side effects of immunomodulating compounds are also conditioned. This aspect is of particular importance since patients who participate in placebo-controlled clinical trials often report side effects similar to the ones experienced by patients receiving full-dose treatment (Enck et al., 2013). For one, this phenomenon may arise due to mere communication of potential adverse effects in the informed consent process (Barsky, Saintfort, Rogers, & Borus, 2002). For another, such effects can also be produced in clinical practice by negative expectations relating to disclosures of possible side effects from treatment (Myers, Cairns, & Singer, 1987). More precisely, the information that a prescribed drug may cause side effects can itself evoke the same side effect without having the pharmacological properties of the drug (Flaten, Simonsen, & Olsen, 1999; Luparello, Leist, Lourie, & Sweet, 1970, Mondaini et al., 2007). Albeit no direct experimental evidence does exist regarding conditioned side effects during behaviorally conditioned immunosuppression, it is very likely that this is not the case. The first clue is given by the fact that immune responses in conditioned immunosuppression are mediated via distinct intracellular mechanisms in T cells compared to the effects on the immune system induced by the drug CsA itself (Riether et al., 2008). This aspect indicates that the main toxic side effect of CsA (nephrotoxicity) is caused by the substance itself (Halloran, 2004; Halloran et al., 1999). However, to definitely clarify this issue further conditioning experiments need to be conducted, elucidating in detail the effect of conditioning of pharmacological properties and side effects of the compounds used as US. Until now this paradigm is restricted to the mode of action of calcineurin inhibitors such as CsA. Thus, it needs to be evaluated whether and to what extent the same mechanisms are also responsible for learned pharmacological responses of immunomodulating drugs with distinct cell signaling pathways.

An even more challenging point regarding the application of placebos as a supportive therapy of pharmacological regimens are ethical limitations. For one, a placebo- or "no treatment," especially in cases of severe disease, is unacceptable to use in sick patients (World Medical Association, 2013). For another, the strongest ethical concern about placebo treatments is the use of deceptive information that is given to the patient (Miller & Colloca, 2009). The only exception would be a case when giving deceptive information prevents harming innocent persons. However, in terms of morality, giving deceptive information is considered as *lying* and thus wrong. To overcome this problem, an authorized deception could be negotiated with the patient before initiating treatment. Such a condition would also strengthen the ethical legitimacy of placebo treatments. Thus, one possibility could be the patients' agreement on interspersed use of placebos during treatment without knowing or notification of which administration contains a placebo (Miller & Kaptchuk, 2008). Alternatively, such an "informed consent" procedure could be modified by briefly delineating that the offered treatment may involve a covert drug-dose reduction (Doering & Rief, 2012; Enck et al., 2013; Schedlowski, Enck, Rief, & Bingel, 2015).

Within the framework of placebo responses in general, two questions arise: (a) who will respond to a "placebo therapy" and (b) are there predictors of placebo responses (Kaptchuk et al., 2008)? This aspect is of particular importance, since quick decisions regarding which person would benefit from a commensurate therapy could be made by distinguishing and thus identifying between "placebo responders" and "placebo nonresponders." Evidence for predictors to distinguish between the two has been identified in the case of conditioned immunosuppression. Here, recent work shows that state anxiety and plasma noradrenaline together with IL-2 concentrations at baseline level predicted almost 60% of the variance in the conditioned response, measured by suppressed IL-2 levels (Ober et al., 2012). Applicability and effectiveness of employing conditioning procedures together with immunopharmacological regimen across various immune-related diseases are not thoroughly investigated yet. Also, little is known about how long placebos may be used to supplement long-term drug intake and whether its effectiveness may diminish or disappear. Thus, further human studies are necessary to define when the use of placebos is effective and in which specific medical conditions. To rule out confounding changes in the efficacy outcome measures (e.g., spontaneous remission, regression to the mean), study protocols as presented in Fig. 2 should comprise three arms: (1) full-dose medication; (2) a partial reinforcement schedule, where the full dose is given 25%–50% of the time and replaced by placebos the other times; and (3) a partial reinforcement control schedule, where the full dose is given 25%–50% without placebo replacement. Moreover, to obtain a meaningful assessment of placebos in immunological diseases, safety, optimization, and feasibility studies are needed to rule out limiting factors such as the irreversibility of a disease, inability to adjust or optimize treatment reduction, or the pharmacokinetic properties of the drug (Colloca et al., 2016).

Nevertheless, in summary, experimental data in rodents demonstrate that learning/conditioning paradigms may become a valuable tool as supportive therapy together with conventional immunopharmacological regimens. Likewise, first observations in human volunteers and patients show that this research has a clear translational relevance and thus offers great potential for therapeutic applications that could maximize the patient's benefit by minimizing unwanted aspects of the treatment such as intensive medication costs or unwanted drug side effects.

ACKNOWLEDGMENTS

This work was funded by a German Research Foundation (DFG) grant (SFB1280/A18) to M.H. and M.S. The authors confirm that this chapter contents have no conflict of interest.

REFERENCES

Acosta, J. I., Thiel, K. J., Sanabria, F., Browning, J. R., & Neisewander, J. L. (2008). Effect of schedule of reinforcement on cue-elicited reinstatement of cocaine-seeking behavior. *Behavioural Pharmacology*, *19*, 129–136.

Ader, R. (1976). Conditioned adrenocortical steroid elevations in the rat. *Journal of Comparative and Physiological Psychology*, *90*, 1156–1163.

Ader, R. (2003). Conditioned immunomodulation: Research needs and directions. *Brain, Behavior, and Immunity*, *17*, 51–57.

Ader, R., & Cohen, N. (1975). Behaviorally conditioned immunosuppression. *Psychosomatic Medicine*, *37*, 333–340.

Ader, R., & Cohen, N. (1982). Behaviorally conditioned immunosuppression and murine systemic lupus erythematosus. *Science*, *215*, 1534–1536.

Ader, R., Mercurio, M. G., Walton, J., James, D., Davis, M., Ojha, V., et al. (2010). Conditioned pharmacotherapeutic effects: A preliminary study. *Psychosomatic Medicine*, *72*, 192–197.

Agren, T. (2014). Human reconsolidation: A reactivation and update. *Brain Research Bulletin*, *105*, 70–82.

Alberini, C. M., & LeDoux, J. E. (2013). Memory reconsolidation. *Current Biology*, *23*, R746–R750.

Albring, A., Wendt, L., Benson, S., Witzke, O., Kribben, A., Engler, H., et al. (2012). Placebo effects on the immune response in humans: The role of learning and expectation. *PLoS One*, *7*, e49477.

Alvarez-Borda, B., Ramirez-Amaya, V., Perez-Montfort, R., & Bermudez-Rattoni, F. (1995). Enhancement of antibody production by a learning paradigm. *Neurobiology of Learning and Memory*, *64*, 103–105.

Banks, W. A. (2005). Blood-brain barrier transport of cytokines: A mechanism for neuropathology. *Current Pharmaceutical Design*, *11*, 973–984.

Barrett, J. E., King, M. G., & Pang, G. (2000). Conditioning rhinitis in allergic humans. *Annals of the New York Academy of Sciences*, *917*, 853–859.

Barsky, A. J., Saintfort, R., Rogers, M. P., & Borus, J. F. (2002). Nonspecific medication side effects and the nocebo phenomenon. *JAMA*, *287*, 622–627.

Batiuk, T. D., & Halloran, P. F. (1997). The downstream consequences of calcineurin inhibition. *Transplantation Proceedings*, *29*, 1239–1240.

Benedetti, F., Carlino, E., & Pollo, A. (2011). How placebos change the patient's brain. *Neuropsychopharmacology*, *36*, 339–354.

Benedetti, F., Pollo, A., Loplano, L., Lanotte, M., Vighetti, S., & Rainero, I. (2003). Conscious expectation and unconscious conditioning in analgesic, motor, and hormonal placebo/nocebo responses. *The Journal of Neuroscience*, *23*, 4315–4323.

Berman, D. E., & Dudai, Y. (2001). Memory extinction, learning anew, and learning the new: Dissociations in the molecular machinery of learning in cortex. *Science*, *291*, 2417–2419.

Bermudez-Rattoni, F. (2004). Molecular mechanisms of taste-recognition memory. *Nature Reviews. Neuroscience*, *5*, 209–217.

Bermudez-Rattoni, F., & McGaugh, J. L. (1991). Insular cortex and amygdala lesions differentially affect acquisition on inhibitory avoidance and conditioned taste aversion. *Brain Research*, *549*, 165–170.

Blalock, J. E., & Smith, E. M. (2007). Conceptual development of the immune system as a sixth sense. *Brain, Behavior, and Immunity*, *21*, 23–33.

Booth, R. J., Petrie, K. J., & Brook, R. J. (1995). Conditioning allergic skin responses in humans: A controlled trial. *Psychosomatic Medicine*, *57*, 492–495.

Bovbjerg, D., Ader, R., & Cohen, N. (1984). Acquisition and extinction of conditioned suppression of a graft-vs-host response in the rat. *The Journal of Immunology*, *132*, 111–113.

Bovbjerg, D., Kim, Y. T., Siskind, G. W., & Weksler, M. E. (1987). Conditioned suppression of plaque-forming cell responses with cyclophosphamide. The role of taste aversion. *Annals of the New York Academy of Sciences*, *496*, 588–594.

Buske-Kirschbaum, A., Kirschbaum, C., Stierle, H., Jabaij, L., & Hellhammer, D. (1994). Conditioned manipulation of natural killer (NK) cells in humans using a discriminative learning protocol. *Biological Psychology*, *38*, 143–155.

Buske-Kirschbaum, A., Kirschbaum, C., Stierle, H., Lehnert, H., & Hellhammer, D. (1992). Conditioned increase of natural killer cell activity (NKCA) in humans. *Psychosomatic Medicine*, *54*, 123–132.

Cohen, N., Ader, R., Green, N., & Bovbjerg, D. (1979). Conditioned suppression of a thymus-independent antibody response. *Psychosomatic Medicine*, *41*, 487–491.

Colloca, L., Enck, P., & DeGrazia, D. (2016). Relieving pain using dose-extending placebos: A scoping review. *Pain*, *157*, 1590–1598.

Colloca, L., & Miller, F. G. (2011). Role of expectations in health. *Current Opinion in Psychiatry*, *24*, 149–155.

Cubero, I., Thiele, T. E., & Bernstein, I. L. (1999). Insular cortex lesions and taste aversion learning: Effects of conditioning method and timing of lesion. *Brain Research*, *839*, 323–330.

Dantzer, R., O'Connor, J. C., Freund, G. G., Johnson, R. W., & Kelley, K. W. (2008). From inflammation to sickness and depression: When the immune system subjugates the brain. *Nature Reviews. Neuroscience*, *9*, 46–56.

de Carvalho Myskiw, J., Furini, C. R., Schmidt, B., Ferreira, F., & Izquierdo, I. (2015). Extinction learning, which consists of the inhibition of retrieval, can be learned without retrieval. *Proceedings of the National Academy of Sciences of the United States of America*, *112*, E230–E233.

Doering, B. K., & Rief, W. (2012). Utilizing placebo mechanisms for dose reduction in pharmacotherapy. *Trends in Pharmacological Sciences*, *33*, 165–172.

Enck, P., Bingel, U., Schedlowski, M., & Rief, W. (2013). The placebo response in medicine: Minimize, maximize or personalize? *Nature Reviews. Drug Discovery*, *12*, 191–204.

Exton, M. S., Bull, D. F., King, M. G., & Husband, A. J. (1995). Modification of body temperature and sleep state using behavioral conditioning. *Physiology & Behavior*, *57*, 723–729.

Exton, M. S., Elfert, A., Jeong, W.-Y., Bull, D. F., Westermann, J., & Schedlowski, M. (2000). Conditioned suppression of contact sensitivity is independent of sympathetic splenic innervation. *The American Journal of Physiology. Endocrinology and Metabolism*, *279*, R1310–R1315.

Exton, M. S., Gierse, C., Meier, B., Mosen, M., Xie, Y. F., Frede, S., et al. (2002). Behaviorally conditioned immunosuppression in the rat is regulated via noradrenaline and β-adrenoceptors. *Journal of Neuroimmunology*, *131*, 21–30.

Exton, M. S., Schult, M., Donath, S., Strubel, T., Nagel, E., Westermann, J., et al. (1998). Behavioral conditioning prolongs heart allograft survival in rats. *Transplantation Proceedings*, *30*, 2033.

Exton, M. S., von Horsten, S., Schult, M., Voge, J., Strubel, T., Donath, S., et al. (1998). Behaviorally conditioned immunosuppression using cyclosporine A: Central nervous system reduces IL-2 production via splenic innervation. *Journal of Neuroimmunology*, *88*, 182–191.

Exton, M. S., Von Horsten, S., Voge, J., Westermann, J., Schult, M., Nagel, E., et al. (1998). Conditioned taste aversion produced by cyclosporine A: Concomitant reduction in lymphoid organ weight and splenocyte proliferation. *Physiology & Behavior*, *63*, 241–247.

Flaten, M. A., Simonsen, T., & Olsen, H. (1999). Drug-related information generates placebo and nocebo responses that modify the drug response. *Psychosomatic Medicine*, *61*, 250–255.

Garcia, J., Kimeldorf, D. J., & Koelling, R. A. (1955). Conditioned aversion to saccharin resulting from exposure to gamma radiation. *Science*, *122*, 157–158.

Garcia, J., Lasiter, P. S., Bermudez-Rattoni, F., & Deems, D. A. (1985). A general theory of aversion learning. *Annals of the New York Academy of Sciences*, *443*, 8–21.

Gauci, M., Husband, A. J., Saxarra, H., & King, M. G. (1994). Pavlovian conditioning of nasal tryptase release in human subjects with allergic rhinitis. *Physiology & Behavior*, *55*, 823–825.

Ghanta, V. K., Hiramoto, N. S., Solvason, H. B., Soong, S. J., & Hiramoto, R. N. (1990). Conditioning: A new approach to immunotherapy. *Cancer Research*, *50*, 4295–4299.

Ghanta, V. K., Hiramoto, R. N., Solvason, H. B., & Spector, N. H. (1985). Neural and environmental influences on neoplasia and conditioning of NK activity. *Journal of Immunology*, *135*, 848s–852s.

Ghanta, V., Hiramoto, R. N., Solvason, B., & Spector, N. H. (1987). Influence of conditioned natural immunity on tumor growth. *Annals of the New York Academy of Sciences, 496*, 637–646.

Ghanta, V. K., Miura, T., Hiramoto, N. S., & Hiramoto, R. N. (1988). Augmentation of natural immunity and regulation of tumor growth by conditioning. *Annals of the New York Academy of Sciences, 521*, 29–42.

Giang, D. W., Goodman, A. D., Schiffer, R. B., Mattson, D. H., Petrie, M., Cohen, N., et al. (1996). Conditioning of cyclophosphamide-induced leukopenia in humans. *The Journal of Neuropsychiatry and Clinical Neurosciences, 8*, 194–201.

Goebel, M. U., Hübell, D., Kou, W., Janssen, O. E., Katsarava, Z., Limmroth, V., et al. (2005). Behavioral conditioning with interferon beta-1a in humans. *Physiology and Behavior, 84*, 807–814.

Goebel, M. U., Meykadeh, N., Kou, W., Schedlowski, M., & Hengge, U. R. (2008). Behavioral conditioning of antihistamine effects in patients with allergic rhinitis. *Psychotherapy and Psychosomatics, 77*, 227–234.

Goebel, M. U., Trebst, A. E., Steiner, J., Xie, Y. F., Exton, M. S., Frede, S., et al. (2002). Behavioral conditioning of immunosuppression is possible in humans. *The FASEB Journal, 16*, 1869–1873.

Goehler, L. E., Erisir, A., & Gaykema, R. P. (2006). Neural-immune interface in the rat area postrema. *Neuroscience, 140*, 1415–1434.

Gorczynski, R. M. (1990). Conditioned enhancement of skin allografts in mice. *Brain, Behavior, and Immunity, 4*, 85–92.

Grigoleit, J. S., Kullmann, J. S., Winkelhaus, A., Engler, H., Wegner, A., Hammes, F., et al. (2012). Single-trial conditioning in a human taste-endotoxin paradigm induces conditioned odor aversion but not cytokine responses. *Brain, Behavior, and Immunity, 26*, 234–238.

Grochowicz, P. M., Schedlowski, M., Husband, A. J., King, M. G., Hibberd, A. D., & Bowen, K. M. (1991). Behavioral conditioning prolongs heart allograft survival in rats. *Brain, Behavior, and Immunity, 5*, 349–356.

Hadamitzky, M., Bosche, K., Wirth, T., Buck, B., Beetz, O., Christians, U., et al. (2016). Memory-updating abrogates extinction of learned immunosuppression. *Brain, Behavior, and Immunity, 52*, 40–48.

Hadamitzky, M., Engler, H., & Schedlowski, M. (2013). Learned immunosuppression: Extinction, renewal, and the challenge of reconsolidation. *Journal of Neuroimmune Pharmacology, 8*, 180–188.

Halloran, P. F. (2004). Immunosuppressive drugs for kidney transplantation. *The New England Journal of Medicine, 351*, 2715–2729.

Halloran, P. F., Helms, L. M., Kung, L., & Noujaim, J. (1999). The temporal profile of calcineurin inhibition by cyclosporine in vivo. *Transplantation, 68*, 1356–1361.

Hiramoto, R., Ghanta, V., Solvason, B., Lorden, J., Hsueh, C. M., Rogers, C., et al. (1993). Identification of specific pathways of communication between the CNS and NK cell system. *Life Sciences, 53*, 527–540.

Husband, A. J., King, M. G., & Brown, R. (1987). Behaviourally conditioned modification of T cell subset ratios in rats. *Immunology Letters, 14*, 91–94.

Janz, L. J., Green-Johnson, J., Murray, L., Vriend, C. Y., Nance, D. M., Greenberg, A. H., et al. (1996). Pavlovian conditioning of LPS-induced responses: Effects on corticosterone, splenic NE, and IL-2 production. *Physiology & Behavior, 59*, 1103–1109.

Justesen, D. R., Braun, E. W., Garrison, R. G., & Pendleton, R. B. (1970). Pharmacological differentiation of allergic and classically conditioned asthma in the guinea pig. *Science, 170*, 864–866.

Kaptchuk, T. J., Kelley, J. M., Deykin, A., Wayne, P. M., Lasagna, L. C., Epstein, I. O., et al. (2008). Do "placebo responders" exist? *Contemporary Clinical Trials, 29*, 587–595.

Kapturczak, M. H., Meier-Kriesche, H. U., & Kaplan, B. (2004). Pharmacology of calcineurin antagonists. *Transplantation Proceedings, 36*, 25S–32S.
Klosterhalfen, S., & Enck, P. (2006). Psychobiology of the placebo response. *Autonomic Neuroscience, 125*, 94–99.
Klosterhalfen, W., & Klosterhalfen, S. (1983). Pavlovian conditioning of immunosuppression modifies adjuvant arthritis in rats. *Behavioral Neuroscience, 97*, 663–666.
Klosterhalfen, S., & Klosterhalfen, W. (1987). Classically conditioned effects of cyclophosphamide on white blood cell counts in rats. *Annals of the New York Academy of Sciences, 496*, 569–577.
Klosterhalfen, S., & Klosterhalfen, W. (1990). Conditioned cyclosporine effects but not conditioned taste aversion in immunized rats. *Behavioral Neuroscience, 104*, 716–724.
Konsman, J. P., Luheshi, G. N., Bluthe, R. M., & Dantzer, R. (2000). The vagus nerve mediates behavioural depression, but not fever, in response to peripheral immune signals; a functional anatomical analysis. *The European Journal of Neuroscience, 12*, 4434–4446.
Leech, J., Mazzone, S. B., & Farrell, M. J. (2012). The effect of placebo conditioning on capsaicin-evoked urge to cough. *Chest, 142*, 951–957.
Longo, D. L., Duffey, P. L., Kopp, W. C., Heyes, M. P., Alvord, W. G., Sharfman, W. H., et al. (1999). Conditioned immune response to interferon-gamma in humans. *Clinical Immunology, 90*, 173–181.
Lückemann, L., Bosche, K., Engler, H., Schwitalla, J. C., Hadamitzky, M., & Schedlowski, M. (2016). Pre-exposure to the unconditioned or conditioned stimulus does not affect learned immunosuppression in rats. *Brain, Behavior, and Immunity, 51*, 252–257.
Lückemann, L., Unteroberdorster, M., Kirchhof, J., Schedlowski, M., & Hadamitzky, M. (2017). Applications and limitations of behaviorally conditioned immunopharmacological responses. *Neurobiology of Learning and Memory, 142*, 91–98.
Luheshi, G. N., Bluthe, R. M., Rushforth, D., Mulcahy, N., Konsman, J. P., Goldbach, M., et al. (2000). Vagotomy attenuates the behavioural but not the pyrogenic effects of interleukin-1 in rats. *Autonomic Neuroscience, 85*, 127–132.
Luparello, T. J., Leist, N., Lourie, C. H., & Sweet, P. (1970). The interaction of psychologic stimuli and pharmacologic agents on airway reactivity in asthmatic subjects. *Psychosomatic Medicine, 32*, 509–513.
Lysle, D. T., Cunnick, J. E., Fowler, H., & Rabin, B. S. (1988). Pavlovian conditioning of shock-induced suppression of lymphocyte reactivity: Acquisition, extinction, and preexposure effects. *Life Sciences, 42*, 2185–2194.
McCaffrey, P. G., Luo, C., Kerppola, T. K., Jain, J., Badalian, T. M., Ho, A. M., et al. (1993). Isolation of the cyclosporin-sensitive T cell transcription factor NFATp. *Science, 262*, 750–754.
Miller, F. G., & Colloca, L. (2009). The legitimacy of placebo treatments in clinical practice: Evidence and ethics. *The American Journal of Bioethics, 9*, 39–47.
Miller, F. G., & Kaptchuk, T. J. (2008). Deception of subjects in neuroscience: An ethical analysis. *The Journal of Neuroscience, 28*, 4841–4843.
Mondaini, N., Gontero, P., Giubilei, G., Lombardi, G., Cai, T., Gavazzi, A., et al. (2007). Finasteride 5 mg and sexual side effects: How many of these are related to a nocebo phenomenon? *The Journal of Sexual Medicine, 4*, 1708–1712.
Myers, M. G., Cairns, J. A., & Singer, J. (1987). The consent form as a possible cause of side effects. *Clinical Pharmacology and Therapeutics, 42*, 250–253.
Nader, K., & Einarsson, E. Ö. (2010). Memory reconsolidation: An update. *Annals of the New York Academy of Sciences, 1191*, 27–41.
Nader, K., Schafe, G. E., & Le Doux, J. E. (2000). Fear memories require protein synthesis in the amygdala for reconsolidation after retrieval. *Nature, 406*, 722–726.

Niemi, M. B., Harting, M., Kou, W., Del Rey, A., Besedovsky, H. O., Schedlowski, M., et al. (2007). Taste-immunosuppression engram: Reinforcement and extinction. *Journal of Neuroimmunology, 188*, 74–79.

Ober, K., Benson, S., Vogelsang, M., Bylica, A., Gunther, D., Witzke, O., et al. (2012). Plasma noradrenaline and state anxiety levels predict placebo response in learned immunosuppression. *Clinical Pharmacology and Therapeutics, 91*, 220–226.

Pacheco-Lopez, G., Doenlen, R., Krugel, U., Arnold, M., Wirth, T., Riether, C., et al. (2013). Neurobehavioural activation during peripheral immunosuppression. *The International Journal of Neuropsychopharmacology, 16*, 137–149.

Pacheco-Lopez, G., Niemi, M. B., Kou, W., Harting, M., Fandrey, J., & Schedlowski, M. (2005). Neural substrates for behaviorally conditioned immunosuppression in the rat. *The Journal of Neuroscience, 25*, 2330–2337.

Pavlov, I. (1927). *Conditioned reflexes: An investigation of the physiological activity of the cerebral cortex.* Cambridge: Oxford University Press.

Price, D. D., Finniss, D. G., & Benedetti, F. (2008). A comprehensive review of the placebo effect: Recent advances and current thought. *Annual Review of Psychology, 59*, 565–590.

Riether, C., Doenlen, R., Pacheco-Lopez, G., Niemi, M. B., Engler, A., Engler, H., et al. (2008). Behavioural conditioning of immune functions: How the central nervous system controls peripheral immune responses by evoking associative learning processes. *Reviews in the Neurosciences, 19*, 1–17.

Riether, C., Kavelaars, A., Wirth, T., Pacheco-Lopez, G., Doenlen, R., Willemen, H., et al. (2011). Stimulation of beta(2)-adrenergic receptors inhibits calcineurin activity in CD4(+) T cells via PKA-AKAP interaction. *Brain, Behavior, and Immunity, 25*, 59–66.

Russell, M., Dark, K. A., Cummins, R. W., Ellman, G., Callaway, E., & Peeke, H. V. (1984). Learned histamine release. *Science, 225*, 733–734.

Sandler, A. D., Glesne, C. E., & Bodfish, J. W. (2010). Conditioned placebo dose reduction: A new treatment in attention-deficit hyperactivity disorder? *Journal of Developmental and Behavioral Pediatrics, 31*, 369–375.

Schedlowski, M., Enck, P., Rief, W., & Bingel, U. (2015). Neuro-bio-behavioral mechanisms of placebo and nocebo responses: Implications for clinical trials and clinical practice. *Pharmacological Reviews, 67*, 697–730.

Schedlowski, M., & Pacheco-Lopez, G. (2010). The learned immune response: Pavlov and beyond. *Brain, Behavior, and Immunity, 24*, 176–185.

Schiller, D., Monfils, M. H., Raio, C. M., Johnson, D. C., Ledoux, J. E., & Phelps, E. A. (2010). Preventing the return of fear in humans using reconsolidation update mechanisms. *Nature, 463*, 49–53.

Siegel, S., Baptista, M. A., Kim, J. A., McDonald, R. V., & Weise-Kelly, L. (2000). Pavlovian psychopharmacology: The associative basis of tolerance. *Experimental and Clinical Psychopharmacology, 8*, 276–293.

Solvason, H. B., Ghanta, V. K., & Hiramoto, R. N. (1988). Conditioned augmentation of natural killer cell activity. Independence from nociceptive effects and dependence on interferon-beta. *Journal of Immunology, 140*, 661–665.

Steinbach, W. J., Reedy, J. L., Cramer, R. A., Jr., Perfect, J. R., & Heitman, J. (2007). Harnessing calcineurin as a novel anti-infective agent against invasive fungal infections. *Nature Reviews. Microbiology, 5*, 418–430.

Stewart-Williams, S., & Podd, J. (2004). The placebo effect: Dissolving the expectancy versus conditioning debate. *Psychological Bulletin, 130*, 324–340.

Tracey, K. J. (2002). The inflammatory reflex. *Nature, 420*, 853–859.

Tracey, K. J. (2010). Understanding immunity requires more than immunology. *Nature Immunology, 11*, 561–564.

Tronson, N. C., & Taylor, J. R. (2007). Molecular mechanisms of memory reconsolidation. *Nature Reviews. Neuroscience, 8*, 262–275.

Vits, S., Cesko, E., Benson, S., Rueckert, A., Hillen, U., Schadendorf, D., et al. (2013). Cognitive factors mediate placebo responses in patients with house dust mite allergy. *PLoS One*, *8*, e79576.

Vits, S., Cesko, E., Enck, P., Hillen, U., Schadendorf, D., & Schedlowski, M. (2011). Behavioural conditioning as the mediator of placebo responses in the immune system. *Philosophical Transactions of the Royal Society of London. Series B, Biological Sciences*, *366*, 1799–1807.

von Horsten, S., Exton, M. S., Schult, M., Nagel, E., Stalp, M., Schweitzer, G., et al. (1998). Behaviorally conditioned effects of cyclosporine A on the immune system of rats: Specific alterations of blood leukocyte numbers and decrease of granulocyte function. *Journal of Neuroimmunology*, *85*, 193–201.

Wayner, E. A., Flannery, G. R., & Singer, G. (1978). Effects of taste aversion conditioning on the primary antibody response to sheep red blood cells and Brucella abortus in the albino rat. *Physiology & Behavior*, *21*, 995–1000.

Wirth, T., Ober, K., Prager, G., Vogelsang, M., Benson, S., Witzke, O., et al. (2011). Repeated recall of learned immunosuppression: Evidence from rats and men. *Brain, Behavior, and Immunity*, *25*, 1444–1451.

World Medical Association. (2013). World Medical Association Declaration of Helsinki: Ethical principles for medical research involving human subjects. *JAMA*, *310*, 2191–2194.

CHAPTER FOUR

Human Pharmacological Conditioning of the Immune and Endocrine System: Challenges and Opportunities

Judith Tekampe[*,†], **Henriët van Middendorp**[*,†], **Fred C.G.J. Sweep**[†], **Sean H.P.P. Roerink**[†], **Ad R.M.M. Hermus**[†], **Andrea W.M. Evers**[*,†,‡,1]

[*]Health, Medical and Neuropsychology Unit, Faculty of Social and Behavioural Sciences, Institute of Psychology, Leiden University, Leiden, The Netherlands
[†]Radboud University Medical Center, Nijmegen, The Netherlands
[‡]Leiden University Medical Center, Leiden, The Netherlands
[1]Corresponding author: e-mail address: a.evers@fsw.leidenuniv.nl

Contents

Abstract

In this chapter, we review recent studies on conditioned pharmacological effects on immune and endocrine responses in humans, and discuss challenges and opportunities for bringing these effects into clinical practice. By altering physiological mechanisms in part independent of pharmacological agents, pharmacological conditioning has high clinical relevance, as illustrated in some patient studies. Methodological challenges for further investigation include broadening the spectrum of opportunities for conditioned pharmacological effects, by investigating conditioning of substances that have not or not often been used before (e.g., corticosteroids) and unraveling mechanisms by which pharmacological responses become conditioned, thereby identifying characteristics that make conditioning designs effective. As an opportunity to optimize external validity, we introduce a design in which the potential of pharmacological conditioning can be pretested in the laboratory. The feasibility of this design is demonstrated by a pilot study.

International Review of Neurobiology, Volume 138
ISSN 0074-7742
https://doi.org/10.1016/bs.irn.2018.01.002

Pharmacological conditioning occurs when administration of an active pharmacological component (unconditioned stimulus, UCS) is repeatedly paired with the occurrence of another stimulus (conditioned stimulus, CS). Once conditioning has taken place, administration of the previously inert CS will elicit a physiological response similar to the response triggered by administration of the UCS. Pharmacological conditioning has important implications for the study of placebo effects (Colloca, 2014), as the effective qualities of the pharmacological stimulus (UCS) are transferred to a previously inert stimulus (CS). Consequently, conditioned pharmacological effects are also called "learned placebo effects" (Pacheco-Lopez, Niemi, Engler, & Schedlowski, 2007; Schedlowski & Pacheco-Lopez, 2010). A key aspect that makes conditioning and other learning mechanisms particularly interesting when compared to other placebo-inducing mechanisms, such as verbal suggestion, is that placebo effects based on conditioning tend to be larger and can be evoked repeatedly (Colloca, 2014).

One of the most widely known examples of pharmacological conditioning is an experiment by Ader and Cohen (1982). In this study, a special breed of mice, which were prone to develop an autoimmune disease similar to lupus erythematosus, were injected with an immunosuppressant drug, cyclophosphamide (UCS), which helps to suppress the symptoms of their autoimmune disease. This injection was repeatedly paired with administration of a sweet-tasting inert saccharine drinking solution (CS). After several pairings of the two stimuli, administering saccharine solution alone triggered a conditioned immunosuppressive response similar to the effect of cyclophosphamide. The conditioned mice also improved on their autoimmune disease (lower mortality rate, slower development of proteinuria) compared to controls. Later, it was discovered that the same principle could be applied in humans (Olness & Ader, 1992). As shown by this example, pharmacological conditioning has high clinical relevance, as it offers the possibility to alter dysfunctional mechanisms in part independent of pharmacological agents, which may eventually lead to lower health-care costs.

For conditioned pharmacological responses to occur, an association between the sensory input of the CS and the peripheral changes caused by the pharmacological agent (UCS) has to be formed in the central nervous system (CNS). In order to orchestrate an unconditioned response (UCR), the CNS must be able to communicate with the peripheral systems in which the response is to occur (Stockhorst, 2005; Vits et al., 2011). As this bidirectional communication between CNS and periphery is a prerequisite for conditioned pharmacological responses, it provides possibilities to study the interaction between the CNS and peripheral functions of the immune and endocrine system.

Most experiments investigating conditioned pharmacological responses follow the same basic procedure (Fig. 1, part 2). In the first phase of the experiment, the acquisition phase (also called learning or association phase), an association is formed between the CS, often a gustatory or olfactory stimulus (e.g., taste or smell) and UCS (e.g., an immunomodulatory drug) by paired administration. Often, this paired administration is repeated several times; however, single-trial conditioning, where the CS and UCS are administered together only once, has also proven to be effective in some cases (Barrett, King, & Pang, 2000; Gauci, Husband, Saxarra, & King, 1994; Fig. 1, part 1). In the second phase of the experiment, the evocation phase (also called test phase), the CS is administered alone and it is measured whether or not a conditioned response occurs. When the conditioned response has been evoked by administration of the CS, it will generally decrease in magnitude with every subsequent administration of the CS. This

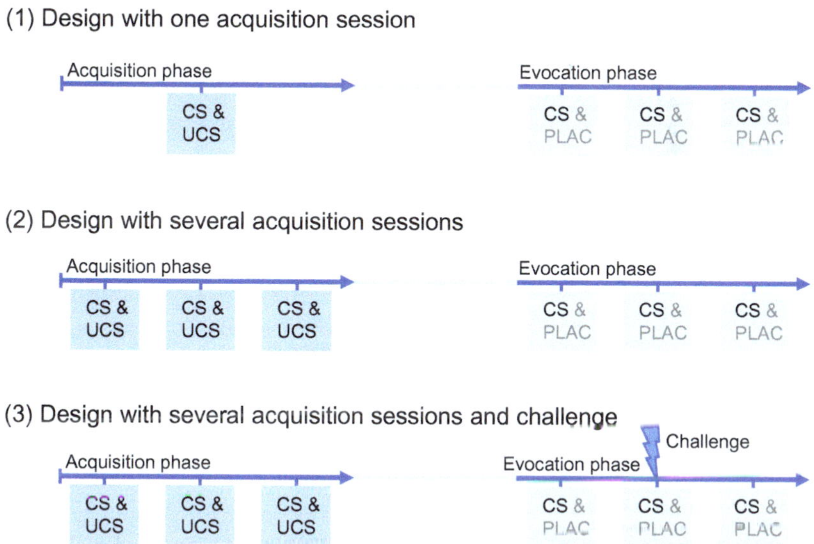

Fig. 1 Examples of different designs used for pharmacological conditioning. The number of expositions to CS and/or UCS and the time interval between expositions and phases may vary between studies. (1) Design with one acquisition session (single-trial conditioning). (2) Design with several acquisition sessions. (3) Design with challenge during the evocation phase to increase external validity. The timing of the challenge may vary between studies. *CS*, conditioned stimulus; *PLAC*, placebo; *UCS*, unconditioned stimulus.

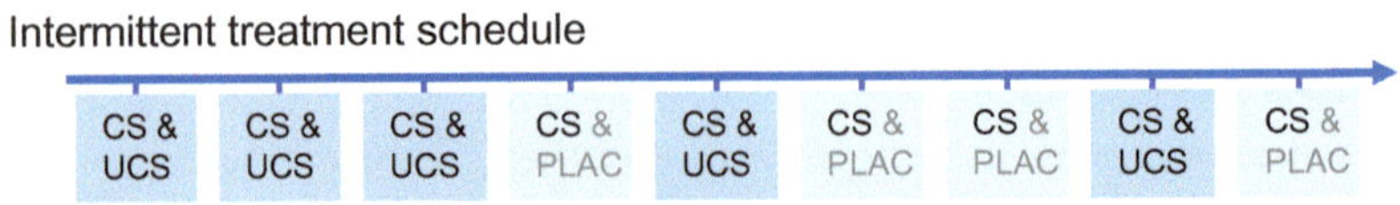

Fig. 2 Intermittent treatment schedule with booster sessions to prevent extinction. This design is used in studies in patient samples. The number of expositions to CS and/or UCS and the time interval between expositions may vary between studies. *CS*, conditioned stimulus; *PLAC*, placebo; *UCS*, unconditioned stimulus.

process is called extinction. One way to prevent extinction (at least to some extent) is to implement an intermittent treatment schedule (e.g., Giang et al., 1996; Longo et al., 1999; Olness & Ader, 1992). In this design (Fig. 2), the CS and UCS are repeatedly administered together to form an association between the two stimuli (similar to an acquisition phase), followed by the CS being administered sometimes alone (similar to the evocation phase in the design described earlier) and sometimes paired with the UCS in a booster session. By paired administration of CS and UCS during the booster session, the association between both stimuli is renewed and the extinction process interrupted.

In this chapter, we briefly review the literature on pharmacological conditioning with the focus on studies in humans to give an overview of what is currently known. We then discuss current challenges and opportunities for clinical application of conditioned pharmacological effects. As part of the solution to some of these challenges, we introduce a design (Fig. 1, part 3) in which the potential of pharmacological conditioning for altering physiological responses involved in pathophysiological processes can be pretested in the laboratory to optimize external validity.

1. PHARMACOLOGICAL CONDITIONING OF IMMUNE AND ENDOCRINE RESPONSES

In the following paragraphs, we will review the recent literature of laboratory studies addressing conditioned pharmacological effects in humans. First we will focus on conditioned immune responses, starting with conditioning in the direction of immunostimulation, followed by immunosuppression and allergic responses. Studies on conditioned endocrine, including glycemic, responses will also be reviewed before finishing this section with a brief discussion of the challenges for future research and clinical application of conditioned pharmacological responses.

1.1 Pharmacological Conditioning of Immune Responses

Research into conditioned pharmacological effects started with the discovery of conditioned immunological responses. First reports of this phenomenon date back to the 1920s or even earlier. In the following years, conditioned responses in the immune system were predominantly investigated in animal studies (see Pacheco-Lopez et al., 2007). In an exemplary study in humans, using a conditioning design with several acquisition and evocation sessions, Buske-Kirschbaum et al. (1992) showed that the increase in the activity of natural killer cells caused by administration of epinephrine could be conditioned. In the acquisition phase of their experiment, they paired a sherbet sweet (CS) with the injection of 0.2 mg of epinephrine (UCS). In the evocation phase, when the CS was administered together with a placebo injection, a conditioned increase in natural killer cell activity (NKCA) was observed. The results of this experiment could only partially be replicated (Kirschbaum et al., 1992). However, more indications for the possibility to condition NKCA by epinephrine and a gustatory CS were found in another study with a discriminant conditioning design. In this study, the administration of one CS, the CS+ (sherbet sweet and white noise), was followed by an injection with the UCS (0.2 mg epinephrine), while the administration of another CS, the CS− (herbal sweet with a specific tone), was followed by no treatment. In the evocation phase, exposure to the CS+ led to an increase in NKCA, while exposure to the CS− had no effect on NKCA (Buske-Kirschbaum, Kirschbaum, Stierle, Jabaij, & Hellhammer, 1994). Taken together, these studies provide relatively strong indications that NKCA can be pharmacologically conditioned in humans.

There are also indications that other proinflammatory immune parameters can be conditioned. One study found higher quinolinic acid and neopterin, and increased CD64 antigen expression after pairing recombinant human interferon gamma (rhIFN-γ, UCS) with the taste of propylene glycol (CS) in an intermittent treatment design (Longo et al., 1999). In another study using a single-trial conditioning design (Fig. 1, part 1), an injection with IFNβ-1a (UCS) was paired with the administration of a distinctively flavored beverage. When the beverage was administered in combination with a placebo injection 8 days later, no conditioned responses were observed (Goebel et al., 2005). Similarly, a study using the same single trial conditioning paradigm with lipopolysaccharide (UCS) and the same distinctively flavored beverage also found no conditioned immunological effects (Grigoleit et al., 2012). Taken together, these studies suggest that conditioning proinflammatory immune reactions by pairing immunostimulant drugs with gustatory stimuli is possible, but more evidence is needed to support these findings. Different conditioning designs can be used to investigate these effects, although single-trial conditioning does not seem to be effective for conditioning proinflammatory immune reactions.

Pharmacological conditioning of immune responses in the direction of immunosuppression has also been investigated. In most studies, the oral administration of the immunosuppressant drug cyclosporine A (CsA, UCS) has been paired with a distinctively tasting beverage (CS) in the acquisition phase. When the CS was administered with a placebo during evocation, this consistently led to a conditioned decrease in production of the cytokine interleukin 2 (Albring et al., 2014; Goebel et al., 2002; Ober et al., 2012; Wirth et al., 2011). Some studies also found conditioned reductions in IFN-γ production (Goebel et al., 2002; Wirth et al., 2011) and lymphocyte proliferation (Goebel et al., 2002). All of these are indicators of conditioned immunosuppression. Through variations in the number of evocation sessions between the studies, indications were found that these conditioned immunosuppressive responses may only become apparent after more than one evocation session (Albring et al., 2012). Once conditioned responses have been established, they are relatively stable, as they have shown only to be extinct after 14 evocations (Albring et al., 2014). Also, this type of conditioning can boost the effect of a subtherapeutic dose of CsA to a level that is clinically relevant (Albring et al., 2014), a finding that hints at the possible opportunities for clinical applications based on conditioned pharmacological effects.

Conditioned allergic responses provide more examples of conditioned pharmacological effects. In two studies, allergens (UCS) were applied via a nose spray containing a distinctive smell (CS) (Barrett et al., 2000; Gauci et al., 1994). When the CS was applied alone (nose spray with the smell, but without the allergen) later, conditioned elevations in histamine (Barrett et al., 2000) and tryptase (Gauci et al., 1994) levels in nasal secretion were found. Interestingly, in contrast to studies in conditioned immunostimulation, conditioned responses occurred even after only one acquisition (single-trial conditioning) and were prone to extinction (Barrett et al., 2000). Nonetheless, conditioned responses became larger with more acquisition trials (Barrett et al., 2000). In another study, the dermal application of allergens (UCS) was paired with a visual CS (the color of the vial containing the allergens) during the acquisition phase. In this experiment, no conditioned responses were found (Booth, Petrie, & Brook, 1995). Although the studies on conditioning with allergens have found effects on histamine and tryptase, which are involved in allergic reactions, none has found effects on allergic symptoms (i.e., congestion of the nose or sneezing). However, Goebel et al. (2008) found that allergic symptoms can be targeted when conditioning antiallergic effects. In this study, the administration of anti-allergic medication (desloratadine, UCS) was paired with a distinctively tasting beverage (CS) during the acquisition phase. When the beverage was later administered in combination with a placebo, a conditioned decrease in basophil activation, a reduced reaction to a skin-prick test, and lower symptom scores were found. In an attempt to replicate these promising results in another study, however, the antiallergic effects of conditioning did not exceed those in the placebo group, but both groups improved in symptoms when compared to a natural history group. This finding suggests that conditioning as well as expectation-based effects may be at play (Vits et al., 2013). Concluding, there is at least preliminary evidence that pharmacological conditioning of immune responses in the direction of immunostimulation and immunosuppression, as well as allergic reactions, is possible in the laboratory.

1.2 Pharmacological Conditioning of Endocrine Responses

Not only immune but also endocrine responses can be affected by pharmacological conditioning. Most studies in this direction focus on conditioned responses in blood glucose levels (glycemic responses). Early reports of indications for conditioned glycemic responses in humans were made in 1959, when Russian experimenters found anticipatory hypoglycemic symptoms in schizophrenic patients receiving insulin shock therapy (Stockhorst et al., 1999). After several animal studies in which conditioned glycemic responses were found (Stockhorst, 2005; Stockhorst, Steingruber, & Scherbaum, 2000), the possibility of conditioned responses in humans received the attention of investigators as possible mechanism behind eating disorders (Overduin & Jansen, 1997). More controlled studies in humans yielded mixed results, finding conditioned responses in blood glucose after conditioning with insulin in some participants but not in others (Fehm-Wolfsdorf, Gnadler, Kern, Klosterhalfen, & Kerner, 1993). In the first placebo-controlled experiment in humans, Stockhorst et al. (1999) showed that by repeatedly pairing an injection of insulin (UCS) with the administration of a distinctive smell (CS), the effect of intravenous insulin on blood glucose became conditioned. In a later study, conditioned effects of insulin on other hormones involved in blood glucose regulation also became apparent (Stockhorst et al., 2004). Interestingly, when intranasal insulin was used as a UCS (Stockhorst, de Fries, Steingrueber, & Scherbaum, 2011), which leads to an increase of the level of insulin in the brain while no changes in blood glucose level occur peripherally, the conditioned response encompasses changes in insulin rather than peripheral blood glucose. This is in contrast to the studies in which intravenous insulin was used as UCS, causing changes in peripheral blood glucose but not in insulin. In these studies, conditioned changes in blood glucose but not in insulin were found (Stockhorst et al., 1999, 2004). This indicates that conditioned pharmacological responses can be very specific, mimicking the bodily reaction the UCS causes (Stockhorst, 2005). Two studies attempted to obtain conditioned changes in blood glucose by using glucose administration as a UCS (Overduin & Jansen, 1997; Stockhorst et al., 2004), which found no conditioned effects. Taken together, these studies provide strong indications that responses involved in blood glucose regulation can be conditioned. Whether these conditioned responses play a role in maladaptive eating behavior and thus eating disorders remains subject to further investigation.

Effects of pharmacological conditioning on other endocrine responses, predominantly corticosterone responses, have been shown in rodents (e.g., Ader, 1976; Buske-Kirschbaum et al., 1996; Janz et al., 1996; Pacheco-Lopez et al., 2004). Conditioning endocrine, specifically hypothalamic–pituitary–adrenal axis (HPA axis) responses, in humans is potentially of great clinical relevance (Sabbioni et al., 1997), as dysfunctions in HPA axis regulation are thought to play a role in various disorders (Fries, Hesse, Hellhammer, & Hellhammer, 2005; Putman & Roelofs, 2011; Yehuda & Seckl, 2011). Only 3 studies so far have investigated conditioned endocrine effects in humans. In the first study, the oral administration of dexamethasone (UCS) was repeatedly paired with the administration of a distinctively tasting beverage (CS) in the acquisition phase. When the CS was administered with a placebo during the evocation phase, indications for a conditioned increase in plasma cortisol were found, but post hoc tests remained inconclusive (Sabbioni et al., 1997). In the second study (Benedetti et al., 2003), sumatriptan was used as a UCS and a conditioned decrease in cortisol and increase in growth hormone were found. Recently, a third study in which corticotropin-releasing hormone was used as UCS was unable to find conditioned responses in cortisol and noradrenaline. Further analysis of the data by a median split revealed indications that participants with high baseline cortisol did show a conditioned increase in cortisol, but participants with low baseline cortisol did not (Petrakova et al., 2017). In sum, there are some indications for conditioned HPA axis responses in humans, but the precise extent and relevance of these responses remains to be elucidated.

2. CHALLENGES

Numerous challenges have to be addressed before the opportunities of pharmacological conditioning are fully unraveled. One of the most important challenges is broadening the spectrum of opportunities for conditioned pharmacological effects, by investigating conditioning of substances that have not or not often been used before (e.g., corticosteroids). Another challenge is the unraveling of the mechanisms by which pharmacological responses become conditioned and thereby identifying characteristics that make conditioning designs effective.

The UCSs used in the studies so far have mostly been immunomodulatory drugs. This is hardly surprising, considering the historical origin of pharmacological conditioning in conditioned immune responses (Pacheco-Lopez et al., 2007). However, the few studies on conditioned endocrine responses show that the potential of pharmacological conditioning is not limited to modulating immune responses. Expanding the focus of laboratory studies from conditioning of the immune system to other regulatory systems is important in revealing more opportunities for clinical application of conditioned pharmacological responses. Conditioning of glycemic responses, for example, might provide insight into the mechanisms involved in onset and maintenance of eating disorders and obesity and eventually may also offer new treatment possibilities. In the same way, conditioned HPA axis responses may bear relevant implications for understanding and ultimately treating stress-related disorders, as dysregulation of the HPA axis seems to play a role in these disorders.

The studies reviewed earlier show differences in the characteristics of the study design potentially related to successful conditioning of pharmacological responses for different UCSs. Regarding the acquisition of conditioned responses, it is generally accepted that studies with more acquisition trials produce conditioned effects more reliably. However, the finding that single-trial conditioning works for some UCSs (allergens) but not others (immunostimulants) seems to indicate that some responses are more readily conditionable than others. This finding corresponds to differences in conditionability of stimuli in fear conditioning studies, where evolutionary factors are thought to play a role (Ohman, 2009). What factors are behind differences in conditionability in pharmacological conditioning remains to be elucidated. Regarding the evocation and ultimately extinction of conditioned pharmacological responses, more differences between various UCSs become apparent. While conditioned immunostimulation and allergic responses emerge at the first evocation, conditioned immunosuppressive responses induced by CsA conditioning may only show after two evocations (Albring et al., 2012). In contrast to allergic responses that show signs of extinction at the second evocation, these immunosuppressive responses remain rather stable across several evocations (Ober et al., 2012; Wirth et al., 2011). A possible explanation for these differences in UCSs may lie in the mechanisms by which an association between CS and UCS is formed in the CNS and a conditioned response is generated by CNS–peripheral interaction. These mechanisms seem to differ between UCSs (Schedlowski & Pacheco-Lopez, 2010). Potentially these underlying mechanisms also differ depending on the population targeted, with on the one hand allergic patients who are familiar with the UCS as well as UCR and on the other hand healthy volunteers for whom UCS as well as UCR is unfamiliar prior to the study (Tekampe, van Middendorp, Meeuwis, et al., 2017). More research into the underlying mechanisms by which conditioning takes place in different populations may provide insight into the design characteristics that make for successful conditioning and hence point to opportunities for effectively bringing conditioned pharmacological responses into clinical practice.

In order to understand the mechanisms involved in conditioning of pharmacological effects, it is important to understand the exact nature of the CS and UCS. The CSs used in the various studies are mostly gustatory or olfactory (taste or smell) stimuli. Forming an association between the ingestion of a certain substance (identified by its taste and/or smell, CS) and its consequences for homeostasis (the effect of the drug, UCR) is considered to have an adaptive value for the survival of an organism (Pacheco-Lopez et al., 2007; Schedlowski & Pacheco-Lopez, 2010). By integrating the two stimuli, the organism is able to learn which substances produce favorable bodily changes and are thus suitable for ingestion and which substances produce changes that threaten homeostasis and should thus be avoided. The latter phenomenon, conditioned taste avoidance, is frequently observed in animal studies on pharmacological conditioning and regarded as one of the outcomes of pharmacological conditioning (Pacheco-Lopez et al., 2007; Schedlowski & Pacheco-Lopez, 2010). This basic learning mechanism is the reason that gustatory or olfactory stimuli seem to be the first choice for pharmacological conditioning. In some studies, the gustatory or olfactory stimuli were combined with auditory (Buske-Kirschbaum et al., 1994) or tactile (Buske-Kirschbaum et al., 1992) stimuli. Another study employed a visual CS, but was unsuccessful in finding conditioned responses (Booth et al., 1995). To what extent the sensory domain of the CS is a determinant for the success of the conditioning procedure remains to be elucidated. Regarding the nature of the UCSs, it is argued that not so much the drug itself but the physiological reaction caused by the drug is the UCS (Woods & Ramsay, 2000). This is important in explaining that some conditioned pharmacological responses, such as immunosuppressive responses, are mimicking the effect of the drug, while others are counterregulatory (Ramsay & Woods, 1997).

From this overview of the research on pharmacological conditioning in humans, it can be concluded that pharmacological conditioning has great potential in affecting immune and endocrine responses. Successfully addressing the challenges discussed earlier will expand our knowledge about the full potential of these responses and, maybe even more importantly, help finding ways to bring conditioned pharmacological responses out of the lab and put them to use in clinical settings.

3. OPPORTUNITIES FOR RESEARCH AND CLINICAL PRACTICE

Conditioned pharmacological effects offer great opportunities for clinical practice. If the effects of drugs can be learned and ultimately produced without the actual presence of the pharmacological agent, this could reduce the amount of medication patients are required to take. As a consequence, pharmacological conditioning might be part of the solution of problems in health-care systems worldwide, such as rising costs or avoidable side effects. When translating findings into opportunities for clinical practice, two main challenges of conditioned pharmacological effects have to be met. The size of the conditioned response has to be clinically relevant (Ader, 2003) and, to be of use for patients requiring medication for longer periods of time, effects of conditioning have to be resistant to extinction (Albring et al., 2014).

Regarding the size of the conditioned responses, pharmacological conditioning in lab research often yields small responses (Tekampe, van Middendorp, Meeuwis, et al., 2017), for example, 10%–15% of the UCR (Stockhorst et al., 1999). It has been argued that this may in part be to maintain homeostasis, as it seems unlikely that bodily systems will produce responses that will threaten their own health (Ader, 2003). However, this does not necessarily mean that conditioned pharmacological responses cannot be large enough to be clinically relevant. A frequently observed example of the strength of conditioned responses is found in conditioned nausea and vomiting, a phenomenon that frequently occurs in cancer patients receiving chemotherapy. In these patients, stimuli such as the hospital environment (CS) that predict the administration of chemotherapy (UCS) have become associated with the immediate negative side effects of the chemotherapy, including nausea and vomiting. Resultantly, only being in the hospital can evoke these symptoms (Stockhorst, Enck, & Klosterhalfen, 2007). Together with studies on conditioned antiallergic effects, reviewed earlier, these findings show that, despite the relatively small conditioned effects found in laboratory studies (Ader, 2003; Tekampe, van Middendorp, Meeuwis, et al., 2017), conditioned pharmacological responses can elicit clinically relevant symptoms. Further studies in patient samples will be needed to unravel the full potential of pharmacological conditioning in eliciting clinically relevant changes in symptoms.

Extinction could potentially be prevented in two ways. Lab research has shown that conditioned immunosuppressive responses can be prevented from extinction by pairing the CS with a subtherapeutic dose of the UCS during the evocation phase instead of administering the CS alone or with a placebo (Albring et al., 2014). Another way of preventing extinction that has already been tested in patient samples is by using a conditioning design entailing an intermittent treatment schedule (Fig. 2), in which booster sessions are used to avoid extinction (Doering & Rief, 2012). In a case study by Olness and Ader (1992), an intermittent treatment design was used to lower the monthly dose of cyclophosphamide in a patient suffering from lupus erythematosus. To this end, the administration of cyclophosphamide (UCS) was paired with the intake of a specific taste and smell (CS). By using an intermittent treatment schedule, the patient improved clinically while receiving only half of the cumulative dose normally necessary to treat the symptoms. However, the nausea induced by cyclophosphamide became associated with the CS as well, causing the patient to suffer from conditioned nausea and ultimately discontinuing the initiated treatment. Preliminary research also suggests that intermittent treatment designs with cyclophosphamide might help lower the cumulative dose needed to treat symptoms of multiple sclerosis (MS). In a pilot study, the monthly cyclophosphamide doses (UCS) of 10 MS patients were paired with the taste of anise syrup (CS)

(Giang et al., 1996). When in 1 month a subtherapeutic dose of cyclophosphamide was administered together with the CS, 8 of the 10 subjects showed at least a small decrease in white blood cell count, which might reflect a conditioned response. In 4 subjects, this decrease after CS alone was of similar magnitude as the decrease caused by cyclophosphamide administration. Regarding side effects, 7 of the 10 included subjects experienced nausea and/or vomiting as side effect of cyclophosphamide administration. Administration of the CS alone led to the same side effects in 4 of the 10 subjects, providing indications that the negative side effects of the UCS may also become conditioned, but are less prevalent than in usual treatment regimes. Intermittent treatment schedules have not only been used to lower the cumulative dose of cyclophosphamide. In patients using a corticosteroid ointment to control symptoms of psoriasis, a conditioning procedure with an intermittent treatment design was tested. Patients on the intermittent treatment schedule received the full dose of corticosteroid treatment 25%–50% of the time and placebo at other times. Compared to a control group receiving the same cumulative dose, the patients in the intermittent treatment group showed lower symptom severity and less relapse (Ader et al., 2010). Other preliminary reports of positive effects of conditioned pharmacological effects by means of an intermittent treatment schedule in patients come from studies in ADHD (Sandler & Bodfish, 2008; Sandler, Glesne, & Bodfish, 2010) and hypertension (Suchman & Ader, 1992). In an ongoing study, the potential effects of conditioning on the achievement of a higher percentage of drug-free clinical remission with the same cumulative dose of active medication by implementing an intermittent treatment schedule with methotrexate and placebo in patients suffering from rheumatoid arthritis are currently investigated (Manaï, van Middendorp, Veldhuijzen, & Evers, 2017).

Concluding from the limited research in clinical samples so far, conditioned pharmacological effects are potentially large enough to be brought into practice by using intermittent treatment schedules in patients who need to use medication for longer periods of time. As in lab research, conditioned pharmacological effects in patient samples are most frequently investigated in immunomodulating drugs. However, as lab studies show that endocrine responses can be conditioned as well, the clinical potential of conditioned pharmacological responses may not be limited to altering immune responses but may also provide opportunities to alter dysfunctional endocrine responses that might be relevant for eating disorders (glycemic responses) or stress-related disorders (HPA axis responses). It has been argued that intermittent treatment schedules might also help to reduce the negative side effects of medication (Goebel et al., 2005). As seen in the case of nausea caused by cyclophosphamide, however, negative symptoms can also become associated with, and hence elicited by, the CS (Olness & Ader, 1992). Still these conditioned side effects may be less frequent than the side effects elicited by administration of cyclophosphamide (Giang et al., 1996).

4. BRIDGING THE GAP BETWEEN LAB AND PRACTICE: INCREASING THE EXTERNAL VALIDITY OF LABORATORY STUDIES BY USING REAL-WORLD CHALLENGES

Research in patient samples may eventually make it possible to bring conditioned pharmacological effects out of the lab and into clinical practice. However, the intermittent treatment protocol used in patient studies entails replacing at least some doses of the active drug with an initially inert CS. This involves the risk that—if conditioning is unsuccessful—the patients do not receive the medication necessary to treat their symptoms and deteriorate clinically. A way to pretest in the lab the potential of pharmacological conditioning for altering physiological responses involved in diseases is offered by an adaptation of the experimental design (Fig. 1, part 3). Based on the classical design with acquisition and evocation phase frequently used in previous studies (e.g., Barrett et al., 2000; Buske-Kirschbaum et al., 1992; Goebel et al., 2002; Sabbioni et al., 1997; Stockhorst et al., 1999), this design includes a relevant real-world challenge of the system in which the conditioned response is to occur during the evocation phase to optimize the external validity. This challenge forms a direct test of the possible clinical relevance of conditioned effects. Examples of relevant challenges of the system in which the conditioned response is to occur are a physiological histamine challenge for conditioned antiallergic responses (Meeuwis, van Middendorp, Veldhuijzen, Pacheco-Lopez, & Evers, 2017), exposure to an immunostimulating vaccination after a psychological training aimed to optimize the immune system (Schakel et al., 2017), or exposure to emotional stimuli, infant crying sounds, and pain during an MRI scan after conditioning with oxytocin (Skvortsova et al., 2016).

To illustrate the feasibility of this design in more detail, a pilot study in 10 healthy participants aiming to condition cortisol levels will be described. Cortisol is a key stress-regulatory parameter, suspected to be involved in the onset and maintenance of various stress-related disorders (Fries et al., 2005; Putman & Roelofs, 2011; Yehuda & Seckl, 2011). If cortisol could be conditioned, this may have clinical implications for these disorders, but conditioned effects under basal circumstances may not generalize to the cortisol stress response, which is particularly important in this regard (Putman & Roelofs, 2011). In a pilot study in 10 participants (Tekampe et al., 2014; Tekampe, van Middendorp, Sweep, et al., 2017), hydrocortisone (UCS) was paired with a distinctively tasting beverage (CS) 3 times on consecutive days during the acquisition phase. In the evocation phase, again consisting of 3 sessions on 3 consecutive days, the CS was administered paired with a placebo. In the first and second evocation session, conditioned effects (e.g., salivary cortisol levels) were measured under basal circumstances, with participants completing undemanding filler tasks between measurements. In the third evocation session, participants were exposed to a psychosocial challenge by means of the Trier Social Stress Test (TSST). Consisting of a 5-min anticipation time, 6-min presentation, and 4-min mental arithmetic task in front of a critical and nonaffirmative panel of sham experts, the TSST reliably induces an HPA axis response, including elevation in cortisol levels (Denson, Spanovic, & Miller, 2009; Dickerson & Kemeny, 2004). This way, conditioned responses on the cortisol response to stress, which is affected in many stress-related disorders, can be investigated. The pilot study showed the feasibility of the design and preliminary results indicated that conditioning with cortisol may affect cortisol levels under basal circumstances as well as in exposure to psychosocial stress. Particularly after exposure to stress, the conditioned participants additionally appeared to have a lower negative affective response compared to subjects in the placebo group (Tekampe et al., 2014; Tekampe, van Middendorp, Sweep, et al., 2017). This might be a first indication that some effects of pharmacological conditioning specifically become apparent by applying challenges during evocation.

In sum, laboratory studies on pharmacological conditioning in which a challenge of the targeted system is implemented during the evocation phase may provide an opportunity to pretest the clinical potential of conditioned effects and to optimize external validity of designs. An important advantage over studies in patients is that this approach does not entail the risk of clinical deterioration of the disease and that it can be applied in the controlled environment of the laboratory.

5. GENERAL CONCLUSION AND DISCUSSION

Pharmacological conditioning has high clinical relevance for altering physiological mechanisms in the immune and endocrine system in part independent of pharmacological agents. As conditioned pharmacological effects illustrate the close interactions between the CNS and peripheral functions (Stockhorst, 2005; Vits et al., 2011), they also are of conceptual interest in understanding placebo effects (Schedlowski & Pacheco-Lopez, 2010). Laboratory research has shown that using pharmacological conditioning in humans has great potential in affecting immune and endocrine responses (Sabbioni et al., 1997; Stockhorst et al., 1999, 2011, 2004). Studies on conditioned glycemic responses have illustrated that conditioned pharmacological responses can be very specific (Stockhorst, 2005; Stockhorst et al., 2011) mimicking not so much the direct effect of the drug, such as elevation of a certain hormone level in the blood, but rather the reaction to the drug, such as a counterregulatory response to reduce the hormone level to a physiological quantity. This finding may contribute to the clinical relevance of conditioned pharmacological effects, as they offer the possibility to target specific responses. Most research has focused on conditioned immune responses (Albring et al., 2014, 2012; Barrett et al., 2000; Booth et al., 1995; Buske-Kirschbaum et al., 1994, 1992; Gauci et al., 1994; Goebel et al., 2005, 2008, 2002; Grigoleit et al., 2012; Kirschbaum et al., 1992; Longo et al., 1999; Ober et al., 2012; Wirth et al., 2011) and despite the potential clinical relevance, there is a lack of studies addressing pharmacological conditioning of endocrine functions (Tekampe, van Middendorp, Meeuwis, et al., 2017). Also, conditioned effects found in the lab are generally small and some conditioned responses, such as allergic responses, are prone to relatively quick extinction. Despite these challenges, a small number of studies have brought pharmacological conditioning into clinical practice by using intermittent treatment schedules with preliminary but positive results (Ader et al., 2010; Giang et al., 1996; Olness & Ader, 1992; Sandler & Bodfish, 2008; Sandler et al., 2010; Suchman & Ader, 1992). Studies addressing pharmacological conditioning can be risky in patient samples, as symptoms may flare up if conditioned responses do not occur. A new way to help bridge the gap between laboratory research and clinical practice was found in an adaptation of the experimental design, implementing a challenge of the system in which the conditioned response is to occur to optimize external validity. This design offers the possibility to pretest the clinical potential of pharmacological conditioning for altering physiological responses in the lab and in healthy participants. A pilot study with this design was conducted, aiming to condition cortisol levels by using hydrocortisone as UCS and challenging the HPA axis by implementing a social stress test during the evocation phase. Preliminary results suggest that some effects of pharmacological conditioning might specifically become apparent by challenging the system that was targeted by conditioning. Future studies including comparable relevant real-life challenges of the conditioned (immune/endocrine) systems offer the opportunity to examine the possible external validity of pharmacological conditioning. If conditioned pharmacological responses can successfully be brought into practice, the ability to transfer the effective qualities of a pharmacological agent to a previously inert stimulus might help reduce the amount of medication patients are required to take and ultimately help lower health-care costs.

REFERENCES

Ader, R. (1976). Conditioned adrenocortical steroid elevations in the rat. *Journal of Comparative and Physiological Psychology, 90*(12), 1156–1163.

Ader, R. (2003). Conditioned immunomodulation: Research needs and directions. *Brain, Behavior, and Immunity, 17*(Suppl 1), S51–S57.

Ader, R., & Cohen, N. (1982). Behaviorally conditioned immunosuppression and murine systemic lupus erythematosus. *Science, 215*(4539), 1534–1536.

Ader, R., Mercurio, M. G., Walton, J., James, D., Davis, M., Ojha, V., et al. (2010). Conditioned pharmacotherapeutic effects: A preliminary study. *Psychosomatic Medicine, 72*(2), 192–197.

Albring, A., Wendt, L., Benson, S., Nissen, S., Yavuz, Z., Engler, H., et al. (2014). Preserving learned immunosuppressive placebo response: Perspectives for clinical application. *Clinical Pharmacology & Therapeutics, 96*(2), 247–255.

Albring, A., Wendt, L., Benson, S., Witzke, O., Kribben, A., Engler, H., et al. (2012). Placebo effects on the immune response in humans: The role of learning and expectation. *PLoS One, 7*(11), e49477.

Barrett, J. E., King, M. G., & Pang, G. (2000). Conditioning rhinitis in allergic humans. *Annals of the New York Academy of Sciences, 917*, 853–859.

Benedetti, F., Pollo, A., Lopiano, L., Lanotte, M., Vighetti, S., & Rainero, I. (2003). Conscious expectation and unconscious conditioning in analgesic, motor, and hormonal placebo/nocebo responses. *Journal of Neuroscience, 23*(10), 4315–4323.

Booth, R. J., Petrie, K. J., & Brook, R. J. (1995). Conditioning allergic skin responses in humans: A controlled trial. *Psychosomatic Medicine, 57*(5), 492–495.

Buske-Kirschbaum, A., Grota, L., Kirschbaum, C., Bienen, T., Moynihan, J., Ader, R., et al. (1996). Conditioned increase in peripheral blood mononuclear cell (PBMC) number and corticosterone secretion in the rat. *Pharmacology Biochemistry and Behavior, 55*(1), 27–32.

Buske-Kirschbaum, A., Kirschbaum, C., Stierle, H., Jabaij, L., & Hellhammer, D. (1994). Conditioned manipulation of natural killer (NK) cells in humans using a discriminative learning protocol. *Biological Psychology, 38*(2–3), 143–155.

Buske-Kirschbaum, A., Kirschbaum, C., Stierle, H., Lehnert, H., & Hellhammer, D. (1992). Conditioned increase of natural killer cell activity (NKCA) in humans. *Psychosomatic Medicine, 54*(2), 123–132.

Colloca, L. (2014). Placebo, nocebo, and learning mechanisms. *Handbook of Experimental Pharmacology, 225*, 17–35.

Denson, T. F., Spanovic, M., & Miller, N. (2009). Cognitive appraisals and emotions predict cortisol and immune responses: A meta-analysis of acute laboratory social stressors and emotion inductions. *Psychological Bulletin, 135*(6), 823–853.

Dickerson, S. S., & Kemeny, M. E. (2004). Acute stressors and cortisol responses: A theoretical integration and synthesis of laboratory research. *Psychological Bulletin, 130*(3), 355–391.

Doering, B. K., & Rief, W. (2012). Utilizing placebo mechanisms for dose reduction in pharmacotherapy. *Trends in Pharmacological Sciences, 33*(3), 165–172.

Fehm-Wolfsdorf, G., Gnadler, M., Kern, W., Klosterhalfen, W., & Kerner, W. (1993). Classically conditioned changes of blood glucose level in humans. *Physiology & Behavior, 54*(1), 155–160.

Fries, E., Hesse, J., Hellhammer, J., & Hellhammer, D. H. (2005). A new view on hypocortisolism. *Psychoneuroendocrinology, 30*(10), 1010–1016.

Gauci, M., Husband, A. J., Saxarra, H., & King, M. G. (1994). Pavlovian conditioning of nasal tryptase release in human subjects with allergic rhinitis. *Physiology & Behavior, 55*(5), 823–825.

Giang, D. W., Goodman, A. D., Schiffer, R. B., Mattson, D. H., Petrie, M., Cohen, N., et al. (1996). Conditioning of cyclophosphamide-induced leukopenia in humans. *Journal of Neuropsychiatry and Clinical Neurosciences, 8*(2), 194–201.

Goebel, M. U., Hubell, D., Kou, W., Janssen, O. E., Katsarava, Z., Limmroth, V., et al. (2005). Behavioral conditioning with interferon beta-1a in humans. *Physiology & Behavior, 84*(5), 807–814.

Goebel, M. U., Meykadeh, N., Kou, W., Schedlowski, M., & Hengge, U. R. (2008). Behavioral conditioning of antihistamine effects in patients with allergic rhinitis. *Psychotherapy and Psychosomatics*, 77(4), 227–234.

Goebel, M. U., Trebst, A. E., Steiner, J., Xie, Y. F., Exton, M. S., Frede, S., et al. (2002). Behavioral conditioning of immunosuppression is possible in humans. *The FASEB Journal, 16*(14), 1869–1873.

Grigoleit, J. S., Kullmann, J. S., Winkelhaus, A., Engler, H., Wegner, A., Hammes, F., et al. (2012). Single-trial conditioning in a human taste-endotoxin paradigm induces conditioned odor aversion but not cytokine responses. *Brain, Behavior, and Immunity, 26*(2), 234–238.

Janz, L. J., Green-Johnson, J., Murray, L., Vriend, C. Y., Nance, D. M., Greenberg, A. H., et al. (1996). Pavlovian conditioning of LPS-induced responses: Effects on corticosterone, splenic NE, and IL-2 production. *Physiology & Behavior, 59*(6), 1103–1109.

Kirschbaum, C., Jabaaij, L., Buske-Kirschbaum, A., Hennig, J., Blom, M., Dorst, K., et al. (1992). Conditioning of drug-induced immunomodulation in human volunteers: A European collaborative study. *British Journal of Clinical Psychology, 31*(Pt. 4), 459–472.

Longo, D. L., Duffey, P. L., Kopp, W. C., Heyes, M. P., Alvord, W. G., Sharfman, W. H., et al. (1999). Conditioned immune response to interferon-gamma in humans. *Clinical Immunology, 90*(2), 173–181.

Manaï, M., van Middendorp, H., Veldhuijzen, D. S., & Evers, A. W. M. (2017). In *Optimizing pharmacotherapeutic effects through pharmacological conditioning with or without an online health training: Study protocol. Paper presented at the ARPH Conference,* Leiden.

Meeuwis, S. H., van Middendorp, H., Veldhuijzen, D. S., Pacheco-Lopez, G., & Evers, A. W. M. (2017). In *Placebo effects on itch: Conditioning of antihistaminergic effects, a trial design. Paper presented at the first official SIPS conference,* Leiden, The Netherlands.

Ober, K., Benson, S., Vogelsang, M., Bylica, A., Gunther, D., Witzke, O., et al. (2012). Plasma noradrenaline and state anxiety levels predict placebo response in learned immunosuppression. *Clinical Pharmacology and Therapeutics, 91*(2), 220–226.

Ohman, A. (2009). Of snakes and faces: an evolutionary perspective on the psychology of fear. *Scandinavian Journal of Psychology, 50*(6), 543–552.

Olness, K., & Ader, R. (1992). Conditioning as an adjunct in the pharmacotherapy of lupus erythematosus. *Journal of Developmental and Behavioral Pediatrics, 13*(2), 124–125.

Overduin, J., & Jansen, A. (1997). Conditioned insulin and blood sugar responses in humans in relation to binge eating. *Physiology & Behavior, 61*(4), 569–575.

Pacheco-Lopez, G., Niemi, M. B., Engler, H., & Schedlowski, M. (2007). Neuro-immune associative learning. In F. Bermúdez-Rattoni (Ed.), *Neural plasticity and memory: From genes to brain imaging.* Boca Raton, FL: CRC Press/Taylor & Francis.

Pacheco-Lopez, G., Niemi, M. B., Kou, W., Harting, M., Del Rey, A., Besedovsky, H. O., et al. (2004). Behavioural endocrine immune-conditioned response is induced by taste and superantigen pairing. *Neuroscience, 129*(3), 555–562.

Petrakova, L., Boy, K., Kugler, M., Benson, S., Engler, H., Moller, L., et al. (2017). Plasma cortisol response cannot be classically conditioned in a taste-endocrine paradigm in humans. *Psychopharmacology, 234*(21), 3249–3257.

Putman, P., & Roelofs, K. (2011). Effects of single cortisol administrations on human affect reviewed: Coping with stress through adaptive regulation of automatic cognitive processing. *Psychoneuroendocrinology, 36*(4), 439–448.

Ramsay, D. S., & Woods, S. C. (1997). Biological consequences of drug administration: Implications for acute and chronic tolerance. *Psychological Review, 104*(1), 170–193.

Sabbioni, M. E., Bovbjerg, D. H., Mathew, S., Sikes, C., Lasley, B., & Stokes, P. E. (1997). Classically conditioned changes in plasma cortisol levels induced by dexamethasone in healthy men. *The FASEB Journal, 11*(14), 1291–1296.

Sandler, A. D., & Bodfish, J. W. (2008). Open-label use of placebos in the treatment of ADHD: A pilot study. *Child: Care, Health and Development, 34*(1), 104–110.

Sandler, A. D., Glesne, C. E., & Bodfish, J. W. (2010). Conditioned placebo dose reduction: A new treatment in attention-deficit hyperactivity disorder? *Journal of Developmental and Behavioral Pediatrics, 31*(5), 369–375.

Schakel, L., Veldhuijzen, D. S., van Middendorp, H., Prins, C., Joosten, S. A., Ottenhoff, T. H. M., et al. (2017). The effects of a psychological intervention directed at optimizing immune function: Study protocol for a randomized controlled trial. *Trials, 18*(1), 243.

Schedlowski, M., & Pacheco-Lopez, G. (2010). The learned immune response: Pavlov and beyond. *Brain, Behavior, and Immunity, 24*, 176–185.

Skvortsova, A., Veldhuijzen, D. S., van Middendorp, H., Pacheco-Lopez, G., van Ijzendoorn, M. H., Bakermans-Kranenburg, M. J., et al. (2016). In *Conditioning of the neuroendocrinesystem: Learned oxytocin responses. Paper presented at the LIBC Day,* Leiden.

Stockhorst, U. (2005). Classical conditioning of endocrine effects. *Current Opinion in Psychiatry, 18*, 181–187.

Stockhorst, U., Enck, P., & Klosterhalfen, S. (2007). Role of classical conditioning in learning gastrointestinal symptoms. *World Journal of Gastroenterology, 13*(25), 3430–3437.

Stockhorst, U., de Fries, D., Steingrueber, H. J., & Scherbaum, W. A. (2011). Unconditioned and conditioned effects of intranasally administered insulin vs placebo in healthy men: A randomised controlled trial. *Diabetologia, 54*(6), 1502–1506.

Stockhorst, U., Gritzmann, E., Klopp, K., Schottenfeld-Naor, Y., Hubinger, A., Berresheim, H. W., et al. (1999). Classical conditioning of insulin effects in healthy humans. *Psychosomatic Medicine, 61*(4), 424–435.

Stockhorst, U., Mahl, N., Krueger, M., Huenig, A., Schottenfeld-Naor, Y., Huebinger, A., et al. (2004). Classical conditioning and conditionability of insulin and glucose effects in healthy humans. *Physiology & Behavior, 81*(3), 375–388.

Stockhorst, U., Steingruber, H. J., & Scherbaum, W. A. (2000). Classically conditioned responses following repeated insulin and glucose administration in humans. *Behavioural Brain Research, 110*(1–2), 143–159.

Suchman, A. L., & Ader, R. (1992). Classic conditioning and placebo effects in crossover studies. *Clinical Pharmacology and Therapeutics, 52*(4), 372–377.

Tekampe, J., van Middendorp, H., Meeuwis, S. H., van Leusden, J. W., Pacheco-Lopez, G., Hermus, A. R., et al. (2017). Conditioning immune and endocrine parameters in humans: A systematic review. *Psychotherapy and Psychosomatics*, *86*(2), 99–107.

Tekampe, J., van Middendorp, H., Sweep, F. C. G. J., Roerink, S. H. P. P., Hermus, A. R. M. M., & Evers, A. W. M. (2014). Pilot study on conditioning cortisol and its psychophysiological effects. *International Journal of Behavioral Medicine*, *21*(S1), S148.

Tekampe, J., van Middendorp, H., Sweep, F. C. G. J., Roerink, S. H. P. P., Hermus, A. R. M. M., & Evers, A. W. M. (2017). In *Conditioning cortisol levels in humans: Design and pilot study. Paper presented at the first official SIPS conference,* Leiden, The Netherlands.

Vits, S., Cesko, E., Benson, S., Rueckert, A., Hillen, U., Schadendorf, D., et al. (2013). Cognitive factors mediate placebo responses in patients with house dust mite allergy. *PLoS One*, *8*(11).

Vits, S., Cesko, E., Enck, P., Hillen, U., Schadendorf, D., & Schedlowski, M. (2011). Behavioural conditioning as the mediator of placebo responses in the immune system. *Philosophical Transactions of the Royal Society of London. Series B, Biological Sciences*, *366*(1572), 1799–1807.

Wirth, T., Ober, K., Prager, G., Vogelsang, M., Benson, S., Witzke, O., et al. (2011). Repeated recall of learned immunosuppression: Evidence from rats and men. *Brain, Behavior, and Immunity*, *25*(7), 1444–1451.

Woods, S. C., & Ramsay, D. S. (2000). Pavlovian influences over food and drug intake. *Behavioural Brain Research*, *110*(1–2), 175–182.

Yehuda, R., & Seckl, J. (2011). Minireview: Stress-related psychiatric disorders with low cortisol levels: A metabolic hypothesis. *Endocrinology*, *152*(12), 4496–4503.

CHAPTER FIVE

Response Expectancy and the Placebo Effect

Irving Kirsch[1]
Harvard Medical School, Boston, MA, United States
[1]Corresponding author: e-mail address: irvkirsch@gmail.com

Contents

Abstract

In this chapter, I review basic tenets of response expectancy theory (Kirsch, 1985), beginning with the important distinction between response expectancies and stimulus expectancies. Although both can affect experience, the effects of response expectancies are stronger and more resistant to extinction than those of stimulus expectancies. Further, response expectancies are especially important to understanding placebo effects. The response expectancy framework is consistent with and has been amplified by the Bayesian model of predictive coding. Clinical implications of these phenomena are exemplified.

INTRODUCTION

There are general agreement and substantial data, indicating that expectancy is central to the placebo effect. Nevertheless, there are two problems with this statement. The first is the use of the term *the* in the phrase *the placebo effect*. As Benedetti (2014) has reminded us, "there is not one single placebo effect, but many" (p. 623). For example, placebo morphine activates opioid receptors (Amanzio & Benedetti, 1999), whereas placebo NSAIDs activate cannabinoid receptors (Benedetti, Amanzio, Rosato, & Blanchard, 2011).

International Review of Neurobiology, Volume 138
ISSN 0074-7742
https://doi.org/10.1016/bs.irn.2018.01.003

Furthermore, the magnitude of the placebo effect varies as a function of factors such as the condition being treated, the color of the placebo, the apparent dose, the brand name, the price, and the mode of administration (Kirsch, 2006). The direction of these differences can easily be predicted by their relationship to intuitive expectancies about effectiveness. For example, many people expect a high price, brand-name medication given at a high dose to be more effective than lower doses of inexpensive generic treatments, and surgery as a more potent intervention than simply taking a pill. These implicit expectancies are matched by differences in the effectiveness of a placebo intervention (Branthwaite & Cooper, 1981; Goetz et al., 2008; Shiv, Carmon, & Ariely, 2005).

The second problem with my opening statement is the implication that expectancy is a unitary construct. It is not. Just as there are different types of placebo effect, so too are there different kinds of expectancy. The distinction between response expectancies and stimulus expectancies is particularly important for understanding placebo effects. In this chapter, I review this distinction and demonstrate its importance in designing studies on placebo. I also examine various dimensions of response expectancy and how they can be used to most effectively promote therapeutic change. Finally, I note the substantial overlap between the framework of response expectancy and that of Bayesian predictive coding.

1. RESPONSE EXPECTANCY VS STIMULUS EXPECTANCY

I first introduced the concept of response expectancy in 1985 as a causal factor in generating placebos effects, hypnosis, anxiety disorders, and the outcome of psychological treatments (Kirsch, 1985). Stimulus expectancies are anticipations of external events. Response expectancies are predictions of one's own nonvolitional responses (i.e., automatic reactions) to events. As applied to expected pain reduction, for example, the expectation that the pain stimulation will be less intense is a stimulus expectancy. In contrast, the expectation that one will feel less pain, despite there being no change in the stimulus intensity, is a response expectancy.[a]

[a] A number of authors have recently suggested that the term *expectation* be used to denote a subset of expectancies, specifically, those that are consciously accessible and therefore verbalizable (Corsi & Colloca, 2017; Kube, Rief, & Glombiewski, 2017). In this chapter, we follow this terminology by restricting the use of the term *expectation* to those expectancies that can be introspected and rated by the individual holding them.

From the example above, it should be clear that response expectancy is closely related to the placebo effect, whereas stimulus expectancies are not. We believe that a drug or procedure has lessened our sensitivity to painful stimuli, not that the pain stimulus has been lowered. But is this distinction important? Perhaps it makes no difference whether the anticipated pain reduction is a stimulus expectancy or a response expectancy. The question is, what differences in outcome are there as a function of believing that the intensity of the stimulus has been reduced, compared to believing that a treatment is making one less sensitive to pain?

With respect to the placebo effect, the most important aspect of response expectancies is their tendencies to be self-confirming (Kirsch, 1985). Non-volitional responses (e.g., pain perception, emotional reactions, and other subjective states and sensations) can be elicited and/or enhanced by the expectancy of their occurrence. Stimulus expectancies can also affect perception (Bugelski & Alampay, 1961), but response expectancies are stronger, more stable, and more resistant to extinction (Schenk, Sprenger, Onat, Colloca, & Büchel, 2017). One reason for this is that the role of expectancies and response sets on perception varies as a function of the ambiguity of the stimulus—the greater the ambiguity of the stimulus, the larger the role of expectancy in the perception of it (Kirsch, 1999). Indeed, studies reported expectancy effects on stimulus perception rely almost entirely on studies of ambiguous stimuli (see target article and commentaries in Firestone & Scholl, 2016). Perceptual expectancies enable more rapid recognition and classification of ambiguous stimuli than would otherwise be possible. The price of speed is accuracy, but the trade-off is often worthwhile. Rapid recognition can have considerable survival value, even if it is sometimes inaccurate. Less harm may be done by mistaking a stick for a rattlesnake than by taking the time to examine a snake with sufficient care to determine reliably that it is indeed dangerous.

Brains evolved because they provide animals with an adaptive advantage. They are "anticipation machines" (Dennett, 1991, p. 178) that allow organisms to interpret and respond their environments effectively and efficiently. According to Kinsbourne (1998), the content of consciousness arises as a pattern of neuronal activity formed by the clash of two colliding waves, an incoming wave from receptors and an outgoing anticipatory wave. Thus, experience "is an inextricable amalgam of represented anticipation and represented perturbation" (Kinsbourne, 1998, p. 241).

In these characterizations of expectancy effects, one can see a close affinity between expectancy theory and Bayesian predictive coding formulations, in which brains are described as "prediction machines" (Clark, 2013, p. 181). Further, in these formulations, terms "prior beliefs," "expectations," and "predictions" are used interchangeably (e.g., Büchel, Geuter, Sprenger, & Eippert, 2014; Edwards, Adams, Brown, Pareés, & Friston, 2012; Wiech, 2016). The Bayesian framework provides computational tools for elucidating how expectancies are changed and how they influence perception.

As in Bayesian approaches, expectancies can operate automatically without conscious awareness (Kirsch & Lynn, 1998, 1999). Perhaps the most compelling example of this involves the behavior of athletes. When catching a ball, for example, the person doing the catch must reach for the spot at which he or she anticipates the ball being, not where it is now, and this must be done with such speed as to preclude conscious deliberation (Reed, McLeod, & Dienes, 2010). Similarly, having developed an expectancy or response set for pain relief following administration of a treatment, that expectancy need not be held in consciousness at the moment at which it influences pain perception. Nevertheless, in many circumstances, people are able to report their beliefs and expectancies when asked to do so, and this enables us to assess response expectancies as predictors of experience and behavior.

To summarize, we are biologically structured to respond rapidly to the environment, and in so doing, we use perceptual templates or expectancies to resolve stimulus ambiguities. Thus, stimulus expectancies alter perceptual responses, especially to ambiguous stimuli. Because internal responses function as highly ambiguous stimuli, expectancies about them (i.e., response expectancies) are self-confirming to a fairly substantial degree. For example, they can be sufficiently self-confirming as to override and reverse the effects of potent chemical agents (e.g., Bingel et al., 2011; Ikemi & Nakagawa, 1962; Wolf, 1950).

2. OVERGENERALIZATION FROM STIMULUS EXPECTANCY TO RESPONSE EXPECTANCY EFFECTS

One of the most common ways of strengthening placebo effects is by means of a conditioning-like procedure (e.g., Kirsch et al., 2014; Montgomery & Kirsch, 1997; Price et al., 1999; Voudouris, Peck, & Coleman, 1985; Watson, El-Deredy, Bentley, Vogt, & Jones, 2006). Typically, a topical placebo is applied to one location but not another. During conditioning trials, stimulus intensity is lowered surreptitiously at the site at which the placebo has been applied. Then, to test the effect of this procedure, stimulus intensity at the "treated" site is increased to the point of being equal to that of the stimulus intensity at the "untreated" site. The difference in pain between the reported pain at the treated site and reported pain at the untreated site during the test trials is the placebo effect.

This procedure produces a response expectancy for lowered pain perception (Montgomery & Kirsch, 1997). It has been shown to be very effective in enhancing placebo analgesia, and the strength of the placebo effect has been found to be highly correlated with the magnitude of the response expectancy. Furthermore, as predicted by response expectancy theory (Kirsch, 1985, 1999), because response expectancies are self-confirming, they are highly resistant to extinction (Colloca & Benedetti, 2006; Montgomery & Kirsch, 1997). Indeed, in the Montgomery and Kirsch study, instead of extinguishing, the placebo effect increased during the extinction period.

That this resistance to extinction is due to the difference between stimulus expectancy and response expectancy has been confirmed in an elegant study by Schenk et al. (2017). They compared the effects of two different conditioning procedures, both involving surreptitious lowering of the pain stimulus intensity during the conditioning phase of the experiment. In both procedures, different colored lights were associated with low and high pain stimulus intensities. Participants in one condition were told that one of the colors would be paired with analgesic treatment by transcutaneous electrical nerve stimulation, thereby inducing a response expectancy. Participants in the other group were told that the colors signaled different intensities of the stimulus, thereby inducing a stimulus expectancy. As predicted, they found that stimulus expectancies, but not response expectancies, were correctly updated during the extinction/test phase. This, in turn, led to greater resistance to extinction of the conditioned hypoalgesia response when conditioning led people to believe that the decrease in pain was due to treatment than when they thought that the stimulus intensity had been lowered.

The distinction between conditioned stimulus expectancies and conditioned response expectancies seems relatively subtle, given the almost complete procedural overlap between the two phenomena. Indeed, I overlooked it in a series of studies on analgesia conditioning on which I was a coauthor (Jensen et al., 2014, 2012), until an anonymous reviewer called it to our attention in the most recent of that series of studies (Jensen, Kirsch, Odmalm, Kaptchuk, & Ingvar, 2015). In those studies, we paired pictures of faces with a surreptitiously lowered intensity pain stimulus that had been presented either subliminally or supraliminally during the test phase (Jensen et al., 2014, 2012) or during both the acquisition and test phases (Jensen et al., 2015). We found that conditioning could produce hypoalgesia regardless of whether the conditional stimuli were presented subliminally or supraliminally during the learning phase, the test phase, or both. The conditioning procedure in these studies was done in what Schenk and colleagues referred to as a stimulus context, rather than a treatment context. As a result, our studies were not directly related to placebo and nocebo effects, as the titles of the first two studies erroneously suggested. This does not diminish the importance of these studies, and it seems likely that the findings regarding subliminal vs supraliminal presentations of the conditional stimuli would be the same in a treatment context as in a stimulus context. Nevertheless, the data reported by Schenk et al. (2017) suggest the advisability of following up on these studies with similar studies framed in a treatment context. These data would confirm whether placebo effects, as well as non-placebo hypoalgesia, can be enhanced by conditioning with subliminal stimuli.

3. DIMENSIONS OF RESPONSE EXPECTANCIES

Response expectancies vary along at least two dimensions: one is the magnitude of the expected response. Another is the strength or certainty with which the expectancy is held (i.e., how likely is it that the response will occur). The magnitude of the expected response can be assessed with questions like "how much pain do you think you will experience after treatment?" It can be manipulated with instructions like "this is a very powerful opioid" vs "this produces a modest reduction in pain." In contrast, the strength of the expectancy can be assessed with questions like "How certain are you that your pain will be reduced after treatment." It can be manipulated with procedures like administering a treatment open label, double blind, or not at all, corresponding to 100%, 50%, or 0% likelihood.

Many studies manipulate and/or assess one or the other dimension of response expectancy (expected magnitude or strength of the expectancy). It is rare for both to be studied in the same research. One of the rare exceptions is a study on placebo caffeine reported by Lynne Weixel and myself in 1988. In that study (Kirsch & Weixel, 1988), we manipulated the strength of the expectancy by leading some participants to believe that they were receiving caffeine (deceptive administration) and leading others to believe that they might or might not be receiving caffeine (double-blind administration), with an additional group of participants not receiving any beverage at all. We manipulated the magnitude of the expectancy by varying the apparent dose of the caffeine that was presumably administered. We did this by 1, 2, 3, 5, or 8 heaping tablespoons of decaffeinated coffee in preparing the beverage.

We predicted that stronger expectancies (deceptive administration) would produce greater changes than weaker expectancies (double-blind administration), and we confirmed this with respect to heart rate and the degree of confidence that caffeine had been consumed (which would be considered a successful manipulation check). We also predicted a quadratic effect as a function of apparent dose. Although the previous research (Ross & Olson, 1982) reported linear effects of small to moderate apparent doses of a placebo, we hypothesized that more extreme apparent doses (e.g., 8 tablespoons of coffee) would strain subjects' credibility, thereby inhibiting the placebo effect. We viewed this as analogous to the negative effect of overselling a treatment. Because the magnitude of the expected response would be very high with extremely large doses, the discrepancy between the anticipated response and the actual response (i.e., the prediction error) would be especially large, leading participants to suspect that they had not received an active drug at all. We observed the predicted quadratic effects of apparent dose of the placebo caffeine on subjective alertness, subjective tension, systolic blood pressure, and participants' ratings of the likelihood that they are been given caffeine, but these effects were found only when the placebo was given with deceptive instructions. With double-blind administration, the curves on all four outcomes were in the opposite direction (i.e., U shaped). The distinction between the strength and the magnitude of response expectancies is mirrored in the Bayesian predictive coding perspective, where what I have called the strength or confidence with which an expectancy is held is referred to as the precision of the prediction (Büchel et al., 2014).

4. MEASURING EXPECTATIONS

Expectation can be measured in various ways, from single expectancy questions to multiitem scales. One common single-item treatment credibility measure is to ask "How effective do you think this treatment will be?" This is unproblematic in conventional randomized controlled trials, because genuine treatments and placebos can often be administered with participants kept blind to which they are receiving. To assess the placebo effect, however, one needs to compare the placebo response to what would happen if the placebo is not given, through the conclusion of a natural history control group. In this case, participants cannot generally be blinded as to whether they are receiving placebos or are in the natural history control condition. Hence, expectations assessed in this manner would only predict outcome within the group that receives the placebo treatment.

An alternative is to phrase expectancy questionnaires to refer to anticipated outcomes without in the same way that the actual outcomes are assessed, with reference to the time at which the outcome will occur rather than the perceived efficacy of the treatment. Thus, one might ask "What do you expect your level of pain intensity to be?" (Vase et al., 2013, p. 1661). Measures of this sort are only weakly correlated with treatment credibility measures, but are much better predictors of outcome than are treatment creditability measures (Kirsch, Tennen, Wickless, Saccone, & Cody, 1983). Therefore, my advice is to measure expectations using a measure that parallels the actual outcome measure being used.

Finally, it is important to note that expectancies are fluid. They change in response to new experiences. Therefore, it can be useful to measure them multiple times in a single study. Vase and colleagues, for example, reported that "expected pain levels accounted for significant and progressively larger amounts of the variance in pain ratings following both active and placebo acupuncture" (Vase et al., 2013, p. 1659).

5. HOW PLACEBOS WORK: AN INTEGRATIVE MODEL

The magnitude of placebo effects has been shown to be affected by verbal information (Amanzio & Benedetti, 1999), conditioning procedures (Montgomery & Kirsch, 1997), observational learning (Colloca & Benedetti, 2009), response expectancies (Kirsch, 1985; Vase, Robinson, Verne, & Price, 2003), and the therapeutic relationship (Kaptchuk et al., 2008). My colleagues and I (Kirsch, Wampold, & Kelley, 2016) have proposed an integrative heuristic model of placebo effects (Fig. 1), based in part on previous models linking these constructs together (Colloca & Miller, 2011).

In our model, the effects of verbal information, conditioning, and observational learning (modeling) are mediated by response expectancy. Benedetti and colleagues, however, reported that while this held for pain, the effect of conditioning on placebo-engendered hormonal responses was independent of expectations (Benedetti et al., 2003). Generalizing from this finding, they hypothesized that mediation by conscious expectations was limited to consciously perceptible response and did not extend to unconscious outcomes. We have included this in our model, although

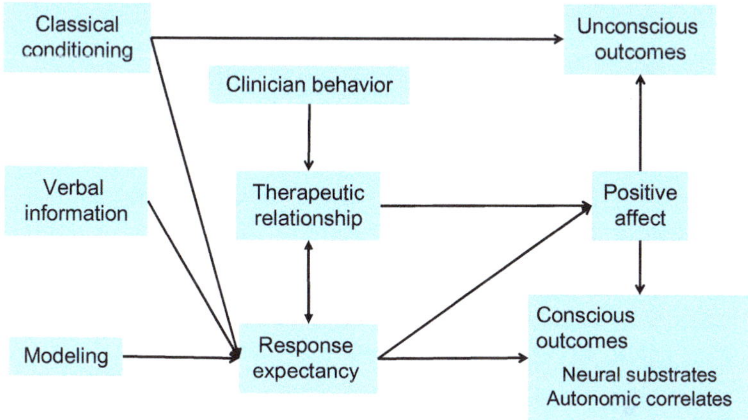

Fig. 1 An integrative heuristic model of placebo effects. *Adapted from Kirsch, I., Wampold, B. E., & Kelley, J. M. (2016). Controlling for the placebo effect in psychotherapy: Noble quest or tilting at windmills?* Psychology of Consciousness: Theory, Research, and Practice, 3*(2), 121–131. doi:10.1037/cns0000065.*

research showing this difference with other conscious vs unconscious responses would be welcome.

Research has shown that clinician behavior indicating warmth (e.g., eye contact) leads to the perception of the clinician as more empathic, warm, and also as more competent (Kraft-Todd et al., 2017). It also leads to better clinical outcomes (Kaptchuk et al., 2008). We postulate a bidirectional causal relationship between response expectancies and the therapeutic relationship. Specifically, trust in a clinician who seems competent and caring is likely to produce expectancies of better outcomes. Reciprocally, the belief that one's condition will improve as a function of treatment is likely to promote a better therapeutic alliance. In general, however, data on the relationship between response expectancies and the therapeutic relationship are lacking.

In addition to affecting each other reciprocally, therapist warmth and the expectation of improvement are likely to produce positive affect, and just as stress might engender negative outcomes, a positive mood might promote positive health outcomes. Evidence for this can be found in the classic "room with a view" study (Ulrich, 1984). Patients who had just had gallbladder surgery were divided into two groups. Those in one group were given rooms with windows looking out over a park-like setting with trees and plants. The other patients were assigned rooms with views of a brick wall. Those given the rooms with the view of trees and plants required significantly less pain medication and were discharged from the hospital sooner than the others. Their nurses were more likely to describe them as doing well and being in good spirits. Patients in the rooms facing the brick wall needed more medication, took longer to be discharged, and were described as upset and needing encouragement. In other words, feeling good psychologically makes you feel good physically. A warm and caring therapeutic relationship feels good. It leads the patient to feel hopeful rather than hopeless. It facilitates an expectation for improvement and that may, at least in part, explain its ability to facilitate healing (Kirsch, 2010).

6. CLINICAL IMPLICATIONS

A good scientific theory is potentially falsifiable (Popper, 1968). If a theory is structured so that no set of observations would prove it to be wrong, it is dubbed a pseudoscience. According to Popper, examples include astrology, psychoanalysis, and Marxism. Conversely, for a response expectancy to be resistant to extinction, it should be as difficult to disconfirm as possible. Indeed, the best placebo communication would not be refutable at all. Here are some examples of verbal information aimed at fortifying the placebo effect:

- Maximizing certainty (precision)
- Minimizing initial magnitude
- Establishing a mindset for gradual change, especially for chronic conditions
- Preparing for setbacks

The first two of these placebogenic communications are based on the distinction between the magnitude of the expected response and strength, confidence, or precision with which the expectancy is held. When a person has a strong expectancy for a relatively small change in response, the expectancy is likely to be confirmed. In contrast, a relatively weak expectancy for a large change in response is more likely to be disconfirmed. Specifically, "the probability of occurrence of a nonvolitional response varies directly with the strength of the expectancy of its occurrence and inversely with the magnitude or difficulty of the expected response. Therefore, a weak expectancy for a large change in subjective response is likely to be disconfirmed, whereas a strong expectancy for a small change is likely to be confirmed, thereby strengthening the expectancy and initiating a cycle for continued change" (Kirsch, 1985, p. 1199).

The timing of the response is an aspect of its expected magnitude that is particularly important for maximizing expectancy effects for chronic conditions. Whereas an expectancy for an initially small change in experience facilitates confirmation of the expectancy, promoting an expectancy for continued improvement provides a means of snowballing this into more powerful therapeutic effects (Kirsch, 1994, 1998). It does so, in part, by facilitating the interpretation of random fluctuations as evidence that the treatment is working, thereby reinforcing and strengthening the expectancy and increasing the likelihood of further change. Finally, cautioning that change is not necessarily linear and may be prone to setbacks can help immunize the person against eventual setbacks by framing them as expected and temporary.

These are a few examples of how communication can facilitate expectancy-induced therapeutic change. For a more detailed account, see *A Clinician's Guide to Expectancy Modification,* which is chapter 7 of *Changing Expectations: A Key to Effective Psychotherapy* (Kirsch, 1990).

REFERENCES

Amanzio, M., & Benedetti, F. (1999). Neuropharmacological dissection of placebo analgesia: Expectation-activated opioid systems versus conditioning-activated specific subsystems. *The Journal of Neuroscience, 19*(1), 484–494.

Benedetti, F. (2014). Placebo effects: From the neurobiological paradigm to translational implications. *Neuron, 84*(3), 623–637.

Benedetti, F., Amanzio, M., Rosato, R., & Blanchard, C. (2011). Nonopioid placebo analgesia is mediated by CB1 cannabinoid receptors. *Nature Medicine, 17*(10), 1228–1230.

Benedetti, F., Pollo, A., Lopiano, L., Lanotte, M., Vighetti, S., & Rainero, I. (2003). Conscious expectation and unconscious conditioning in analgesic, motor, and hormonal placebo/nocebo responses. *Journal of Neuroscience, 23*(10), 4315–4323.

Bingel, U., Wanigasekera, V., Wiech, K., Mhuircheartaigh, R. N., Lee, M. C., Ploner, M., et al. (2011). The effect of treatment expectation on drug efficacy: Imaging the analgesic benefit of the opioid remifentanil. *Science Translational Medicine, 3*(70), 70ra14.

Branthwaite, A., & Cooper, P. (1981). Analgesic effects of branding in treatment of headaches. *British Medical Journal (Clinical Research Ed), 282*(6276), 1576–1578.

Büchel, C., Geuter, S., Sprenger, C., & Eippert, F. (2014). Placebo analgesia: A predictive coding perspective. *Neuron, 81*(6), 1223–1239.

Bugelski, B. R., & Alampay, D. A. (1961). The role of frequency in developing perceptual sets. *Canadian Journal of Psychology/Revue Canadienne de Psychologie Expérimentale, 15*(4), 205.

Clark, A. (2013). Whatever next? Predictive brains, situated agents, and the future of cognitive science. *Behavioral and Brain Sciences, 36*(3), 181–204.

Colloca, L., & Benedetti, F. (2006). How prior experience shapes placebo analgesia. *Pain, 124*(1), 126–133.

Colloca, L., & Benedetti, F. (2009). Placebo analgesia induced by social observational learning. *Pain, 144*(1–2), 28–34. https://doi.org/10.1016/j.pain.2009.01.033.

Colloca, L., & Miller, F. G. (2011). How placebo responses are formed: A learning perspective. *Philosophical Transactions of the Royal Society, B: Biological Sciences, 366*(1572), 1859–1869. https://doi.org/10.1098/rstb.2010.0398.

Corsi, N., & Colloca, L. (2017). Placebo and nocebo effects: The advantage of measuring expectations and psychological factors. *Frontiers in Psychology, 8*, 308. https://doi.org/10.3389/fpsyg.2017.00308.

Dennett, D. C. (1991). *Consciousness explained.* New York: Little Brown & Co.

Edwards, M. J., Adams, R. A., Brown, H., Pareés, I., & Friston, K. J. (2012). A Bayesian account of 'hysteria'. *Brain, 135*(11), 3495–3512.

Firestone, C., & Scholl, B. J. (2016). Cognition does not affect perception: Evaluating the evidence for "top-down" effects. *Behavioral and Brain Sciences 39*, e229. https://doi.org/10.1017/S0140525X15000965.

Goetz, C. G., Wuu, J., McDermott, M. P., Adler, C. H., Fahn, S., Freed, C. R., et al. (2008). Placebo response in Parkinson's disease: Comparisons among 11 trials covering medical and surgical interventions. *Movement Disorders, 5*, 690–699.

Ikemi, Y., & Nakagawa, S. (1962). A psychosomatic study of contagious dermatitis. *Kyoshu Journal of Medical Science, 13*, 335–350.

Jensen, K. B., Kaptchuk, T. J., Chen, X., Kirsch, I., Ingvar, M., Gollub, R. L., et al. (2014). A neural mechanism for nonconscious activation of conditioned placebo and nocebo responses. *Cerebral Cortex, 25*, 3903–3910. https://doi.org/10.1093/cercor/bhu275.

Jensen, K. B., Kaptchuk, T. J., Kirsch, I., Raicek, J., Lindstrom, K. M., Berna, C., et al. (2012). Nonconscious activation of placebo and nocebo pain responses. *Proceedings of the National Academy of Sciences of the United States of America, 109*(39), 15959–15964. https://doi.org/10.1073/pnas.1202056109.

Jensen, K. B., Kirsch, I., Odmalm, S., Kaptchuk, T. J., & Ingvar, M. (2015). Classical conditioning of analgesic and hyperalgesic pain responses without conscious awareness. *Proceedings of the National Academy of Sciences of the United States of America, 112*, 7863–7867. https://doi.org/10.1073/pnas.1504567112.

Kaptchuk, T. J., Kelley, J. M., Conboy, L. A., Davis, R. B., Kerr, C. E., Jacobson, E. E., et al. (2008). Components of placebo effect: Randomised controlled trial in patients with irritable bowel syndrome. *BMJ, 336*(7651), 999–1003. https://doi.org/10.1136/bmj.39524.439618.25.

Kinsbourne, M. (1998). Unity and diversity in the human brain: Evidence from injury. *Daedalus, 127*(2), 233–256.

Kirsch, I. (1985). Response expectancy as a determinant of experience and behavior. *American Psychologist, 40*(11), 1189–1202.

Kirsch, I. (1990). *Changing expectations: A key to effective psychotherapy*. Pacific Grove, CA: Brooks/Cole.

Kirsch, I. (1994). Clinical hypnosis as a nondeceptive placebo—Empirically derived techniques. *American Journal of Clinical Hypnosis, 37*(2), 95–105.

Kirsch, I. (1998). Clinical hypnosis as a nondeceptive placebo. In I. C. A. Kirsch, et al. (Ed.) *Clinical hypnosis and self regulation: Cognitive behavioral perspectives* (pp. 211–225). Washington, DC: American Psychological Association.

Kirsch, I. (1999). Response expectancy: An introduction. In I. Kirsch (Ed.), *How expectancies shape experience* (pp. 3–13). Washington DC: American Psychological Association.

Kirsch, I. (2006). Placebo: The role of expectancies in the generation and alleviation of illness. In P. Halligan & A. Mansel (Eds.), *The power of belief: Psychosocial influence on illness, disability and medicine* (pp. 55–67). Oxford: Oxford University Press.

Kirsch, I. (2010). *The emperor's new drugs: Exploding the antidepressant myth*. New York: Basic Books.

Kirsch, I., Kong, J., Sadler, P., Spaeth, R., Cook, A., Kaptchuk, T. J., et al. (2014). Expectancy and conditioning in placebo analgesia: Separate or connected processes? *Psychology of Consciousness: Theory, Research, and Practice, 1*(1), 51–59. https://doi.org/10.1037/cns0000007.

Kirsch, I., & Lynn, S. J. (1998). Hypnotic involuntariness and the automaticity of everyday life. In I. Kirsch, A. Capafons, E. Cardeña, & S. Amigó (Eds.), *Clinical hypnosis and self regulation: Cognitive behavioral perspectives* (pp. 49–72). Washington, DC: American Psychological Association.

Kirsch, I., & Lynn, S. J. (1999). Automaticity in clinical psychology. *American Psychologist, 54*(7), 504–515.

Kirsch, I., Tennen, H., Wickless, C., Saccone, A. J., & Cody, S. (1983). The role of expectancy in fear reduction. *Behavior Therapy*, *14*(4), 520–533. https://doi.org/10.1016/S0005-7894(83)80075-6.

Kirsch, I., Wampold, B. E., & Kelley, J. M. (2016). Controlling for the placebo effect in psychotherapy: Noble quest or tilting at windmills? *Psychology of Consciousness: Theory, Research, and Practice*, *3*(2), 121–131. https://doi.org/10.1037/cns0000065.

Kirsch, I., & Weixel, L. J. (1988). Double-blind versus deceptive administration of a placebo. *Behavioral Neuroscience*, *102*(2), 319–323.

Kraft-Todd, G. T., Reinero, D. A., Kelley, J. M., Heberlein, A. S., Baer, L., & Riess, H. (2017). Empathic nonverbal behavior increases ratings of both warmth and competence in a medical context. *PLoS One*, *12*(5), e0177758. https://doi.org/10.1371/journal.pone.0177758.

Kube, T., Rief, W., & Glombiewski, J. A. (2017). On the maintenance of expectations in major depression—Investigating a neglected phenomenon. *Frontiers in Psychology*, *8*, 9. https://doi.org/10.3389/fpsyg.2017.00009.

Montgomery, G. H., & Kirsch, I. (1997). Classical conditioning and the placebo effect. *Pain*, *72*(1–2), 107–113.

Popper, K. R. (1968). *Conjectures and refutations: The growth of scientific knowledge*. New York: Harper & Row.

Price, D. D., Milling, L. S., Kirsch, I., Duff, A., Montgomery, G. H., & Nicholls, S. S. (1999). An analysis of factors that contribute to the magnitude of placebo analgesia in an experimental paradigm. *Pain*, *83*(2), 147–156.

Reed, N., McLeod, P., & Dienes, Z. (2010). Implicit knowledge and motor skill: What people who know how to catch don't know. *Consciousness and Cognition*, *19*(1), 63–76. https://doi.org/10.1016/j.concog.2009.07.006.

Ross, M., & Olson, J. K. (1982). Placebo effects in medical research and practice. In J. R. Eiser (Ed.), *Social psychology and behavioral medicine* (pp. 441–458). New York: Wiley.

Schenk, L. A., Sprenger, C., Onat, S., Colloca, L., & Büchel, C. (2017). Suppression of striatal prediction errors by the prefrontal cortex in placebo hypoalgesia. *The Journal of Neuroscience*, *37*(40), 9715–9723. https://doi.org/10.1523/jneurosci.1101-17.2017.

Shiv, B., Carmon, Z., & Ariely, D. (2005). Placebo effects of marketing actions: Consumers may get what they pay for. *Journal of Marketing Research*, *42*(4), 383–393. https://doi.org/10.1509/jmkr.2005.42.4.383.

Ulrich, R. S. (1984). View through a window may influence recovery from surgery. *Science*, *224*(4647), 420–421.

Vase, L., Baram, S., Takakura, N., Yajima, H., Takayama, M., Kaptchuk, T. J., et al. (2013). Specifying the non-specific components of acupuncture analgesia. *Pain*, *54*(9), 1659–1667. https://doi.org/10.1016/j.pain.2013.05.008

Vase, L., Robinson, M. E., Verne, G. N., & Price, D. D. (2003). The contributions of suggestion, desire, and expectation to placebo effects in irritable bowel syndrome patients: An empirical investigation. *Pain*, *105*(1), 17–25. https://doi.org/10.1016/S0304-3959(03)00073-3.

Voudouris, N. J., Peck, C. L., & Coleman, G. (1985). Conditioned placebo responses. *Journal of Personality and Social Psychology*, *48*, 47–53.

Watson, A., El-Deredy, W., Bentley, D. E., Vogt, B. A., & Jones, A. K. P. (2006). Categories of placebo response in the absence of site-specific expectation of analgesia. *Pain*, *126*, 115–122.

Wiech, K. (2016). Deconstructing the sensation of pain: The influence of cognitive processes on pain perception. *Science*, *354*(6312), 584–587. https://doi.org/10.1126/science.aaf8934.

Wolf, S. (1950). Effects of suggestion and conditioning on the action of chemical agents in human subjects—The pharmacology of placebos. *Journal of Clinical Investigation*, *29*, 100–109.

CHAPTER SIX

A Functional-Cognitive Perspective on the Relation Between Conditioning and Placebo Research

Jan De Houwer[1]
Ghent University, Ghent, Belgium
[1]Corresponding author: e-mail address: jan.dehouwer@ugent.be

Contents

Abstract

From a functional-cognitive perspective, conditioning is relevant for placebo research because it (a) highlights stimulus pairings (e.g., taking a pill that contains an active pharmacological substance causing a drop in blood pressure) as a potential environmental cause of creating or boosting placebo effects (e.g., reduction in blood pressure after taking a similar pill that no longer contains the active substance) and (b) orients researchers to potential mental mechanisms that might underlie those effects (e.g., the formation of associations or propositions). After describing the functional-cognitive perspective on conditioning, I provide a brief overview of three generations of conditioning theories (stimulus–response, stimulus–stimulus, and propositional theories) and evaluate different ways in which conditioning and placebo research can be related. Finally, I discuss the implications of the functional-cognitive perspective on conditioning for the status of the placebo phenomenon.

International Review of Neurobiology, Volume 138
ISSN 0074-7742
https://doi.org/10.1016/bs.irn.2018.01.007

Extensive empirical evidence supports the conclusion that conditioning plays a crucial role in establishing and boosting placebo effects (e.g., Colloca & Miller, 2011). In this chapter, I put forward a functional-cognitive perspective on the relation between conditioning and placebo effects. This perspective is built on a definition of conditioning that does not refer to any explanatory mental mechanisms such as expectations or associations. By separating conditioning as a to-be-explained effect from the mental processes that might explain this effect, new light can be shed on different claims about conditioning and its role in placebo effects. In the first section, I describe what it means to define conditioning as an effect rather than a mental process. Next I present a brief overview of various ideas about the mental processes that might explain conditioning effects. Afterward, I discuss the implications for placebo research.

1. CONDITIONING AS AN EFFECT

In a recent paper, my colleagues and I defined learning as ontogenetic adaptation, that is, as the impact of regularities in the environment on behavior during the lifetime of the organism (see De Houwer, Barnes-Holmes, & Moors, 2013; De Houwer & Hughes, 2017, for more details). From this perspective, different types of learning can be distinguished on the basis of the type of regularity that changes behavior. Nonassociative types of learning encompass the effects of regularities in the presence of one stimulus. For instance, habituation refers to the decline in intensity of an original response that occurs as the result of the repeated presentation of the response evoking stimulus. The term conditioning, on the other hand, refers to associative types of learning in which a change in behavior is due to regularities in the presence of two events. In this chapter, I focus on classical conditioning (also known as Pavlovian conditioning), in which behavior changes are a function of the pairing of two stimuli, namely a conditioned stimulus (CS) and an unconditioned stimulus (US). For instance, dogs that initially do not salivate upon hearing a buzzer start salivating in response to the buzzer (CS) after it was paired with food (US) (Pavlov, 1927).

This definition of conditioning is functional in nature; that is, it defines conditioning as a function that maps regularities in the environment onto changes in behavior. Put differently, it defines conditioning as an effect, that is, as the impact of CS–US pairings on behavior (also see Eelen, 1980;

Stewart-Williams & Podd, 2004). As such, the claim that a particular change in behavior is an instance of classical conditioning implies a functional explanation of that change in behavior. More specifically, it implies that the change in behavior (e.g., increased salivation in response to the buzzer) is due to the pairings of stimuli (e.g., buzzer and food) rather than other environmental events (e.g., the mere repeated presentation of food). Crucially, a functional definition of conditioning does not entail any assumption about the mental processes via which CS–US pairings influence behavior. Instead, it allows for all kinds of mental explanations of the conditioning phenomenon and therefore does not limit conditioning to a subclass of effects that is supposedly mediated by one particular mechanism. In fact, the theoretical freedom offered by functional definitions of psychological phenomena such as conditioning is one of the important strength of those types of definitions. It allows one to adopt a functional-cognitive framework that combines the strength of functional (e.g., Skinnerian) approaches directed at explaining behavior in term of environmental events and cognitive approaches aimed at explaining psychological phenomena in terms of mental mechanisms (see De Houwer, 2011; Hughes, De Houwer, & Perugini, 2016, for more details).

Rather than accepting a particular mental explanation of conditioning in an a priori manner by incorporating it in the definition of conditioning, theories about the mechanisms that mediate conditioning effects can be substantiated only by examining the moderators of conditioning, that is, the variables that influence its presence and magnitude (De Houwer, 2011; Hughes et al., 2016). A good theoretical model of conditioning is one that can explain the effect of known moderators (high heuristic value) and correctly predicts the effects of yet untested moderators (high predictive value). Unfortunately, like most other humans, scientists are not immune to the logical fallacy known as "affirming the consequent." Hence, when they observe that the prediction of a theoretical model is confirmed, they often conclude too quickly that the model is correct. This fallacy sometimes even results in scientists conflating a phenomenon with one possible explanation of that phenomenon. In fact, as will become clear in the next section of this chapter, much of the confusion surrounding the conditioning literature resulted from this kind of conflation. In order to clear up some of this confusion, I will list and evaluate a number of theoretical models of classical conditioning while keeping those explanations clearly separated from the to-be-explained phenomenon.

2. A BRIEF HISTORY OF CONDITIONING MODELS

2.1 First Generation: S–R Association Formation Models

Conditioning effects were initially explained in terms of the automatic formation of stimulus–response (S–R) associations (e.g., Byrne & Bates, 2006). Before the CS and US are paired, the US (e.g., food) already evokes a certain unconditioned response (UR; e.g., salivation). A prototypical S–R model postulates that, as the result of pairing the CS and US, the CS co-occurs with the UR, which is a necessary and sufficient condition for the formation of an association between the CS and UR. As the S–R associations gradually grow in strength with each CS–US pairing, the presentation of the CS also starts to evoke the UR, which is now called the conditioned response (CR; e.g., salivation upon hearing the buzzer). From this perspective, conditioning is simply about allowing neutral stimuli (CSs) to evoke novel responses that are initially evoked only by the stimuli it is paired with (USs). This process is assumed to be (a) low level in that it does not depend on (conscious) cognitive processes and (b) automatic because it can occur even under conditions that typically hamper cognitive processing (e.g., lack of awareness, time, attention, or intention).

Although this type of model was popular in the 1940s and 1950s, it was abandoned by most learning researchers from the 1960s onward. In a high-profile review article published 30 years ago, Rescorla (1988) reproduced a number of descriptions of this prototypical S–R model as they were provided in psychology textbooks at that time. He noted that "these quotations will certainly sound so familiar that many readers may wonder what is wrong with them. I want to suggest that the answer is "almost everything." … [they] come from … the reflex tradition … which sees conditioning as a kind of low-level mechanical process in which control over a response is passed from one stimulus to another" (pp. 151–152, parentheses added). Rescorla then pointed out that "the prevalent modern view [is] that conditioning involves the learning of relations among events … [this] is not a stupid process by which the organism willy-nilly forms associations between any two stimuli that happen to co-occur. Rather, the organism is better seen as an information seeker using logical and perceptual relations among events, along with its own preconceptions, to form a sophisticated representation of its world. Indeed, in teaching undergraduates, I favor an analogy between animals showing Pavlovian conditioning and scientists identifying the cause of a phenomenon" (pp. 153–154, parentheses added).

Rescorla's conclusions (Rescorla, 1988) are based on a wide range of findings that contradict the predictions of S–R models of conditioning. For instance, it is abundantly clear that the CR (i.e., the new response to the CS that occurs as the result of CS–US pairings) can be fundamentally different from the UR (i.e., the response evoked by the US). Just think of a tone that is repeatedly followed by a painful electric shock. Whereas the shock evokes the experience of pain as an UR, the tone will not evoke pain but rather fear as a CR. Fear is not part of the UR but is induced by the anticipation of the US. Although it is difficult to exclude the possibility that not a single instance of conditioning is due to the formation of S–R associations, at least in humans, there is surprisingly little strong evidence for the operation of these processes (see Hogarth & Troisi, 2015; Moors, Boddez, & De Houwer, 2017).

2.2 Second Generation: S–S Association Formation Models

The next generation of conditioning models remained true to the mechanism of association formation but integrated it within a cognitive approach in at least three ways. First, the second-generation models postulated that associations were formed between cognitive representations of stimuli, that is, nodes in memory that encode information about the sensory properties of those stimuli (e.g., Wagner, 1981). Because of this assumption, these types of models are often referred to as stimulus–stimulus (S–S) models. Second, these models typically assume that CS–US pairings have an impact on S–S associations only when the cognitive conditions are right. For instance, the pairing of CS and US will lead to an association between the CS and US representation only to the extent that the US (Rescorla & Wagner, 1972) and CS (e.g., Mackintosh, 1975) are processed attentively. Third, it is often assumed that S–S associations can influence cognitive states. More specifically, the presentation of a CS can result in an expectancy that the US will be presented, which in turn leads to CRs that prepare the organism for the arrival of the US. Because preparatory responses are often different from URs (e.g., fear of an impending shock vs pain following an actually shock), S–S models can account for the observation that CRs and URs are often fundamentally different.

Many different S–S models have been proposed over the years (see Bouton, 2016, for a recent review). Although they share the assumption that conditioning is mediated by S–S associations in memory, they differ with regard to the (cognitive) conditions under which associations are formed, the nature of the representations that are associated, and the conditions under which associations influence behavior and thinking. It is safe to say that S–S models are so dominant in the current conditioning literature, that it seems almost self-evident to say that conditioning is mediated by S–S association formation, much like it was self-evident in the 1950s to say that conditioning is due to S–R association formation. In part, this evolution is due to the many successes of S–S models in explaining and predicting the effects of moderators of conditioning (see Bouton, 2016, for a review).

Nevertheless, there are reasons to believe that also S–S models do not provide an adequate account of conditioning. Hence, conditioning phenomena cannot simply be equated with an S–S association formation mechanism. In the milestone review paper that I referred to earlier, Rescorla (1988, p. 154) likened conditioning to "scientists identifying the cause of a phenomenon." The S–S association formation models that have been proposed until now, however, are a far cry from models that could underlie scientific discovery. Although S–S models undoubtedly assign a bigger role to cognitive processes than S–R models, they cling on to the idea of an essentially stupid association formation mechanism (i.e., what fires together wires together) that is made "smart" only by calling to arms cognitive processes that modulate the activity of the association formation mechanism and by allowing it to produce expectations. However, current S–S models lack the ability to encode important aspects of relational information, which render them incapable of accounting for higher order cognitive processes such as analogical reasoning and scientific discovery. As Gentner (2016, p. 651) recently pointed out, "simple associative processes ... can tell us that cow is strongly associated with calf and also with milk, but they cannot record the nature of the relation. Using a purely associative process, we would not be able to discern that the relation between mare and colt is more like that between cow and calf than that between cow and milk."

The fact that current S–S models have little to say about scientific discovery only discredits them as models of conditioning to the extent that conditioning is actually akin to scientific discovery, as Rescorla (1988) claimed. Although surprising from the perspective of S–R and S–S models, evidence is indeed accumulating in support of Rescorla's (1988) proposal. Let us consider the impact of additivity on blocking. Blocking refers to the observation that conditioned responding to CS1 after CS1+CS2–US trials (i.e., CS1 and CS2 are presented together and followed by the US) is weakened or eliminated when CS2 is first paired with the US (i.e., CS2–US followed by CS1+CS2–US) compared to when CS2 is never paired with the US (e.g., only CS1+CS2–US). Typically, blocking is attributed to the fact that

CS2 interferes with the formation of the CS1–US association (e.g., because CS2 reduces attention to the US on CS1+CS2–US trials; Rescorla & Wagner, 1972). In fact, the phenomenon of blocking is widely considered as one of the cornerstones on which S–S models of conditioning are built (Rescorla, 1988). My colleagues and I have, however, showed that blocking is often weak and highly parameter dependent (e.g., Beckers, Miller, De Houwer, & Urushihara, 2006; De Houwer, Beckers, & Glautier, 2002; Maes et al., 2016). Based on these findings, we proposed that blocking depends on an inferential reasoning process that allows participants to discount CS1 as a potential cause of the US (De Houwer, Beckers, & Vandorpe, 2005). More specifically, blocking occurs when participants can verify that CS1 does not add anything to the causal effect that CS2 has on the US. For instance, whereas blocking is strong when CSs are described as potential causes of the US, it is weak or absent when there are reasons to doubt that the CSs have causal (and thus additive) effects on the US (e.g., by telling them the CS is mere effects or predictors of the US; see De Houwer et al., 2002; Waldmann & Holyoak, 1992). Likewise, when the context implies nonadditive effects of causal CSs on US intensity (e.g., when USs have their maximal intensity on CS2–US trials and ceiling effects therefore prevent participants from determining whether CS1 adds anything to the effect of CS2), blocking effects are weak or absent, both in humans (De Houwer et al., 2002) and in rats (Beckers et al., 2006). In sum, there are reasons to assume that Rescorla (1988) was right in assuming that conditioning has much in common with scientific discovery. Hence, S–S models are fundamentally flawed as models of conditioning.

2.3 Third Generation: Propositional Models

In response to growing evidence for the role of higher order reasoning processes in conditioning, a new type of model was proposed that drops the idea of association formation altogether. Instead, these models postulate that conditioning is mediated by the formation of propositions (De Houwer, 2009; Mitchell, De Houwer, & Lovibond, 2009). Unlike simple associations, propositions can specify the precise way in which events are related. For instance, the propositions "CS causes US" and "CS predicts US" both relate the CS to the US but differ in the way that the CS and US are said to be related. Because they contain relational information and therefore have a truth value (Lagnado, Waldmann, Hagmayer, & Sloman, 2007), they allow for inferences (i.e., the construction of new propositions on the basis of existing ones) and can thus account for the evidence supporting the role of inferences in blocking and other conditioning phenomena (see the previous section). They are also compatible with the well-known finding that even instructions about CS–US relations (e.g., "the tone will be followed by a shock") result in CRs (e.g., tone evokes fear; Cook & Harris, 1937). From the perspective of propositional models, actual CS–US pairings are just one source of propositions about CS–US relations, next to other sources such as instructions, observations, and inferences (e.g., Mitchell et al., 2009). Moreover, propositional models fit well with the observation that conditioning effects in humans typically require awareness of the CS–US relation (see Lovibond & Shanks, 2002, for a review). In sum, there are good reasons to believe that at least part of the conditioning effects that have been described in the literature are mediated by propositional processes. As such, propositional models are compatible with the idea that conditioning has much in common with higher order cognitive phenomena such as problem solving and scientific discovery. In fact, my colleagues and I have recently argued that conditioning in humans is a symbolic phenomenon that is fundamentally similar to learning via instructions (De Houwer & Hughes, 2016). Symbols (defined broadly as stimuli that refer to other events) are not necessarily verbal but also encompass certain gestures (as in sign language) or other movements (e.g., the wink of an eye). We proposed that also spatiotemporal events such as the pairing of a CS and US can function as symbols (also see Hayes, Barnes-Holmes, & Roche, 2001). From this perspective, CS–US pairings are a symbolic cue that signals the type of relation between the CS and US, much like a sentence such as "the CS predicts the US" provides information about the CS–US relation.

2.4 Multiple Process Models

Regardless of the merits of the symbolic perspective on conditioning, it is now widely accepted that higher order cognitive processes mediate at least some instances of conditioning. Many learning researchers, however, cling on to the idea that there are also instances of conditioning that are mediated by S–R association formation (based on the famous aliens in the original Star Trek universe, I sometimes refer to these researchers as first-generation Klingons) or S–S association formation (i.e., next-generation Klingons; e.g., McLaren et al., 2014). As such, these researchers defend a multiple process perspective according to which conditioning effects can result from different types of mechanisms (e.g., McLaren et al., 2014). Although such a perspective seems attractive at first sight, it also has important downsides (see Boddez, De Houwer, & Beckers, 2017; Mitchell et al., 2009, for a discussion). Most importantly, its merits depend on the extent that specific and plausible assumptions are made about how different types of processes interact (e.g., when which process will underlie conditioning effects). Without these assumptions, multiple process models (i.e., models which assume that different instances of conditioning might be mediated by different mental processes; e.g., McLaren et al., 2014) offer little more than post hoc explanations and a false sense of understanding.

3. DIFFERENT WAYS IN WHICH CONDITIONING MIGHT BE INVOLVED IN THE PLACEBO EFFECT

3.1 S–R Association Formation as a Source of Placebo Effects

So what does all of this tell us about the way in which conditioning might contribute to placebo effects? When placebo researchers claim that conditioning processes contribute to placebo effects, they sometimes seem to use the term "conditioning" to refer to the process of S–R association formation. For instance, in a review paper on placebo-induced relief from nausea, Quinn and Colagiuri (2015, p. 450) wrote that "Conditioning-based models of placebo responding propose that contextual features, such as the treatment setting or ritual of administration, associated with the nauseogenic drug or procedure function as conditioned stimuli (CS). These stimuli become associated with the source of nausea, or unconditioned stimulus (US), and hence the nausea itself, the unconditioned response (UR). In turn, these contextual features could themselves elicit nausea and vomiting, the conditioned response (CR)." Sometimes, conditioning is contrasted with expectancies as being different potential sources of placebo effects, with the added assumption that both mechanisms operate under different conditions. For instance, Benedetti et al. (2003, p. 4315) argued that "placebo responses are mediated by conditioning when unconscious physiological functions such as hormonal secretion are involved, whereas they are mediated by expectation when conscious physiological processes such as pain and motor performance come into play."

It should be clear that this perspective on the relation between conditioning and placebo effects is not only limiting but also implausible. It is limiting in the sense that conditioning can be much more than simply the operation of that one mechanism. It is implausible in that many findings question the validity of S–R models of conditioning (Rescorla, 1988) and very little evidence has been found for the operation of the S–R mechanism in humans (e.g., see Hogarth & Troisi, 2015, for a recent review questioning the role of S–R learning in addiction). If there is so little evidence for S–R association formation processes in humans, then maybe those S–R processes also do not play such a big role in placebo effects. In any case, it is important not to simply equate claims about the importance of conditioning for placebo effects with claims about the importance of S–R association formation processes for placebo effects.

3.2 S–S Association Formation as a Source of Placebo Effects

Regardless of whether some instances of conditioning are mediated by an S–R mechanism, it is now widely accepted that conditioning cannot simply be contrasted with cognitive accounts of placebo effects. Most notably, the current emphasis on the role of expectancies in placebo effects (e.g., Colloca & Miller, 2011) is perfectly compatible not only with the notion of conditioning as an effect but also with both S–S and propositional models of conditioning effects (Crombez, Baeyens, Vansteenwegen, & Hermans, 1997; Kirsch, 1985; Stewart-Williams & Podd, 2004). As Colloca and Miller (2011, p. 1864) argued, "it is necessary to overcome any strict dichotomy between conditioning and expectation mechanisms, as the former involves information processing by which a subject anticipates (i.e. expects) a future event." Note, however, that placebo researchers should also be wary of equating conditioning with S–S association formation. Neither S–R association formation nor S–S association formation is necessarily the only mechanism via which stimulus pairings can produce placebo effects. Hence, claims of the relevance of conditioning (defined as the impact of stimulus pairings on behavior) for placebo research can go beyond claims of the relevance of S–S association formation for placebo research.

3.3 Propositions About Stimulus Relations as a Source of Placebo Effects

Although S–S models of conditioning provide an interesting perspective on the potential role of conditioning in placebo effects, I believe that this perspective on conditioning is still too limiting.[a] If conditioning is defined

[a] Proponents of S–S models of course do not deny that elements other than stimulus pairings (e.g., instructions, context) are important for placebo effects (e.g., Colloca & Miller, 2011). However, defining conditioning in terms of S–S association formation does imply that *stimulus pairings* have an impact only because they result in the formation of S–S associations.

functionally as the impact of stimulus pairings on behavior, then one can also consider the possibility that stimulus pairings are not just causes of S S associations that produce expectancies. From a propositional perspective, stimulus pairings produce full-blown propositional beliefs about the way in which events in the world are related. It might well be that these propositional beliefs underlie the expectancies that produce placebo effects. There are a number of reasons for why such a propositional perspective on the role of conditioning is preferable to an S–S perspective. First, one could argue that expectancies are themselves propositional beliefs (e.g., the belief that an event is imminent). Rather than assuming a mysterious transformation from nonpropositional associations to propositional expectancies, it seems more plausible to assume that expectancies result from other propositional beliefs, including beliefs about the pairing of stimuli (e.g., the proposition that the CS repeatedly precedes the US which might be formed after experiencing the pairing of CS and US). In fact, there is strong evidence to support the conclusion that CS–US pairings result in US expectancies only if people have conscious beliefs about the CS–US pairings (see Lovibond & Shanks, 2002, for a review; but see Jensen, Kirsch, Odmalm, Kaptchuk, & Ingvar, 2015, for evidence of unconscious conditioning in the context of placebo research). Second, a propositional perspective fits well with the finding that conditioning, instructions, and prior beliefs closely interact in determining placebo effects (Colloca & Miller, 2011; Quinn & Colagiuri, 2015). For instance, placebo instructions seem to be much more effective when they are combined with initial conditioning trials in which the treatment is paired with an active pharmacological agent (e.g., Quinn & Colagiuri, 2015). The idea that conditioning experiences (i.e., stimulus pairings) produce propositional beliefs about stimulus relations implies that those beliefs can easily be related to and thus interact with other propositional beliefs that are based on instructions or prior experiences. In sum, compared to S–S models, propositional models of conditioning provide a much richer framework for understanding how conditioning experiences relate to expectancies.

Propositional models also highlight the fact that stimulus pairings can lead to much more than expectancies about the presence of stimuli. Most importantly, they also result in beliefs about the nature of the CS–US *relation* (e.g., CS predicts the US, CS causes the US, CS is an effect of the US, CS sometimes co-occurs with the US, etc.). Although expectancies might indeed be particularly important in guiding behaviors, some data suggest that propositions about CS–US relations can have an effect even in the absence of

(US) expectancies. For instance, research on evaluative conditioning (i.e., changes in liking due to stimulus pairings) suggests that conditioned preferences linger on even after CSs have been extinguished as predictors of the US, thus showing conditioned responding in the absence of US expectancies (e.g., Hermans, Vansteenwegen, Crombez, Baeyens, & Eelen, 2002; see Hofmann, De Houwer, Perugini, Baeyens, & Crombez, 2010, for a review). Likewise, consider the idea of drinking water from toilet bucket that has been perfectly disinfected. People are uncomfortable with doing so even though they have no reason to expect that there will be any discomfort or risk involved (Rozin, Millman, & Nemeroff, 1986). Examples like these might shed new light on the so-called open placebo effects. Beneficial effects of placebo treatments are sometimes observed even when participants are informed that the placebo treatment does not contain any active pharmacological substance (e.g., Kaptchuk et al., 2010). In these cases, the mere fact of entertaining a proposition that relates a treatment with an outcome might produce that outcome even when there is no expectancy that the treatment will work. Although this idea is highly speculative at this point in time and the merit of propositional models for placebo research does not hinge upon the veracity of this idea, it is worth keeping in mind that there is more to psychology than expectations.

3.4 Stimulus Pairings as a Source of Placebo Effects

Despite the merits of a propositional perspective on the role of conditioning in placebo effects, it should be clear that even this perspective is too limiting. At least in principle, it is possible that different conditioning effects are mediated by different mediating mechanisms, even mechanisms that have not yet been considered. Rather than continuing to make the mistake of equating the phenomenon of conditioning with whatever mechanism is in vogue at a certain period in time, it is better to think of conditioning as an effect when considering its relevance for placebo research (also see Stewart-Williams & Podd, 2004).

More specifically, there is merit in examining whether placebo effects qualify as instances of conditioning. Such a claim implies that the placebo effect involves a change in behavior (in a broad sense that includes motoric, physiological, and neurological responses) that is due to the pairing of stimuli. Take the prototypical example of a pill that initially contains an active pharmacological component that produces a drop in blood pressure. After someone has taken the active pill on multiple occasions, a similar pill without the pharmacological component might also produce a drop in blood pressure. The latter effect could be defined as a placebo effect that is an instance of classical conditioning. If one allows for the possibility that stimulus pairings exert an impact on behavior that occurs a long time after those stimulus pairings, one could even argue that most placebo effects are instances of conditioning. Indeed, effects of many placebo treatments might hinge on having previous experiences in which similar treatments were paired with active components that led to beneficial effects (Colloca & Miller, 2011).

The claim that a particular placebo effect is an instance of conditioning has explanatory value in that it elucidates the environmental causes of the observed change in behavior (De Houwer, 2011; Hughes et al., 2016). For instance, explaining the effect of a placebo pill in terms of conditioning implies that the pill has its effects because similar pills previously went together with a certain active pharmacological substance. Such functional explanations are not trivial because they eliminate other possible functional explanations (e.g., the mere repeated presentation of the pharmacological substance) and imply that the placebo effects might be similar to other conditioning effects in terms of its moderators (e.g., the fact that exposure to the placebo pill prior to receiving an active pill might reduce the placebo effect). Moreover, although the claim that placebo effects are instances of conditioning does not commit itself to a particular explanatory mechanism, it also has implications for mechanistic accounts. Most importantly, it orients researchers to a set of possible mechanisms that *might* produce the placebo effect (e.g., the formation of S–R associations, S–S associations, or propositions). Once it has been demonstrated that the placebo effect qualifies as an instance of conditioning (i.e., that the observed change is due to stimulus pairings), further studies can be conducted to examine which conditioning mechanism is most likely to mediate the effect.

In sum, a functional cognitive perspective on the role of conditioning in placebo research can help to reveal the environmental causes and potential moderators of placebo effects, while it also facilitates the discovery of mechanistic explanations of those effects. This is why I believe that research about the relation between conditioning and placebo effects is best conceptualized in terms of the contribution of stimulus pairings to those effects (see De Houwer & Hughes, 2016, for a similar conclusion and a more in-depth discussion about the nature of evaluative conditioning, that is, the conditioning of preferences).

4. WHAT ARE PLACEBO EFFECTS?

Having said this, the idea that placebo effects might qualify as instances of conditioning raises a number of important questions about the nature and usefulness of the concept of placebo effects. Matters are still relatively straightforward when placebo treatments are defined as "substances, given in the guise of active medication, but which in fact have no pharmacological effect on the condition being treated" (Kirsch, 1985, p. 238). Based on such a definition, stimulus pairings could be seen as one reason for why substances (e.g., placebo sugar pills) can have effects even when they do not contain a pharmacological substance that can have that effect. It becomes more complicated, however, when the concept of placebo is extended to include procedures (e.g., sham operations) in addition to substances (e.g., sugar pills). Procedures could be referred to as placebo procedures when there is no known physical mechanism via which those procedures can exert their effect. Such a definition would, however, relegate all psychological interventions (e.g., psychotherapy) to the domain of placebo effects (Stewart-Williams & Podd, 2004). One could extend the definition of placebo procedures in such a way that it requires also the absence of any known mechanism via which the procedure could exert its effect, including psychological mechanisms. Apart from the fact that this definition would render it difficult to establish placebo effects (because it requires one to exclude the impact of every possible physical and psychological mechanism) and implies that the boundaries of placebo effects change over time (when new physical and psychological mechanisms are discovered), it ignores scientific explanations that are not situated at the mechanistic level. Most crucially within the context of this chapter, knowing that a placebo effect is an instance of conditioning reveals the functional cause of the observed change in behavior (it is due to stimulus pairings) regardless of whether the mediating mechanism (e.g., association formation) is known. Because also functional explanations reveal the causes of behavior, it seems improper to refer to functionally explained behavior as a placebo effect. To put it differently, why would some conditioning effects be referred to as placebo effects (e.g., the reduction in blood pressure due to a pill that no longer contains the active substance it used to contain) when others are not (e.g., the increase in salivation in response to a tone that results from prior tone-food pairings)?

Referring to certain conditioning effects as placebo effects also adds little because it does not provide additional insights in either the environmental causes or mental mechanisms involved in those effects. The concept of placebo effects not only lacks explanatory value, but it also does not refer to a clearly delineated set of phenomena given that it is defined primarily in negative terms, that is, in terms of what does not count as a placebo effect (e.g., effects of pharmacological substances). One radical solution would therefore be to drop the placebo concept altogether. Regardless of the eventual fate of the placebo concept, progress in understanding placebo-related phenomena is bound to depend on a continuing close integration of placebo research with other (health) psychological research on environmental factors and mental mechanisms that determine (health related) behavior, including conditioning research.

ACKNOWLEDGMENTS

The preparation of this chapter was made possible by Methusalem Grant BOF16/MET_V/002 of Ghent University. I thank Geert Crombez for helpful discussions and feedback on a first draft of this chapter.

REFERENCES

Beckers, T., Miller, R. R., De Houwer, J., & Urushihara, K. (2006). Reasoning rats: Forward blocking in Pavlovian animal conditioning is sensitive to constraints of causal inference. *Journal of Experimental Psychology: General, 135*, 92–102.

Benedetti, F., Pollo, A., Lopiano, L., Lanotte, M., Vighetti, S., & Rainero, I. (2003). Conscious expectation and unconscious conditioning in analgesic, motor, and hormonal placebo/nocebo responses. *Journal of Neuroscience, 23*, 4315–4323.

Boddez, Y., De Houwer, J., & Beckers, T. (2017). The inferential reasoning theory of causal learning: Towards a multi-process propositional account. In M. R. Waldmann (Ed.), *The Oxford handbook of causal reasoning*. Oxford: Oxford University Press.

Bouton, M. E. (2016). *Learning and behavior: A contemporary synthesis* (2nd ed.). Sunderland, MA: Sinauer Associates.

Byrne, R. W., & Bates, L. A. (2006). Why are animals cognitive? *Current Biology, 16*, 445–448.

Colloca, L., & Miller, F. G. (2011). How placebo responses are formed: A learning perspective. *Philosophical Transactions of the Royal Society B, 366*, 1859–1869.

Cook, S. W., & Harris, R. E. (1937). The verbal conditioning of the galvanic skin reflex. *Journal of Experimental Psychology, 21*, 202–210. https://doi.org/10.1037/h0063197.

Crombez, G., Baeyens, F., Vansteenwegen, D., & Hermans, D. (1997). Psychofarmaca en gedragstherapie: een klassieke analyse. *Gedragstherapie, 30*, 135–162.

De Houwer, J. (2009). The propositional approach to associative learning as an alternative for association formation models. *Learning & Behavior, 37*, 1–20.

De Houwer, J. (2011). Why the cognitive approach in psychology would profit from a functional approach and vice versa. *Perspectives on Psychological Science, 6*, 202–209.

De Houwer, J., Barnes-Holmes, D., & Moors, A. (2013). What is learning? On the nature and merits of a functional definition of learning. *Psychonomic Bulletin & Review, 20*, 631–642. https://doi.org/10.3758/s13423-013-0386-3.

De Houwer, J., Beckers, T., & Glautier, S. (2002). Outcome and cue properties modulate blocking. *Quarterly Journal of Experimental Psychology, 55A*, 965–985.

De Houwer, J., Beckers, T., & Vandorpe, S. (2005). Evidence for the role of higher-order reasoning processes in cue competition and other learning phenomena. *Learning & Behavior, 33*, 239–249.

De Houwer, J., & Hughes, S. (2016). Evaluative conditioning as a symbolic phenomenon: On the relation between evaluative conditioning, evaluative conditioning via instructions, and persuasion. *Social Cognition, 34*, 480–494.

De Houwer, J., & Hughes, S. (2017). Environmental regularities as a concept for carving up the realm of learning research: Implications for relational frame theory. *Journal of Contextual Behavioral Science, 6*, 343–346.

Eelen, P. (1980). Klassieke conditioning: Klassiek en toch modern [Classical conditioning: Classic but nevertheless modern]. Liber Amicorum Prof. In J. R. Nuttin (Ed.), *Gedrag, dynamische relatie en betekeniswereld [Behavior, dynamic relation, and world of meaning]*. Leuven, Belgium: Universitaire Pers Leuven.

Gentner, D. (2016). Language as cognitive toolkit: How language supports relational thought. *American Psychologist, 71*, 650–657.

Hayes, S. C., Barnes-Holmes, D., & Roche, B. (Eds.), (2001). *Relational frame theory: A post-Skinnerian account of human language and cognition.* New York: Plenum Press.

Hermans, D., Vansteenwegen, D., Crombez, G., Baeyens, F., & Eelen, P. (2002). Expectancy-learning and evaluative learning in human classical conditioning: Affective priming as an indirect and unobtrusive measure of conditioned stimulus valence. *Behaviour Research and Therapy, 40*, 217–234.

Hofmann, W., De Houwer, J., Perugini, M., Baeyens, F., & Crombez, G. (2010). Evaluative conditioning in humans: A meta-analysis. *Psychological Bulletin, 136*, 390–421.

Hogarth, L., & Troisi, J. R. (2015). A hierarchical instrumental decision theory of nicotine dependence. *Current Topics in Behavioral Neurosciences, 23*, 165–191. https://doi.org/10.1007/978-3-319-13665-3_7.

Hughes, S., De Houwer, J., & Perugini, M. (2016). The functional-cognitive framework for psychological research: Controversies and resolutions. *International Journal of Psychology, 51*, 4–14.

Jensen, K., Kirsch, I., Odmalm, S., Kaptchuk, T. J., & Ingvar, M. (2015). Classical conditioning of analgesic and hyperalgesic pain responses without conscious awareness. *Proceedings of the National Academy of Sciences of the United States of America, 112*, 7863–7867.

Kaptchuk, T. J., Friedlander, E., Kelley, J. M., Sanchez, M. N., Kokkotou, E., Singer, J. P., et al. (2010). Placebos without deception: A randomized controlled trial in irritable bowel syndrome. *PLoS One, 5*(12), e15591.

Kirsch, I. (1985). The logical consequences of the common-factor definition of the term placebo. *American Psychologist, 40*, 237–238.

Lagnado, D. A., Waldmann, M. R., Hagmayer, Y., & Sloman, S. A. (2007). Beyond covariation: Cues to causal structure. In A. Gopnik & L. Schulz (Eds.), *Causal learning: Psychology, philosophy, and computation* (pp. 154–172). Oxford: Oxford University Press.

Lovibond, P. F., & Shanks, D. R. (2002). The role of awareness in Pavlovian conditioning: Empirical evidence and theoretical implications. *Journal of Experimental Psychology: Animal Behavior Processes, 28*, 3–26.

Mackintosh, N. J. (1975). A theory of attention: Variation in the associability of stimuli with reinforcement. *Psychological Review, 82*, 276–298.

Maes, E., Boddez, Y., Alfei, J. M., Krypotos, A. M., D'Hooge, R., De Houwer, J., et al. (2016). The elusive nature of the blocking effect: 15 failures to replicate. *Journal of Experimental Psychology: General, 145*, e49–e71.

McLaren, I. P. L., Forrest, C. L. D., McLaren, R. P., Jones, F. W., Aitken, M. R. F., & Mackintosh, N. J. (2014). Associations and propositions: The case for a dual-process account of learning in humans. *Neurobiology of Learning and Memory, 108*, 185–195.

Mitchell, C. J., De Houwer, J., & Lovibond, P. F. (2009). The propositional nature of human associative learning. *Behavioral and Brain Sciences*, *32*, 183–198.

Moors, A., Boddez, Y., & De Houwer, J. (2017). The power of goal-directed processes in the causation of emotional and other actions. *Emotion Review*, *9*, 310–318.

Pavlov, I. P. (1927). *Conditioned reflexes*. London: Oxford University Press.

Quinn, V., & Colagiuri, B. (2015). Placebo interventions for nausea: A systematic review. *Annals of Behavioral Medicine*, *49*, 449–462.

Rescorla, R. A. (1988). Pavlovian conditioning: It's not what you think it is. *American Psychologist*, *43*, 151–160.

Rescorla, R. A., & Wagner, A. R. (1972). A theory of Pavlovian conditioning: Variations in the effectiveness of reinforcement and nonreinforcement. In A. H. Black & W. F. Prokasy (Eds.), *Classical conditioning II* (pp. 64–99). Appleton-Century-Crofts.

Rozin, P., Millman, L., & Nemeroff, C. (1986). Operation of the laws of sympathetic magic in disgust and other domains. *Journal of Personality and Social Psychology*, *50*, 703–712.

Stewart-Williams, S., & Podd, J. (2004). The placebo effect: Dissolving the expectancy versus conditioning debate. *Psychological Bulletin*, *130*, 324–340. https://doi.org/10.1037/0033-2909.130.2.324.

Wagner, A. R. (1981). SOP: A model of automatic memory processing in animal behavior. In N. E. Spear & R. R. Miller (Eds.), *Information processing in animals: Conditioned inhibition* (pp. 223–266). Hillsdale, NJ: Lawrence Erlbaum Associates.

Waldmann, M. R., & Holyoak, K. J. (1992). Predictive and diagnostic learning within causal models: Asymmetries in cue competition. *Journal of Experimental Psychology: General*, *121*, 222–236.

CHAPTER SEVEN

The Application of Persuasion Theory to Placebo Effects

Andrew L. Geers*[,1], Pablo Briñol[†], Erin A. Vogel[‡], Olivia Aspiras*, Fawn C. Caplandies*, Richard E. Petty[§]
*University of Toledo, Toledo, OH, United States
[†]Universidad Autónoma de Madrid, Madrid, Spain
[‡]University of California, San Francisco, San Francisco, CA, United States
[§]Ohio State University, Columbus, OH, United States
[1]Corresponding author: e-mail address: andrew.geers@utoledo.edu

Contents

Abstract

Placebo effects, or positive outcomes resulting from expectations about a treatment, are powerful components of modern medical care. In this chapter, we suggest that our understanding of placebo effects may benefit from more explicitly connecting this phenomenon to the existing empirical psychological literature on persuasion. Persuasion typically involves an attempt to bring about a change in beliefs or attitudes as a result of providing information on a topic. We begin by providing a brief overview of the psychological literature on placebo effects. We then point to connections between this literature and research on persuasive communication. Although some links have been made, these initial connections have predominantly relied on classic theories of persuasion rather than on more contemporary and comprehensive models. Next, we describe a modern theory of persuasion that may facilitate the study of placebo effects and analyze two issues pertinent to the literature on placebo effects from the lens of this model. Specifically, we consider how and when characteristics of a practitioner (e.g., variables such as perceptions of a practitioner's confidence or competence) can influence the magnitude of placebo effects, and how modern persuasion theory can help in understanding the durability of placebo effects over time. We conclude that examining placebo effects as an outcome of persuasive communication would be a fruitful line of future research.

International Review of Neurobiology, Volume 138
ISSN 0074-7742
https://doi.org/10.1016/bs.irn.2018.01.004

The placebo effect concept evolved during the prescientific era of medicine as a term for positive subjective changes resulting from the administration of inert medical treatments to patients (Shapiro & Shapiro, 1997). Early discussion of this concept generally considered placebos useful for placating patients but not for producing medically relevant change. In the middle of the 20th century, it became scientific practice to administer placebo treatments to participants in randomized controlled trials as a methodological safeguard against bias (De Craen, Kaptchuk, Tijssen, & Kleijnen, 1999). Following the emergence of this practice, the term placebo effect became synonymous with nuisance variance that needed to be extracted from clinical trials to validate genuine medical treatments. In these early perspectives, placebo effects were viewed as a form of unwanted psychological bias evoked by inert treatments. The notion that placebo effects could be incorporated in clinical care was considered unethical and inconsistent with a scientifically derived biomedical model of health care, which focused on uncovering physiological mechanisms uncontaminated by the subjectivity of the mind (Miller & Kaptchuk, 2008).

In contrast, a contemporary view is that placebo effects in healing are not artifacts of the past and are not simply psychological bias. There is substantial evidence that the success of many modern drugs and therapies benefits from placebo effects (Benedetti, 2014; Colloca, Flaten, & Meissner, 2013; Kirsch, 1999). Further, over the past two decades, a rigorous empirical literature has evolved demonstrating placebo effects across many medical domains, types of treatments, and outcomes. Based on this emerging literature, placebo effects are being reconceptualized as complex, beneficial, and scientifically measurable changes that are a direct result of the treatment administration process (Colagiuri, Schenk, Kessler, Dorsey, & Colloca, 2015; Colloca & Benedetti, 2005). Researchers now view placebo effects as an important, and potentially mutable, component of medical care and are searching for techniques that can enhance them (Enck, Bingel, Schedlowski, & Rief, 2013).

Although the value of placebo effects in medicine is gaining recognition, much about this phenomenon has yet to be explained and conceptualized. In this chapter, we suggest that an understanding of placebo effects can be improved by more explicitly connecting this phenomenon to the existing empirical literature on persuasion. That is, in many respects, placebo effects can be conceptualized as a consequence of persuasive communication. Unlike *compliance* and other forms of influence that typically involve a situation in which a behavioral response is sought in response to a request to do something, *persuasion* typically involves an attempt to bring about a change in beliefs or attitudes as a result of providing information on a topic (e.g., delivering a message on the benefits of a new exercise routine; Briñol & Petty, 2012). The connection between persuasive communication and placebo effects is apparent in the exemplar case of prescribing placebos in medical care, when a practitioner prescribes an inert treatment with the suggestion of benefits to a patient. Here, we see the four central elements of the persuasive communication process (McGuire, 1969) at play: the message factor (i.e., suggested benefits of treatment), the source factor (i.e., medical practitioner), the recipient factor (i.e., the patient), and the context in which persuasion takes place (i.e., a medical office). Even in less prototypical examples, such as when a consumer selects a pain reliever at a drug store, many of the key elements of persuasion remain. For example, the store-bought drug will have packaging features that suggest the effectiveness of the treatment (i.e., message factor) and aspects of the consumer, such as mood and personality, can also influence how that message is evaluated (i.e., recipient factor).

Importantly, each of the four classic elements of persuasive communication (McGuire, 1969) has received attention in the literature on placebo effects. For example, researchers have found that variations in the characteristics or descriptions of the treatment can change the magnitude of placebo effects. These features include the color, size, taste, and cost of a drug (e.g., De Craen, Roos, De Vries, & Kleijnen, 1996; Moerman, 2002; Wright, da Costa Hernandez, Sundar, Dinsmore, & Kardes, 2013). Similarly, differences in the characteristics of the medical practitioner, the source of the treatment message, are also important determinants of placebo effects (Howe, Goyer, & Crum, 2017; Kaptchuk et al., 2008). Although research has shown that placebo effects vary based on the characteristics of the treatment message, source recipient, and context, these studies are generally considered in isolation and have not been integrated into a single organizational theoretical perspective. Considering these characteristics as elements involved in the process of persuasive communication would allow researchers to connect this work to established theories and principles that have been generated from decades of research on the psychology of persuasion. Further, through this integration, it may be possible to better understand the psychological processes by which variables, such as characteristics of a doctor, influence placebo effects (Geers & Miller, 2014).

In this chapter, we provide a brief review of the literature on the psychology of placebo effects. Next, we describe connections between this literature and persuasive communication. To illustrate the potential benefits of considering placebo effects from the vantage point of persuasion, we then explain a modern and influential theory of persuasion, the elaboration likelihood model (ELM; Petty & Briñol, 2012; Petty & Cacioppo, 1986). By focusing on the fundamental psychological mechanisms underlying persuasion, the ELM has served to organize the many different theories, outcomes, and variables relevant to persuasion across diverse domains. In this regard, the ELM provides a potent theoretical platform for understanding a variety of aspects of placebo effects. Further, models such as the ELM afford predictions regarding additional facets of placebo effects, such as their directionality, durability, and likelihood of altering subsequent behaviors. After describing this model that has guided much of the persuasion research in the last decades (Petty & Briñol, 2012; Petty & Cacioppo, 1986), we present several examples to illustrate possible ways in which the ELM could enhance our understanding of findings in the placebo effect literature.

1. RESEARCH ON PLACEBO EFFECTS

Placebo effects are the result of psychobiological processes that are activated during the treatment context and can be separated from the physical properties of the treatment itself (Colloca & Benedetti, 2005). Self-fulfilling expectations and classical conditioning are the primary psychological mechanisms for placebo effects that have been examined thus far. In terms of expectations, placebo effects appear to be driven by an individual's belief that receiving a treatment will result in a particular response or outcome (Kirsch, 1999). When considering classical conditioning, placebo effects can be conditioned responses to a treatment, with active medications as the unconditioned stimuli, and the methods or techniques used to administer or accompany treatments as the conditioned stimuli (Stewart-Williams & Podd, 2004).

A standard procedure to experimentally test for placebo effects is to alter a treatment efficacy message given to one group of participants (e.g., this pill will reduce your pain) and then compare the responses of this group to those of another group not given this efficacy message. To isolate the impact of this expectation of outcome manipulation on treatment effects, inert treatments are typically administered to all participants. Studies of this kind offer evidence that the treatment message provided to patients is central in determining the illness experience and the symptomatic expressions of disease (Benedetti, 2014; Colloca et al., 2013; Kirsch, 1999). Placebo effects have been found across cultures, age groups, and with samples ranging from healthy children to those with severe health conditions, such as neuropathic pain patients (e.g., Petersen et al., 2012; Weimer et al., 2013; Zimba, 1993). Placebo effects can cause clinically significant effects and in some cases the benefits appear comparable to those provided by standard treatments, most notably in the domains of pain relief and depression (Benedetti, 2014; Colloca et al., 2013).

Changes induced by placebo manipulations have been observed on a wide variety of assessment tools. Placebo effects can manifest on subjective measures of pain, depression, mood, headache, self-esteem, anxiety, relaxation, sleep quality, confidence in memory, and drug cravings (e.g., Caplandies, Colagiuri, Helfer, & Geers, 2017; Greenwald, Spangenberg, Pratkanis, & Eskenazi, 1991; Mills, Boakes, & Colagiuri, 2016). Placebo effects have also been exhibited on measures of cognitive processing such as reaction time, word generation, recall, recognition, Stroop interference, attention, and implicit learning (e.g., Colagiuri, Livesey, & Harris, 2011; Draganich & Erdal, 2014; Slama, Caspar, Gevers, & Cleeremans, 2013; Weger & Loughnan, 2013). Placebo effects can also occur on behavioral measures, including pain tolerance, completion of puzzle tasks, reduced sleep latency, approach of snakes and electric shocks, amount of alcohol consumed, talking time by socially anxious individuals, speed and endurance in aerobic athletes, weightlifting by competitive lifters, and motor performance in patients with Parkinson's disease (e.g., Benedetti et al., 2003; Fillmore & Vogel-Sprott, 1992; Shiv, Carmon, & Ariely, 2005). Placebo treatments may also influence subsequent decisions and actions, such as later purchasing behavior, number of cigarettes smoked, acts of aggression, and reductions in opioid analgesics used by pain patients (e.g., Berkowitz & Thome, 1987; Doering & Rief, 2012). Further, placebo manipulations have altered physiological parameters including blood pressure, pulse rate, sexual arousal, heart rate variability, salivary cortisol, startle eyeblink reflex, electrodermal activity, beta-band frequency during sleep, and bronchoconstriction in asthmatics (e.g., Benedetti, 2014; Darragh, Vanderboor, Booth, Sollers, & Consedine, 2015). Finally, there is also now strong neurobiological evidence linking placebo effects to complex mechanisms in the brain and spinal cord, indicating that the effects are not just the result of subjective biases (for a review, see Wager & Atlas, 2015).

Important for the present context, many psychosocial variables moderate placebo effects and these variables can be grouped into the four central elements of the persuasive communication process noted earlier: the message, source, recipient, and context. As previously noted, many features of the treatment message alter the magnitude of placebo effects, including the color of a pill, the size of a pill, the taste of a pill, the intrusiveness of the administration induction, and the cost of a drug. Moreover, a variety of studies indicate that more "impressive" medical administration procedures (e.g., injections, acupuncture) tend to yield stronger placebo effects than less impressive procedures (e.g., oral placebos; Brunoni, Lopes, Kaptchuk, & Fregni, 2009; de Craen, Tijssen, de Gans, & Kleijnen, 2000; Kaptchuk et al., 2006; Linde, Niemann, & Meissner, 2010). As another illustration, consider a series of experiments by Shiv et al. (2005). In three different studies, these researchers manipulated the cost of a sham energy drink that was purported to increase mental acuity. Across all studies, participants who paid more for the placebo energy drink performed better on a subsequent series of puzzle tasks than those who paid less. Similar moderating effects have been found with the other three elements of persuasive communication (e.g., Geers et al., 2013). Thus, variations in the four elements of persuasion alter the strength of placebo effects.

It should also be noted that expectations for treatment outcomes can be negative as well as positive. Nocebo effects refer to negatively valenced placebo effects that result from the anticipation of unpleasant treatment outcomes or side effects. The term nocebo was coined by Kennedy (1961) and early on was discussed in relation to sociocultural reports of voodoo death. As with placebo effects, contemporary research finds nocebo effects across symptom domains and in both clinical and experimental studies (Colloca, 2017), and the effects are altered by communication factors such as whether the treatment is presented as name brand or generic (Faasse, Cundy, Gamble, & Petrie, 2013). Although nocebo and placebo effects may involve some distinct features, nocebo effects can be considered a variant of the broader placebo effect phenomenon. Keeping with this perspective, much of what is discussed herein pertains to both placebo and nocebo effects.

Finally, it should be noted that placebo effects are not restricted to the administration of inactive treatments. Rather, expectations regarding treatment efficacy can alter the efficacy of *active* treatments as well (Colloca, Lopiano, Lanotte, & Benedetti, 2004). For example, studies employing the balanced placebo design experimentally cross treatment expectation manipulations (e.g., explicitly stating whether a capsule contains caffeine or not) with active treatment manipulations (e.g., administering caffeine pills or not). Many studies using this design find that treatment expectations can independently increase the success of active treatments (see Dinnerstein & Halm, 1970; Hull & Bond, 1986). These findings reinforce the notion that the processes responsible for placebo effects do not only impact the outcomes of inert treatments. Rather, these processes are at play with active treatments as well, and placebo effects can be construed as a component of virtually any treatment effect. From this perspective, placebo effect research has implications for health treatments and interventions that have both strong and weak effects.

Although the research literature on placebo effects has uncovered many important findings, much remains unknown. For instance, in studies where placebo effects arise, the effect sizes can vary substantially (Petersen et al., 2014; Vase, Petersen, Riley, & Price, 2009), and in some cases expectation manipulations produce result that are in contrast to the expectation (Ross & Olson, 1981). For example, Storms and Nisbett (1970) found that insomniacs actually fell asleep faster when they were given a placebo pill that purportedly increased arousal, as compared to insomniacs not given a placebo pill. Further, little is currently known about the persistence of placebo effects over time, and there is much still unknown regarding the psychological processes involved in placebo effects (Geers & Miller, 2014). In particular, in many reports, the discussion of psychological processes begins and ends by evoking the mechanisms of expectations or conditioning. We suggest that existing research on persuasive communication may be able to reconcile what otherwise might look like conflicting findings in placebo research and offer greater insight into the psychological variables involved in these effects.

2. PLACEBO EFFECTS AND PERSUASIVE COMMUNICATION

Messages pertaining to treatment efficacy (e.g., this drug will reduce your pain) are a ubiquitous feature of standard health care. Treatment efficacy messages can be conveyed either verbally (e.g., from a clinical provider) or nonverbally (e.g., print packaging material, online advertisements, social observation). Like other messages, treatment efficacy messages are, in essence, persuasive communications. These messages can change recipients' beliefs about treatment efficacy or evaluations of treatments, which, in turn, can cause placebo effects. As such, to understand how and when treatment messages result in placebo effects, it is valuable to understand the processes of persuasion leading to belief and evaluation change.

It is important to note that, in the present context, the term persuasion does not imply an extensive or deliberate attempt at changing views. That is, when we refer to persuasion, we are not contending that placebo effects manifest only when medical practitioners deliberatively and actively try to change patient responses. Rather, persuasion is used here as a broad term for communication variables and processes that govern belief, attitude, and behavioral change (Petty & Briñol, 2012). Thus, the variables involved in persuasion can be complex or simple, consciously or nonconsciously perceived, relevant or irrelevant to an individual's current concerns, and deliberatively or unintentionally applied to modify evaluations and beliefs. It is theorized here that changes in these evaluations and beliefs can lead to corresponding changes in placebo effects.

Notably, this is not the first instance in which principles of persuasion have been proposed to help explain placebo effects. For example, Kirsch (1999) suggests that hypnosis and placebo effects result from the same underlying mechanism—expectations. Further, Wickless and Kirsch (1989) surmised that the expectation responsible for both hypnosis and placebo effects results from verbal persuasion and direct experience. In their research, Wickless and Kirsch found evidence that hypnotizability was increased by a verbal persuasive message as well as direct experience, which can be considered another type of persuasion—self-generated persuasion.

Another connection was made by Frank (1961) in his influential book, *Persuasion and Healing*. Frank posited that most therapies share similar ritualistic elements that serve to enhance the well-being of patients by providing hope and relief from distress. Furthermore, Frank surmised that the beneficial effects of treatments, such as psychotherapy, can result from the ceremonial actions and features of the clinical encounter that serve as persuasive influences. These influences were thought to improve patient outcomes through placebo effects. In this account, a single variable (e.g., wearing a white lab coat) is viewed as leading to greater persuasion (e.g., positive attitude toward healing). This approach is similar to early empirical approaches to the study of persuasion, which postulated a unidirectional effect of one variable on a given outcome (see Petty & Briñol, 2012). According to this main effect view, any one variable (e.g., source expertise, emotional state of the recipient) is likely to have just one effect on persuasion—either enhancing or reducing it.

A third link between placebo effects and the persuasion literature was drawn by Liberman (1962), who relied upon the "Yale" approach. The Yale approach was one of the earliest and most influential general theories of persuasion in the modern era and was based on learning theory principles (Hovland, Janis, & Kelley, 1953). This approach held that anything that facilitated attending to, comprehending, and learning the contents of a persuasive message would be beneficial for attitude change, and anything that would disrupt those learning processes would be detrimental. For example, distracting someone from the message was predicted to reduce persuasion because it would interfere with learning. Similarly, providing a person with a credible source would increase persuasion because it would motivate people to learn the message (see Petty & Cacioppo, 1986, for a review).

When applied to placebo effects, these early accounts may suggest, for example, that a doctor with higher confidence should always obtain placebo effects of a greater magnitude than less confident doctors, due to enhanced patient learning (i.e., people would attend more to a confident doctor and thus learn more). Although single effect and single process assumptions of this kind have merits and were useful for early studies of persuasion, extensive research has revealed important limitations. For instance, it has been shown that any one variable (e.g., an expert confident source, a positive mood of the recipient) can increase persuasion in some instances but reduce it in others (for a discussion, see Briñol & Petty, 2012), and learning is not the only (or even the most important) mechanism of persuasion. Moreover, attitude changes are sometimes relatively durable and impactful (e.g., when they guide behavior) but other times rather transitory and inconsequential. Finally, research reveals that persuasion is not always the result of the same psychological mechanism, but rather many different mechanisms can be responsible for attitude change and these mechanisms can have different short- and long-term outcomes. Because of these and other challenges, modern theories of persuasion have become more complex and have been specifically devised to account for multiple effects, processes, and consequences.

3. MODERN MODELS OF PERSUASION

In the 1970s, evidence mounted that even seemingly simple psychological variables such as expert sources or positive moods could produce multiple and opposite effects, and that multiple processes could underlie these diverse outcomes. As a result, the 1980s saw the emergence of several multifaceted theories in social psychology designed to accommodate these disparate findings. In the persuasion literature specifically, the ELM (Petty & Briñol, 2012, 2014; Petty & Cacioppo, 1986) and the heuristic-systematic model (HSM; Chaiken, Liberman, & Eagly, 1989) advanced and articulated multiple processes by which variables could affect evaluative judgments in different situations.

Next, we outline one of these contemporary persuasion models, the ELM. In the present context, is it notable that the ELM was forged in response to inconsistencies in the attitude change literature. Beyond understanding change about issues and objects, the ELM has proven to be broadly applicable and has been able to accommodate virtually any change in judgment caused by social and contextual factors, including internal evaluations such as self-worth judgments (self-esteem), changes in personal identity, and changes in mood and affect (see Wegener & Petty, 1996). Our focus here is on how persuasion processes can explain changes in the evaluative judgments pertaining to symptomology that underlie placebo effects. A central idea is that when treatment messages are evaluated more favorably (i.e., the persuasion is more effective), placebo effects will be stronger.

4. MAIN POSTULATES OF THE ELM

In this section, we describe the main postulates of the ELM. For additional and more detailed descriptions, see Petty and Cacioppo (1986), Petty and Briñol (2012, 2014).

First, according to the ELM, evaluations can change through a high degree of thought or a relatively low degree of thought. That is, persuasion occurs along an "elaboration continuum" that ranges from low (investing little thought in the persuasive message) to high (investing a great deal of thought in the persuasive message). Notably, the ELM holds that there are numerous specific processes of change that operate along this continuum. For example, mental heuristics (e.g., "there are many arguments so it must be good") require relatively little thought, and therefore operate at the low end of the continuum. Alternatively, deliberative attributions and expectancy-value assessments (e.g., evaluating each argument for its merit rather than counting them) tend to require higher degrees of thought and thus operate at the higher end of the continuum. Low-elaboration processes are collectively referred to as following the *peripheral route* to persuasion, whereas high-elaboration processes are collectively referred to as following the *central route* to persuasion.

The ELM also postulates that whether people engage in the peripheral or the central route to persuasion depends on their ability and motivation to think in the context in which persuasion occurs. For instance, distractions, rapid speech, and little domain knowledge can undermine one's ability to think about a message (Petty & Cacioppo, 1986). Other variables, such as message repetition, general intelligence, and high message clarity, can aid one's ability to process a message. Whether the peripheral or central route is used also depends on the perceiver's motivation to process the message. Perhaps the greatest determinant of motivation is the perceived *personal relevance* of the information. When the personal relevance of a message is perceived to be high, individuals scrutinize the evidence more carefully such that if the evidence is found to be strong, more evaluative change results, but if the evidence is found to be weak, less evaluative change occurs (Petty & Cacioppo, 1979). Conversely, when motivation to think is low, individuals are less influenced by argument quality and more influenced by simple cues in the message that take less cognitive effort to process, such as the length of the message and whether a message is endorsed by an attractive or unattractive source.

In one prototypical early study of the ELM illustrating this point, Petty, Cacioppo, and Schumann (1983) varied participants' interest in an advertisement for the "Edge razor" by informing them that they would select from brands of razors (high relevance) or brands of toothpaste (low relevance) as compensation for participating in the study. Subsequently, participants were exposed to a razor advertisement containing either strong (e.g., "In direct comparison tests, the Edge blade gave twice as many close shaves as its nearest competitor") or weak (e.g., "In direct comparison tests, the Edge blade gave no more nicks or cuts than its competition") arguments that were presented by a famous celebrity or an ordinary citizen. Petty and colleagues found larger argument quality effects (i.e., more persuasion for the strong than weak arguments) when the razor advertisement was high as opposed to low in personal relevance but a larger impact of the celebrity endorser when the ad was low as opposed to high in personal relevance. This and other studies have supported the ELM proposal that high and low thinking individuals follow different routes to persuasion.

In the context of placebo effects, many variables could determine whether the peripheral or central route is followed. For example, fear and anxiety could undermine a patient's ability to evaluate a treatment efficacy message, and involvement in treatment decision-making may increase interest in a treatment and raise motivation (e.g., Geers et al., 2013).

Next, the ELM postulates that the route to persuasion determines how consequential the evaluation change is. Specifically, evaluations made through the central route of effortful information processing tend to persist over time, resist attempts at change, and have consequences for other judgments and behavior (Petty, Haugtvedt, & Smith, 1995). This postulate is referred to as the elaboration—strength hypothesis.

As an illustration of this hypothesis, consider how the extent of elaboration can be applied to understanding long-term changes from psychotherapy interventions. In one experiment targeting poor body image, Gascó, Briñol, and Horcajo (2010) changed participants' self-evaluations through a persuasive procedure involving high elaboration (self-persuasion) or through a less engaging procedure involving low elaboration (passive exposure; cf. Janis & King, 1954). Participants in the high-elaboration condition were requested to actively generate reasons why they liked their bodies, whereas those in the low-elaboration condition simply read positive affirmations taken from a program on eating disorders (Rosen, 1997). As expected, participants in both the high- and low-elaboration groups showed more favorable attitudes toward their bodies than those in the control group. More interestingly, although the treatments were equally effective in changing attitudes, the strength associated with those attitudes was significantly different depending on the amount of thinking involved in the process of change. Specifically, attitudes were stronger in the high-elaboration condition than low. Participants in the high-elaboration condition perceived their attitudes to be more valid and more resistant to change than those in the low-elaboration condition. In other words, the treatment had a stronger effect at higher levels of thought. In line with the elaboration-strength hypothesis, it could be surmised that placebo expectations formed through higher levels of elaboration should last longer, be more resistant to change, and guide behavior more than those formed through lower levels of elaboration.

In addition to elaboration, the ELM incorporates another key process called *validation* (Petty, Briñol, & Tormala, 2002). Unlike the mechanisms of change described thus far, which focus on primary or first-order cognition (i.e., amount and direction of thoughts), validation emphasizes secondary or metacognition (i.e., thoughts about our own thoughts). The key notion of validation is that simply generating thoughts is not sufficient for the thoughts to impact judgment. Rather, one must also have sufficient confidence in one's thoughts (cognitive validation) or feel sufficiently good about them (affective validation) for the thoughts to have an impact. Thoughts that are not perceived as valid or that are disliked are mentally discarded.

In an initial set of studies, Petty et al. (2002) provided evidence for a causal link between thought confidence and judgment. In one study, participants were exposed to a message containing very strong arguments (pretested to elicit positive thoughts) or very weak arguments (pretested to elicit mostly negative thoughts) in favor of a campus proposal. Following the message, participants listed their thoughts about the proposal and were then asked to think about past situations in which they experienced confidence or doubt. Those who articulated past instances of confidence became more certain of the validity of their recently generated thoughts to the message compared to those who reflected upon instances of doubt. That is, the feeling of confidence stemming from the randomly assigned memory exercise was misattributed to the thoughts recently generated regarding the persuasive message. This differential thought confidence then polarized attitudes toward the proposal, such that exposure to strong arguments led to very positive attitudes and exposure to weak arguments resulted in very negative attitudes.

There are many aspects of the source (e.g., similarity, status), the message (e.g., matching, quality), the recipient (e.g., bodily responses, emotions), and the situation (e.g., consensus) that can influence the extent of validation following the generation of message-relevant thoughts (Briñol & Petty, 2009a). As an illustration, consider a study by Brinol, Petty, Valle, Rucker, and Becerra (2007). They presented participants with either strong or weak arguments in favor of a proposal advocating for mandatory vaccination. This manipulation of argument quality prompted participants to generate either mostly positive (i.e., when arguments were strong) or mostly negative (i.e., when arguments were weak) thoughts about the new proposal. After participants generated their thoughts about the arguments, but before providing their attitudes toward the vaccination idea, they were instructed to recall either two incidents in their lives in which they had power over another person (high-power condition), or in which someone else had power over them (low-power condition). Relative to the low-power condition, those induced to feel powerful reported greater confidence in their thoughts about the mandatory vaccination. Presumably, they experienced strong feelings of confidence from the power manipulation, and they misattributed that confidence to their thoughts about the message. Consistent with validation, attitudes were polarized in the direction of the thoughts generated when power was high as compared to low. Therefore, generating message-relevant thoughts increased persuasion for strong arguments and decreased persuasion for weak arguments when participants experienced validation (i.e., feelings of confidence). This research illustrates that people may not only need to have the right expectations for placebo effects to work, but also need to have enough confidence and liking regarding their thoughts about the placebo to translate them into relevant judgments and actions.

In sum, the self-validation logic suggests that people often look for ways to validate whatever mental contents have been activated, including both positive and negative treatment expectations. Importantly, increasing perceived validity in positive (placebo) expectations will increase placebo effects, whereas increasing perceived validity in negative (nocebo) expectations will increase nocebo effects. Therefore, researchers and practitioners need to consider not only the direction of the expectations but also the validity associated with them.

5. MULTIPLE PROCESSES: BEYOND ELABORATION AND VALIDATION

A valuable element of the ELM is that it organizes the many specific processes by which variables (e.g., personality factors, the expertise of the communicator, current mood state) affect evaluative judgments into a finite set of five processes that operate at different points along the elaboration continuum (for a discussion, see Petty & Briñol, 2014). For example, the ELM postulates that variables such as the attractiveness of the source of a message can affect how much a person thinks about the message, thereby altering their location along the elaboration continuum. However, if circumstances have already conspired to place the person at the low end of the thinking continuum, then the variable can serve as a simple cue, affecting judgments in a direction that is consistent with its valence (e.g., an attractive source would lead to positive persuasion outcomes). On the other hand, if the person is at the high end of the continuum, the variable can affect judgments through one of three more deliberative processes. Specifically, the variable can be examined as an argument (e.g., does the fact that the source is attractive provide some relevant evidence about the merit of what he/she is advocating?), the variable can affect the valence of the thoughts that come to mind (e.g., exposure to an attractive source can make positive thoughts more salient and motivate positive thinking), and the variable can affect a structural feature of the thoughts generated (e.g., an attractive source could make one's thoughts more likeable or held with greater confidence).

The influence of all communications variables (source, message, recipient, and context) can operate through any of these five processes depending on the circumstances. Thus, the ELM describes five different roles that communication variables can have in altering persuasion: variables can (1) serve as simple cues (peripheral route), (2) alter one's location on the elaboration likelihood continuum (determining whether the peripheral or central route operates), (3) be examined as an argument (central route), (4) affect the valence of the thoughts that come to mind (central route), and (5) affect the metacognitive confidence and the liking toward the thoughts generated (central route).

In terms of placebo effects, this multiprocess perspective can help to clarify when a placebo effect is most likely to arise and which type of process is likely responsible. To illustrate this, consider the use of attractive, expensive-looking packaging for a new pain medication. The general finding could be that the attractive cover of the placebo treatment reduces pain to a greater extent compared to an unattractive container. But what is the process by which this effect occurs? The ELM postulates how high- vs low-elaboration processes may alter treatment outcomes, as well as indicating the conditions under which the different processes should manifest. For example, the attractive packaging could serve as a peripheral cue that associates the treatment with positive feelings (simple learning effect) or the attractive packaging could lead to a simple expectation that "if it looks good it is bound to work." At higher levels of elaboration, more thoughtful processes are anticipated to underlie apparently equivalent placebo effects. For example, when an individual is actively reflecting on the treatment, the attractive packaging could increase the positivity of the thoughts generated with regard to the treatment and even increase the confidence and liking associated with those thoughts. In turn, these favorable thoughts held with validity could lead to stronger treatment efficacy expectations, and promote strong and long-lasting placebo effects.

6. THE ELM AND PLACEBO EFFECTS

With the ELM outlined, we now revisit the placebo effect phenomenon. As may be readily apparent from the explanation of the ELM, the model provides a rich and diverse lens from which to study and understand placebo effects. In this section, we focus on two illustrative questions that arise from the placebo effect literature and provide answers for them drawn from the ELM. First, we discuss how the ELM could be used to explain when and how characteristics of a practitioner can alter placebo effects. Second, we discuss how the ELM can help to explain the duration of placebo effects.

7. WHEN AND WHY DO CHARACTERISTICS OF A PROVIDER ALTER PLACEBO EFFECTS?

As expectations for treatment success are typically induced by another person (e.g., a healthcare provider), there has been great interest in the possibility that the therapeutic encounter and characteristics of a provider can trigger or modify placebo effects (e.g., Benedetti, 2013; Bensing & Verheul, 2010; Brody & Brody, 2000; Frank, 1961). Characteristics of the provider encompass a broad array of potentially placebogenic elements, including amount of communication, gestures and postures, facial expressions, eye gaze, tone of voice, touch, adherence to medical rituals, conversational rapport, and patient perceptions of physician empathy, confidence, competence, social power, and status (e.g., Di Blasi, Harkness, Ernst, Georgiou, & Kleijnen, 2001; Kaptchuk et al., 2008; Vase, Nørskov, Petersen, & Price, 2011).

A variety of evidence suggests that practitioner characteristics can alter placebo effects. For example, in a study by Kaptchuk et al. (2008), patients with irritable bowel syndrome were randomly assigned to a wait list group, a placebo acupuncture group, or a group that received placebo acupuncture with a supportive practitioner. The supportive practitioners were instructed to listen to participants and to be warm, confident, and thoughtful. The results revealed that participants with the supportive practitioner reported more improvement 3 weeks later than the other two groups. Although the findings of this experiment are in-line with the position that practitioner characteristics alter placebo effects, other factors also differed across conditions, including the amount of time spent with patients. More recently, Howe et al. (2017) directly manipulated practitioner warmth and competence and found that these specific characteristics of a practitioner can strengthen placebo effects.

Although aspects of the practitioner can alter placebo effects, studies examining attributes of the practitioner have not always found them to work (e.g., Di Blasi et al., 2001; Schnittker & Liang, 2006), raising questions about the conditions under which practitioner attributes (e.g., perceived confidence) manifest. Further, even in situations when aspects of the practitioner promote placebo effects, it remains unclear why these effects occur. Thus, much is still unknown about how and when practitioner characteristics alter placebo effects.

As previously noted, the ELM is useful in that it describes five different roles that communication variables can have in altering persuasion. This multiprocess perspective can clarify when a factor, such as a provider characteristic, is likely to increase placebo effects and which type of psychological process is likely responsible. To illustrate this, consider the confidence of a doctor when they are presenting a placebo treatment for pain. The general finding could be that a placebo treatment reduces pain to a greater extent when the doctor is perceived as confident. But what is the process by which this effect occurs? As described earlier, the ELM postulates five processes that may influence treatment outcomes and the conditions under which they occur. For example, in low elaboration situations, the doctor's confidence could lead to the adoption of the simple heuristic that "if the doctor is confident, this treatment must work."

Importantly, displays of confidence are not always hypothesized to cause placebo effects through this sort of low effortful process. Instead, when individuals are able and motivated to engage in high levels of elaboration, more thoughtful processes are anticipated to underlie placebo effects. For example, the level of confidence could alter the valence of thoughts that comes to the patient's mind. That is, when the patient is engaged in higher levels of thought, they may actively reflect on the doctor's display of confidence. This confidence could increase thought positivity, which, in turn, could lead to stronger treatment efficacy expectations and subsequent placebo effects. Alternatively, if the doctor is perceived as expressing low levels of confidence, this could lead to the generation of negative thoughts, which could ultimately undercut a placebo effect.

As noted, under high elaboration conditions, other processes beyond biasing the direction of thoughts are possible. For example, the confidence expressed by the doctor can be misattributed to the thoughts generated by the patient, validating whatever comes to mind at the time. If thoughts are positive, increasing confidence will increase placebo effects, but if the patient's thoughts are negative, increasing confidence will decrease placebo effects. This metacognitive process of thought validation is more likely to operate not only when patients think about their thoughts but also when they consider validating information following (or during) thought generation.

When the level of thinking is moderate (rather than high or low) and the patient considers the doctor's confidence before processing the message, the doctor's confidence could affect the *extent* of thinking. Consider the case in which a patient is passively listening to a doctor explaining a medical treatment. The patient may be attending to the doctor, but also devoting some thought to other events (such as issues of payment for the medical visit).

If during this interaction the patient detects that the doctor is not very confident in the treatment, this could prompt an increase in thinking on the part of the patient. This increased thinking could reduce the effectiveness of the doctor's treatment efficacy message by high elaborative processes if the message is weak, but increase the persuasive effect if the message is strong. Of course, if the confidence of the doctor is distracting or is interpreted as a sign that everything is going well, making the patient feel pleasant and safe, then it can decrease (rather than increase) elaboration (see Briñol & Petty, 2009b, for a review on source factors in persuasion).

In sum, source characteristics—such as the confidence of a doctor—may enhance or undermine placebo effects through multiple processes depending on elaboration and timing conditions. Importantly, however, the long-term consequences of source characteristic variables are likely to differ depending on whether persuasion occurred with relatively high or low amounts of thinking. As we describe next, when patients experience placebo effects through high thinking, the placebo effects would be expected to be more durable and more likely to influence subsequent decisions and behavior.

8. WHEN AND WHY DO PLACEBO EFFECTS ENDURE OVER TIME?

One key question regarding placebo effects is, "how long do they last?" In medical care, this question has important implications. Some have suggested that placebo effects from expectations are likely to degrade quickly over time as compared to active treatment effects. However, there are a few examples in the empirical literature of placebo effects lingering for many weeks and even months. As much of the experimental literature on placebo effects comes from single session studies, this durability issue has not yet received a great deal of attention. Consequently, at present, there is an insufficient research base and theory from which to predict how and when treatment expectations lead to more long-lasting and durable placebo effects (Geers & Miller, 2014). Clarifying when placebo effects endure would have valuable implications for medical care.

As described earlier, the ELM proposes that the amount of elaboration embedded in the processes of change is consequential for short- and long-term change. Therefore, a valuable feature of the ELM approach to placebo effects is that the model leads to predictions regarding the longevity, durability, and behavioral implications of placebo effects. Specifically, the model proposes that placebo effects brought about through high-elaboration processes should be more persistent, resistant, and predictive of behavior change than those brought about through of low-elaboration processes. Furthermore, placebo effects forged through high-elaboration processes should also be more accessible and held with greater confidence. In summary, based on the elaboration—strength hypothesis of the ELM, it can be surmised that placebo effects emerging from higher levels of thought should last longer than ones resulting from lower levels of thought. Moreover, placebo effects that result from high thinking should be more resistant to contrary internal information (e.g., physical sensations) and external information (e.g., social observations) that does not support a placebo expectation. Thus, the ELM provides a novel platform for testing the strength of placebo effects.

Do placebo effects formed through higher elaboration last longer than ones formed through lower levels of elaboration? Currently, there are no direct data relevant to this hypothesis. However, one recent experiment concerning the effect of expectations on postexercise feelings provides some suggestive evidence (Helfer, Elhai, & Geers, 2015). In this experiment, one group of participants was presented with information indicating that engaging in exercise results in positive mood states. A control group, in contrast, learned about the specific qualities of an exercise bicycle. Orthogonal to this expectation manipulation was an elaboration manipulation. Half of the participants were asked to further think about the information they were provided. Specifically, the high-elaboration participants were asked to write down personal examples that illustrate the idea that exercise improves mood. The low-elaboration participants did not engage in this additional writing. All participants then completed 10 min of light-intensity exercise on a stationary bicycle in the laboratory. Afterward, participants were given the opportunity to take part in a 2 week follow up survey about exercising. The data revealed that participants who were told that exercising elevates mood states displayed a more positive mood state immediately after the exercise than the no-expectation participants, regardless of amount of elaboration. Thus, in the short-term, high and low elaboration resulted in similar immediate changes in feelings and thoughts (a placebo-like effect). However, at the 2-week follow-up, it was found that only the high-elaboration group reported a more positive mood state following exercise. Consistent with the ELM, the results indicate that greater message elaboration at the time of the initial treatment evaluation increased the duration of the expectation message's effect on mood.

The research just described provides initial evidence that level of message elaboration can alter the duration of expectation effects on mood states. Future research should explore this prediction in standard placebo paradigms, ideally including physiological outcome measures. Further, it will be valuable to test other indicators of the elaboration-strength hypothesis. For example, based on the ELM, it can be predicted that when individuals elaborate on placebo expectations given with strong messages, they will be more likely to alter many other outcome variables such as purchasing behavior, treatment adherence, and making treatment recommendations to others. Additionally, placebo effects developed in this manner may be more resistant to internal physical cues and sensations that do not support the placebo effect.

9. SUMMARY AND CONCLUSIONS

In this chapter, we reviewed both the placebo and persuasion literature and suggested a new conceptualizing of placebo effects as consequences of persuasive communication. Placebo effects often arise when treatment-relevant information is transmitted from a source, in a specific context, and changes a recipient's responses. To illustrate the potential of this integration, we outlined a modern theoretical model of persuasion, the ELM, and suggested ways this model can link to the literature on placebo effects. From this account, placebo effects can occur through both high and low levels of mental elaboration, and motivation (e.g., personal relevance) and ability (e.g., distraction, knowledge) are critical determinants of the extent of elaboration. The model also emphasizes the role of thought confidence and thought liking as other critical factor that influence the development of placebo effects through metacognitive processes. Additionally, when placebo effects arise through high-elaboration processes, they are predicted to be more persistent, resistant, and directive of behavior than if they are brought about through low-elaboration processes. Our review also emphasized the metacognitive process of validation, which influences the use of the thoughts that determine placebo effects. A critical aspect of the ELM is that it identifies five different roles that any single communication variable can have in altering placebo effects. We believe that the ELM provides a theoretically rich framework for future research regarding the psychological processes underlying placebo effects.

REFERENCES

Benedetti, F. (2013). Placebo and the new physiology of the doctor-patient relationship. *Physiological Reviews*, *93*, 1207–1246.

Benedetti, F. (2014). *Placebo effects: Understanding the mechanisms in health and disease* (2nd ed.). Oxford: Oxford University Press.

Benedetti, F., Pollo, A., Lopiano, L., Lanotte, M., Vighetti, S., & Rainero, I. (2003). Conscious expectation and unconscious conditioning in analgesic, motor, and hormonal placebo/nocebo responses. *Journal of Neuroscience*, *23*, 4315–4323.

Bensing, J. M., & Verheul, W. (2010). The silent healer: The role of communication in placebo effects. *Patient Education and Counseling*, *80*, 293–299.

Berkowitz, L., & Thome, P. R. (1987). Pain expectation, negative affect, and angry aggression. *Motivation and Emotion*, *11*, 183–193.

Briñol, P., & Petty, R. E. (2009a). Persuasion: Insights from the self-validation hypothesis. *Advances in Experimental Social Psychology*, *41*, 69–118.

Briñol, P., & Petty, R. E. (2009b). Source factors in persuasion: A self-validation approach. *European Review of Social Psychology*, *20*, 49–96.

Briñol, P., & Petty, R. E. (2012). The history of attitudes and persuasion research. In A. Kruglanski & W. Stroebe (Eds.), *Handbook of the history of social psychology* (pp. 175–212). New York: Psychology Press.

Brinol, P., Petty, R. E., Valle, C., Rucker, D. D., & Becerra, A. (2007). The effects of message recipients' power before and after persuasion: A self-validation analysis. *Journal of Personality and Social Psychology*, *93*, 1040–1053.

Brody, H., & Brody, D. (2000). *The placebo response*. New York: HarperCollins Publishers.

Brunoni, A. R., Lopes, M., Kaptchuk, T. J., & Fregni, F. (2009). Placebo response of non-pharmacological and pharmacological trials in major depression: A systematic review and meta-analysis. *PLoS One*, *4*, e4824.

Caplandies, F., Colagiuri, B., Helfer, S. G., & Geers, A. L. (2017). Effect type but not attribute framing alters nocebo headaches in an experimental paradigm. *Psychology of Consciousness: Theory, Research, and Practice*, *4*, 259–273.

Chaiken, S., Liberman, A., & Eagly, A. H. (1989). Heuristic and systematic processing within and beyond the persuasion context. In J. S. Uleman & J. A. Bargh (Eds.), *Unintended thought: Limits of awareness, attention, and control* (pp. 212–252). New York: Guilford.

Colagiuri, B., Livesey, E. J., & Harris, J. A. (2011). Can expectancies produce placebo effects for implicit learning? *Psychonomic Bulletin & Review*, *18*, 399–405.

Colagiuri, B., Schenk, L. A., Kessler, M. D., Dorsey, S. G., & Colloca, L. (2015). The placebo effect: From concepts to genes. *Neuroscience*, *307*, 171–190.

Colloca, L. (2017). Nocebo effects can make you feel pain. *Science*, *358*, 44.

Colloca, L., & Benedetti, F. (2005). Placebos and painkillers: Is mind as real as matter? *Nature Reviews Neuroscience*, *6*, 545–552.

Colloca, L., Flaten, A. M., & Meissner, K. (Eds.), (2013). *Placebo and pain: From bench to bedside*. San Diego, CA: Academic Press.

Colloca, L., Lopiano, L., Lanotte, M., & Benedetti, F. (2004). Overt versus covert treatment for pain, anxiety, and Parkinson's disease. *The Lancet. Neurology*, *3*, 679–684.

Darragh, M., Vanderboor, T., Booth, R. J., Sollers, J. J., & Consedine, N. S. (2015). Placebo 'serotonin' increases heart rate variability in recovery from psychosocial stress. *Physiology and Behavior*, *145*, 45–49.

De Craen, A. J., Kaptchuk, T. J., Tijssen, J. G., & Kleijnen, J. (1999). Placebos and placebo effects in medicine: Historical overview. *Journal of the Royal Society of Medicine*, *92*, 511–515.

De Craen, A. J., Roos, P. J., De Vries, A. L., & Kleijnen, J. (1996). Effect of colour of drugs: Systematic review of perceived effect of drugs and of their effectiveness. *British Medical Journal*, *313*, 1624–1626.

de Craen, A. J., Tijssen, J. G. P., de Gans, J., & Kleijnen, J. (2000). Placebo effect in the acute treatment of migraine: Subcutaneous placebos are better than oral placebos. *Journal of Neurology*, *247*, 183–188.

Di Blasi, Z., Harkness, E., Ernst, E., Georgiou, A., & Kleijnen, J. (2001). Influence of context effects on health outcomes: A systematic review. *The Lancet*, *357*, 757–762.

Dinnerstein, A. J., & Halm, J. (1970). Modification of placebo effects by means of drugs: Effects of aspirin and placebos on self-rated moods. *Journal of Abnormal Psychology*, *75*, 308–314.

Doering, B. K., & Rief, W. (2012). Utilizing placebo mechanisms for dose reduction in pharmacotherapy. *Trends in Pharmacological Sciences*, *33*, 165–172.

Draganich, C., & Erdal, K. (2014). Placebo sleep affects cognitive functioning. *Journal of Experimental Psychology: Learning, Memory, and Cognition*, *40*, 857–864.

Enck, P., Bingel, U., Schedlowski, M., & Rief, W. (2013). The placebo response in medicine: Minimize, maximize or personalize? *Nature Reviews Drug Discovery*, *12*, 191–204.

Faasse, K., Cundy, T., Gamble, G., & Petrie, K. J. (2013). The effect of an apparent change to a branded or generic medication on drug effectiveness and side effects. *Psychosomatic Medicine*, *75*, 90–96.

Fillmore, M., & Vogel-Sprott, M. (1992). Expected effect of caffeine on motor performance predicts the type of response to placebo. *Psychopharmacology*, *106*, 209–214.

Frank, J. (1961). *Persuasion and healing: A comparative study of psychotherapy*. Baltimore: Johns Hopkins University Press.

Gascó, M., Briñol, P., & Horcajo, J. (2010). Cambio de actitudes hacia la imagen corporal: El efecto de la elaboración sobre la fuerza de las actitudes. *Psicothema*, *22*, 71–76.

Geers, A. L., & Miller, F. G. (2014). Understanding and translating the knowledge about placebo effects: The contribution of psychology. *Current Opinion in Psychiatry*, *27*, 326–331.

Geers, A. L., Rose, J. P., Fowler, S. L., Rasinski, H. M., Brown, J. A., & Helfer, S. G. (2013). Why does choice enhance treatment effectiveness? Using placebo treatments to demonstrate the role of personal control. *Journal of Personality and Social Psychology*, *105*, 549–566.

Greenwald, A. G., Spangenberg, E. R., Pratkanis, A. R., & Eskenazi, J. (1991). Double-blind tests of subliminal self-help audiotapes. *Psychological Science*, *2*, 119–122.

Helfer, S. G., Elhai, J., & Geers, A. L. (2015). Affect and exercise: Positive affective expectations can increase post-exercise mood and exercise intentions. *Annals of Behavioral Medicine*, *49*, 269–279.

Hovland, C. I., Janis, I. L., & Kelley, H. H. (1953). *Communication and persuasion; psychological studies of opinion change*. New Haven, CT: Yale University Press.

Howe, L. C., Goyer, J. P., & Crum, A. J. (2017). Harnessing the placebo effect: Exploring the influence of physician characteristics on placebo response. *Health Psychology*, *36*, 1074–1082.

Hull, J. G., & Bond, C. F. (1986). Social and behavioral consequences of alcohol consumption and expectancy: A meta-analysis. *Psychological Bulletin*, *99*, 347–360.

Janis, I. L., & King, B. T. (1954). The influence of role playing on opinion change. *The Journal of Abnormal and Social Psychology*, *49*, 211–218.

Kaptchuk, T. J., Kelley, J. M., Conboy, L. A., Davis, R. B., Kerr, C. E., Jacobson, E. E., et al. (2008). Components of placebo effect: Randomised controlled trial in patients with irritable bowel syndrome. *British Medical Journal*, *336*, 999–1003.

Kaptchuk, T. J., Stason, W. B., Davis, R. B., Legedza, A. R., Schnyer, R. N., Kerr, C. E., et al. (2006). Sham device v inert pill: Randomised controlled trial of two placebo treatments. *British Medical Journal*, *332*(7538), 391–397.

Kennedy, W. P. (1961). The nocebo reaction. *Medical World*, *912*, 203–205.

Kirsch, I. (Ed.), (1999). *How expectancies shape experience.* Washington, DC: American Psychological Association.
Liberman, R. (1962). An analysis of the placebo phenomenon. *Journal of Chronic Diseases, 15*, 761–783.
Linde, K., Niemann, K., & Meissner, K. (2010). Are sham acupuncture interventions more effective than (other) placebos? A re-analysis of data from the Cochrane review on placebo effects. *Complementary Medicine Research, 17*, 259–264.
McGuire, W. J. (1969). The nature of attitudes and attitude change. In E. Aronson & L. Gardner (Eds.), *The handbook of social psychology* (pp. 136–314). Reading, MA: Addison-Wesley.
Miller, F. G., & Kaptchuk, T. J. (2008). The power of context: Reconceptualizing the placebo effect. *Journal of the Royal Society of Medicine, 101*, 222–225.
Mills, L., Boakes, R. A., & Colagiuri, B. (2016). Placebo caffeine reduces withdrawal in abstinent coffee drinkers. *Journal of Psychopharmacology, 30*, 388–394.
Moerman, D. (2002). *Meaning, medicine, and the 'placebo effect'.* Cambridge, UK: Cambridge University Press.
Petersen, G. L., Finnerup, N., Colloca, L., Amanzio, M., Price, D. D., Jensen, T., et al. (2014). The magnitude of nocebo effects in pain: A meta-analysis. *Pain, 155*, 1426–1434.
Petersen, G. L., Finnerup, N. B., Nørskov, K. N., Grosen, K., Pilegaard, H. K., Benedetti, F., et al. (2012). Placebo manipulations reduce hyperalgesia in neuropathic pain. *Pain, 153*, 1292–1300.
Petty, R. E., & Briñol, P. (2012). The elaboration likelihood model. In P. A. M. Van Lange, A. Kruglanski, & E. T. Higgins (Eds.), *Handbook of theories of social psychology* (pp. 224–245). London, England: Sage.
Petty, R. E., & Briñol, P. (2014). The elaboration likelihood and metacognitive models of attitudes. In J. W. Sherman, B. Gawronski, & Y. Trope (Eds.), *Dual-process theories of the social mind* (pp. 172–187). New York: NY: Guilford Press.
Petty, R. E., Briñol, P., & Tormala, Z. L. (2002). Thought confidence as a determinant of persuasion: The self-validation hypothesis. *Journal of Personality and Social Psychology, 82*, 722–741.
Petty, R. E., & Cacioppo, J. T. (1979). Issue involvement can increase or decrease persuasion by enhancing message-relevant cognitive responses. *Journal of Personality and Social Psychology, 37*, 1915–1926.
Petty, R. E., & Cacioppo, J. T. (1986). The elaboration likelihood model of persuasion. *Advances in Experimental Social Psychology, 19*, 123–205.
Petty, R. E., Cacioppo, J. T., & Schumann, D. (1983). Central and peripheral routes to advertising effectiveness: The moderating role of involvement. *Journal of Consumer Research, 10*, 135–146.
Petty, R. E., Haugtvedt, C. P., & Smith, S. M. (1995). Elaboration as a determinant of attitude strength: Creating attitudes that are persistent, resistant, and predictive of behavior. In R. E. Petty & J. A. Krosnick (Eds.), *Attitude strength: Antecedents and consequences* (pp. 93–130). New York: NY: Psychology Press.
Rosen, J. C. (1997). Cognitive-behavioral body image therapy. In D. M. Garner & P. E. Garfinke (Eds.), *Handbook of treatment for eating disorders* (2nd ed., pp. 188–201). The Guilford Press.
Ross, M., & Olson, J. M. (1981). An expectancy-attribution model of the effects of placebos. *Psychological Review, 88*, 408–437.
Schnittker, J., & Liang, K. (2006). The promise and limits of racial/ethnic concordance in physician-patient interaction. *Journal of Health Politics, Policy and Law, 31*, 811–838.
Shapiro, A. K., & Shapiro, E. (1997). *The powerful placebo effect.* Baltimore, MA: John Hopkins University Press.

Shiv, B., Carmon, Z., & Ariely, D. (2005). Placebo effects of marketing actions: Consumers may get what they pay for. *Journal of Marketing Research*, *42*, 383–393.
Slama, H., Caspar, E. A., Gevers, W., & Cleeremans, A. (2013). Placebo-suggestion modulates conflict resolution in the Stroop task. *PLoS One*, *8*, e75701.
Stewart-Williams, S., & Podd, J. (2004). The placebo effect: Dissolving the expectancy versus conditioning debate. *Psychological Bulletin*, *130*, 324–340.
Storms, M. D., & Nisbett, R. E. (1970). Insomnia and the attribution process. *Journal of Personality and Social Psychology*, *16*, 319–328.
Vase, L., Nørskov, K. N., Petersen, G. L., & Price, D. D. (2011). Patients' direct experiences as central elements of placebo analgesia. *Philosophical transactions of the Royal Society of London. Series B, Biological Sciences*, *366*, 1913–1921.
Vase, L., Petersen, G. L., Riley, J. L., & Price, D. D. (2009). Factors contributing to large analgesic effects in placebo mechanism studies conducted between 2002 and 2007. *Pain*, *145*, 36–44.
Wager, T. D., & Atlas, L. Y. (2015). The neuroscience of placebo effects: Connecting context, learning and health. *Nature Reviews. Neuroscience*, *16*, 403–418.
Wegener, D. T., & Petty, R. E. (1996). Effects of mood on persuasion processes: Enhancing, reducing, and biasing scrutiny of attitude-relevant information. In L. L. Martin & A. Tesser (Eds.), *Striving and feeling: Interactions among goals, affect, and self-regulation* (pp. 329–362). Hillsdale, NJ: Lawrence Erlbaum Associates.
Weger, U. W., & Loughnan, S. (2013). Mobilizing unused resources: Using the placebo concept to enhance cognitive performance. *The Quarterly Journal of Experimental Psychology*, *66*, 23–28.
Weimer, K., Gulewitsch, M. D., Schlarb, A. A., Schwille-Kiuntke, J., Klosterhalfen, S., & Enck, P. (2013). Placebo effects in children: A review. *Pediatric Research*, *74*, 96–102.
Wickless, C., & Kirsch, I. (1989). Effects of verbal and experiential expectancy manipulations on hypnotic susceptibility. *Journal of Personality and Social Psychology*, *57*, 762–768.
Wright, S. A., da Costa Hernandez, J. M., Sundar, A., Dinsmore, J., & Kardes, F. R. (2013). If it tastes bad it must be good: Consumer naïve theories and the marketing placebo effect. *International Journal of Research in Marketing*, *30*, 197–198.
Zimba, C. G. (1993). An experimental study of the placebo effect in African traditional medicine. *Behavioral Medicine*, (3), 101–109.

CHAPTER EIGHT

Mindsets Matter: A New Framework for Harnessing the Placebo Effect in Modern Medicine

Sean R. Zion[1], Alia J. Crum
Stanford University, Stanford, CA, United States
[1]Corresponding author: e-mail address: szion@stanford.edu

Contents

Abstract

The clinical utility of the placebo effect has long hinged on physicians deceptively administering an objective placebo treatment to their patients. However, the power of the placebo does not reside in the sham treatment itself; rather, it comes from the psychosocial forces that surround the patient and the treatment. To this end, we propose a new framework for understanding and leveraging the placebo effect in clinical care. In outlining this framework, we first present the placebo effect as a neurobiological effect that is evoked by psychological processes. Next, we argue that along with implicit learning and expectation formation, mindsets are a key psychological process involved in the placebo effect. Finally, we illustrate the critical role of the social environment and treatment context in shaping these psychological processes. In doing so, we offer a guide for how the placebo effect can be understood, harnessed, and leveraged in the practice of modern medicine.

International Review of Neurobiology, Volume 138
ISSN 0074-7742
https://doi.org/10.1016/bs.irn.2018.02.002

1. INTRODUCTION

Henry Beecher, a World War II army medic, is credited with bringing about the modern study of the placebo effect. Upon exhausting his supply of morphine, Beecher continued treating his wounded patients with saline solution while reassuring them of the pain-relieving power of the infusion. His observation that patients tended to improve with the sham treatment and comforting words led to the publication of "The Powerful Placebo," which reported on the magnitude of the placebo effect across 15 clinical trials. Beecher's analysis revealed that over 35% of patients experienced therapeutic benefit from placebo treatments (Beecher, 1955). His finding intrigued a generation of scientists and laid the groundwork for decades of subsequent research on the placebo effect.

Since Beecher's time, the placebo effect has taken on new meaning depending on the context in which it is studied. In randomized, double-blind clinical drug trials, new drugs and treatments are compared to a placebo control. This gold standard for pharmaceutical research and development enables researchers to quantify the efficacy of the treatment as it compares to the effect of the placebo. The ultimate goal of these trials is to demonstrate the benefit of the active treatment beyond that of the placebo, and consequently, much of clinical research aims to subtract out or do away with the placebo effect (Crum, Leibowitz, & Verghese, 2017). Even when a significant difference between the active treatment and the placebo is found, any effect of the placebo is later usually forgotten. Despite the fact that the placebo effect yields clinically significant benefits in clinical trials for well over half of all medical conditions—including pain, depression, Parkinson's disease, anxiety disorders, cardiovascular disorders, and immunological diseases (de la Fuente-Fernandez & Stoessl, 2002; Goebel, Meykadeh, Kou, Schedlowski, & Hengge, 2008; Levine, Gordon, Smith, & Fields, 1981; Petrovic et al., 2005; Pollo, Vighetti, Rainero, & Benedetti, 2003; Shetty, Friedman, Kieburtz, Marshall, & Oakes, 1999; Walsh, Seidman, Sysko, & Gould, 2002)—the symbolic effect of randomized placebo-controlled trials is that placebos are seen either as an adversary that competes with the drug of interest or as an irrelevant variable to be discounted.

However, a handful of researchers have explored the placebo effect in its own right. Inspired to understand the strength of and mechanisms behind the placebo effect, psychologists and neurobiologists have designed studies to investigate how a placebo alone can produce measurable benefit. These studies consider the placebo as the *primary variable of interest*, not merely as a control. Laboratory experiments have demonstrated the ability to evoke placebo responses through classical conditioning paradigms and through the manipulation of expectations (Montgomery & Kirsch, 1997). These paradigms reveal that placebo administration drives changes in both endogenous opioid and nonopioid neurotransmitter systems and modulates metabolic activity in many regions of the brain (Amanzio & Benedetti, 1999; Benedetti, 2008; Levine & Gordon, 1984).

While both clinical trials and direct placebo research have made great strides in illuminating the power and mechanisms of placebo effects, relatively little research has been devoted to understanding how placebo effects may be leveraged in clinical practice without the deceptive administration of an actual placebo treatment. In an effort to make progress in the clinical applications of the placebo effect, we offer a new framework for understanding the nature and clinical utility of this effect in the practice of medicine. This framework builds upon the wealth of existing clinical, neurobiological, and psychological research on the placebo effect and extends it by explaining how the components underlying placebo effects operate in clinical practice, without an objective placebo. We argue that the placebo effect is: (1) an integral component of the overall treatment effect in medicine, (2) a neurobiological effect that is evoked by specific psychological processes, which are shaped by social and environmental factors, and (3) a variable that can be harnessed, personalized, and maximized in the practice of medicine without the use of inert placebo pills or sham treatments. Through this conceptualization of the placebo effect, we can see that the placebo effect is neither a nuisance nor a mystery. It is a real and powerful effect that can be utilized in the practice of medicine by understanding and leveraging the psychosocial forces that surround a medical treatment.

In the sections that follow, we review the components of the treatment effect and explore how the effects of the drug and the placebo can be disentangled. We then describe the psychological processes that drive the neurobiological mechanisms that underlie the placebo effect. Finally, we discuss the social and contextual factors that inform and shape these psychological processes, offering a roadmap for how they can be harnessed in the practice of medicine.

2. THE TREATMENT EFFECT: DRUG PLUS PLACEBO EFFECTS

The drug effect and the placebo effect have long been considered separate and often competing entities, as evidenced by the placebo's aforementioned role in randomized control trials. What are these two effects and how can they be disentangled? The drug effect is the quantifiable change in disease processes that result from the pharmacological or physical properties of an active treatment—often a medication. These medications evoke clinical change by either: (a) simulating or stimulating normal biological processes that occur in the body or inhibiting processes that contribute to disease, or (b) blocking critical processes in microorganisms inhabiting the body. But these mechanisms are responsible for only part of a treatment's total effect. Placebo effects also occur as an inherent part of all active medical treatments, and the effect of the drug and the effect of the placebo work together to produce the total treatment effect (Fig. 1). From the patient's perspective, the components that contribute to the treatment effect are not particularly relevant—improvement from the drug or improvement from the psychosocial factors that drive the placebo effect is improvement nonetheless. However, for the health care practitioner, it is incredibly important to understand the components that drive the placebo effect and how they might interact with the drug effect.

The placebo effect can be considered in terms of its biological, psychological, and social/contextual components (Table 1). Driven by neurobiological

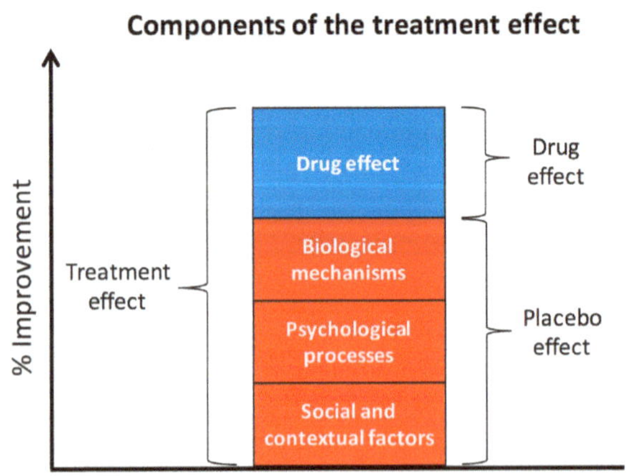

Fig. 1 The drug effect and the placebo effect contribute to the total effect of treatment. These components drive the total improvement a patient experiences from a medical treatment.

Table 1 The Components of the Placebo Effect: Neurobiological Mechanisms (The Body's Healing Properties and Neurophysiology), Psychological Processes (Implicit Learning, Expectations, and Mindsets), and Social/Contextual Factors (Social Environment and Treatment Context)

Components of the Placebo Effect

Biological mechanisms	
Body's healing properties	Biological properties of the body that facilitate healing, including homeostatic mechanisms, immune, and inflammatory responses. These contribute to the natural history of a disease, but can also be targets of placebo effects
Neurophysiology	Dopamine, endogenous opioids, and endocannabinoids are three of the major neurotransmitter systems implicated in moderating the placebo effect
Psychological processes	
Implicit learning	The nonconscious acquisition of knowledge. Classical conditioning, a form of implicit learning, is implicated in certain instances of the placebo effect
Expectations	A belief about the future based on a prediction of what is most likely to happen. Expectations underlie certain instances of the placebo effect and drive neurobiological mechanisms
Mindsets	A lens or frame of mind that orients an individual to a particular set of beliefs, associations, and expectations, and functions to guide attentional and motivational processes
Social and contextual factors	
Development and culture	Our caregivers and social environment influence the psychological processes that underlie the placebo effect. These processes are continuously shaped throughout life by the ideas, institutions, and interactions that constitute the culture in which we live
Patient-provider relationship	The patient–provider relationship shapes the mindsets a patient holds about health, illness, and treatments, and affects the quality of care a patient receives. This relationship is influenced by the warmth and competence of the provider and is further shaped by characteristics like empathy and trust

Continued

Table 1 The Components of the Placebo Effect: Neurobiological Mechanisms (The Body's Healing Properties and Neurophysiology), Psychological Processes (Implicit Learning, Expectations, and Mindsets), and Social/Contextual Factors (Social Environment and Treatment Context)—cont'd

Components of the Placebo Effect

Observational learning and social influence	Learning through direct observation of others undergoing treatment (i.e., other patients) as well as interactions with individuals who yield influence over the patient (i.e., physicians and nurses) both may powerfully drive placebo effects
Treatment characteristics	The specific characteristics of the treatment that is provided to the patient. This includes factors like the shape, color, and branding of the treatment, the method of administration, and the physical environment in which the treatment is administered

mechanisms, the placebo effect recruits the involvement of disease-specific biological and neurotransmitter systems, such as components of the immune system and the endogenous opioid system. These biological mechanisms are evoked and modulated by conscious and nonconscious psychological processes, including implicit learning, expectations, and mindsets. Psychological processes, in turn, are shaped by the social environment and treatment context. As such, a social or environmental factor—like a knowledgeable and understanding physician who the patient trusts—can shape a patient's mindset about a disease or treatment, which can in turn evoke a biological change and subsequent healing response. In the remaining sections, we unpack these biological, psychological, and social elements underlying placebo effects that form the foundation of the treatment effect for all drugs and therapies.

3. PSYCHOLOGICAL PROCESSES ACTIVATE NEUROBIOLOGICAL MECHANISMS

Placebo effects are marked by neurobiological underpinnings (Wager & Atlas, 2015), which are activated by psychological processes. Two such processes have received the majority of research in this domain: nonconscious implicit learning, such as classical conditioning, and conscious expectations (Finniss, Kaptchuk, Miller, & Benedetti, 2010; Price et al., 1999). These mechanisms are neither mutually exclusive nor the only two mechanisms through which the placebo effect is thought to operate

(Stewart-Williams & Podd, 2004). Other mechanisms, like our mindsets, also play an important role in the placebo effect, but have received less attention. In the next sections, we review the existing literature on the role of implicit learning, expectations, and mindsets as they relate to the placebo effect in patients undergoing medical treatment. In particular, we review how these psychological elements trigger the neurobiological processes that lead to the measurable changes we refer to collectively as the placebo effect.

3.1 Implicit Learning

Implicit learning is a process by which information is learned outside of conscious awareness (Frensch & Rünger, 2003). It is a process of detecting associations within an environment and storing this information in the form of abstract representations (Seger, 1994). Classical conditioning, a form of implicit learning, underlies certain instances of the placebo effect. Early evidence for the role of implicit learning in the placebo effect came from animal studies (Ader & Cohen, 1982, 1993), versions of which were later replicated in humans (Goebel et al., 2002). Many of these studies repeatedly paired a neutral stimulus—the placebic vehicle (i.e., syringe or capsule)—with an unconditioned stimulus (i.e., the drug inside the syringe or capsule). Other experimental techniques in humans have also been used to tease apart the mechanism of conditioned placebo responses. For example, the use of "surreptitious reduction" paradigms in which a placebo treatment is paired with the hidden reduction of a painful stimulus also provides evidence for conditioned responses in studies of placebo analgesia (Voudouris, Peck, & Coleman, 1989, 1990).

Through similar processes, symbols and rituals within the medical context become associated with healing. Being directed to the exam room, having temperature, blood pressure, and heart rate measurements taken, and waiting patiently for the physician may all serve as situational cues that become implicitly associated with healing. Over time, these contextual cues are repeatedly paired with active medical treatments. Eventually, exposure to these cues alone may evoke conditioned responses in patients. Outside of the doctor's office, positive or negative experiences with active treatments may lead to associative links between treatment characteristics and outcomes. A child given bright pink, bubble-gum flavored liquid penicillin each time he or she has a bacterial infection will come to associate the perceptual characteristics of this medication with the subsequent healing response. Indeed, multiple sclerosis patients who received cyclophosphamide (an immune suppressant) paired with a flavored syrup later displayed drug-consistent immune responses to the flavored syrup alone (Giang et al., 1996).

How does the nonconscious formation of associative links between healing symbols and treatment outcomes evoke quantifiable biological changes in the patient? Implicit learning recruits disease-related processes in the body to change objective biological markers. This has been demonstrated in studies of conditioned immune and allergic responses in both humans and animals. For example, conditioning rats with cyclosporine A, an immunosuppressive drug, resulted in sympathetic nervous system-mediated immune suppression even when the drug was absent (Exton et al., 2002). In humans, pairing the same immunosuppressant drug, cyclosporine A, with a flavored beverage reduced lymphocyte cell count, cytokine release, and expression of mRNA when the flavored beverage was later given on its own (Goebel et al., 2002). Side effects of active treatments also appear to be conditioned alongside the intended treatment effects. In a study by Benedetti and colleagues, subjects were given a powerful opioid analgesic medication that was subsequently replaced with a placebo treatment as part of a conditioning paradigm. Subjects not only experienced the main effects of the opioid when given the placebo, but they also exhibited respiratory depression, a common side effect of opioids (Benedetti, Amanzio, Baldi, Casadio, & Maggi, 1999).

In the practice of medicine, some benefit is derived from years of associating positive outcomes with white coats, pills, and exam rooms. While this may paint a picture of a patient mindlessly linking stimuli and response (Wickramasekera, 1980), implicit learning processes often induce positive responses by reinforcing patient expectations (Rescorla, 1988), as most conditioning paradigms in humans inherently manipulate expectations to some degree (Benedetti, Pollo, et al., 2003; Montgomery & Kirsch, 1997). Indeed, implicit learning is not required for placebo effects. Thus, while implicit learning may work directly, it may also operate by influencing our conscious expectations.

3.2 Expectancy

Expectations are beliefs about the nature and likelihood of future states. The expectation of a specific outcome can elicit cognitive, emotional, and behavioral changes that increase the likelihood of that event occurring (Kirsch, 1985; Montgomery & Kirsch, 1997). Expectations have been shown to be one mechanism driving many instances of the placebo effect.

Studies employing an open/hidden experimental design allow researchers to disentangle the contribution of expectations from drug effects. In this paradigm, a physician administers a drug either in full view of the patient (open condition) or hidden from the patient (hidden condition). Open administration mimics the conditions of routine medical practice, while hidden administration removes the external factors that contribute to the formation of expectations. The difference between these two conditions reflects the impact of the patients' psychological processes, like expectations, that can elicit real and quantifiable effects. Critically, this paradigm allows the role of expectations to be quantified without actually administering a placebo treatment. Studies employing open/hidden designs have found that medical treatments given covertly are less effective than those given openly, highlighting the contribution of expectations to the efficacy of treatments (Colloca, Lopiano, Lanotte, & Benedetti, 2004). This has been observed in numerous clinical conditions, including pain, anxiety, and Parkinson's disease (Benedetti, Maggi, et al., 2003; Levine & Gordon, 1984; Pollo et al., 2002).

Placebo analgesia has also been used as a paradigm to demonstrate the effect of positive or negative expectations on treatment efficacy (Price, Finniss, & Benedetti, 2008; Price et al., 1999). Compared to a no-expectation condition, patients who expected the potent analgesic remifentanil to work well experienced twice the analgesic effect, while those who held negative expectations experienced no analgesia (Bingel et al., 2011). These subjective effects corresponded with significant changes in the endogenous pain modulatory system. Imaging data from various clinical populations further demonstrate the mechanisms by which expectations evoke neurobiological responses in patients. In a study of patients with irritable bowel syndrome, verbal suggestions of pain relief produced clinically significant placebo effects. This placebo analgesia corresponded with reduced activity in the thalamus, somatosensory cortices, insula, and anterior cingulate and increased activity was noted in the rostral portion of the anterior cingulate, the amygdala, and the periaqueductal gray (Price et al., 2008). Furthermore, in a revealing study employing an open/hidden paradigm to quantify the magnitude of expectations on postsurgical dental pain, injecting saline in full view of the patient reduced pain at a magnitude equal to 6–8 mg of morphine (Levine & Gordon, 1984; Levine, Gordon, Bornstein, & Fields, 1979). In other words, when patients were not aware they were receiving treatment, and thus did not expect to receive benefit, it was as if they had been given 6–8 mg *less* morphine than they actually had.

The effect of expectation on pain is powerful, and it is therefore critical for physicians to be aware of how they are inducing and shaping patient expectations. Physicians may even have the ability to shape expectations and evoke placebo effects in patients who have formed maladaptive or harmful associative links between certain treatments and poor outcomes. For example, a 2003 study demonstrated that manipulating expectations was effective in overriding negative responses to pharmacological preconditioning in a placebo analgesia paradigm (Benedetti, Pollo, et al., 2003). However, in the clinical context, expectations are not always induced intentionally or explicitly. They are shaped by patients' mindsets, the social environment, and the treatment context in which a medical intervention occurs.

3.3 Mindsets

Mindsets are lenses or frames of mind that orient individuals to particular sets of associations and expectations (Crum, Salovey, & Achor, 2013). Mindsets help individuals make sense of complex information by offering them simple schematics about themselves and objects in their world. For patients, mindsets provide a scaffolding for understanding the broad nature of illnesses and treatments. While expectations and mindsets are intimately connected, they are not the same thing. Expectations are specific beliefs about future events. Mindsets are a more general psychological construal that orient an individual to a number of mindset-consistent expectations. For example, the mindset that "cancer is a catastrophe" may be associated with a number of different expectations such as "the treatment will be painful and keep me from the things I enjoy" or "I will not be able to cope with this" that are beyond the more specific expectations of believing a treatment will or will not work. Thus, understanding broader mindsets is also important for understanding the impact of illnesses and treatments.

Mindsets guide patients' attentional and motivational processes and affect both subjective and objective measures of health and well-being (Crum & Zuckerman, 2007). This has been documented in studies of stress, diet, and exercise, in which mindsets were found to affect both psychological states and markers of physical health, including blood pressure, weight loss, cortisol response, and hormone secretion (Crum, Corbin, Brownell, & Salovey, 2011; Crum & Langer, 2007; Crum et al., 2013). Research is beginning to shed light on mindsets about health and disease and their subsequent impact on patients, providers, and the health care system (Crum et al., 2017).

Mindsets can be intentionally and adaptively changed through targeted interventions. These interventions bridge the gap between traditional psychosocial interventions that primarily focus on improving subjective measures of well-being and medical interventions that are often evaluated exclusively in terms of their impact on physiological measures of health. For example, when exposed to information about the positive aspects of stress—that it can enhance immune function and boost cognitive performance—individuals with high stress finance jobs adopted a "stress is enhancing" mindset that shaped subsequent work performance (Crum et al., 2013). Similarly, hotel employees who were taught that their work provides a sufficient amount of daily physical activity, showed improvement on vital measures of health, without evidence of a corresponding behavior change (Crum & Langer, 2007). This research has also demonstrated that effective interventions can be short, simple, and inexpensive. So while patients may come to health care with their own preexisting mindsets, the malleability of mindsets suggests that physicians can intentionally shape their patients' mindsets. Helping a patient develop the mindset that their disease is manageable (as opposed to a catastrophe), for instance, may impact patient expectations about the course of their illness, the nature and occurrence of symptoms, and the efficacy of treatments. Rather than simply shaping expectations (e.g., this drug will work), physicians may be able to help their patients form more adaptive mindsets that elicit multiple downstream effects.

How do mindsets and expectations relate to one another in the clinical context? A patient in pain, for instance, may have the specific expectation that a treatment will relieve their discomfort. However, this expectation may hinge on the broader mindset that their illness is manageable. These mindsets and the expectations they influence activate distinct brain regions associated with pain, anxiety, and reward (Benedetti, Carlino, & Pollo, 2011; Bingel et al., 2011; Zubieta & Stohler, 2009). They also affect the function of the peripheral nervous system and its downstream target organs and modulate the activity of the immune and endocrine systems (Crum et al., 2011, Pollo et al., 2003). While specific expectations facilitate placebo effects in experimental paradigms, mindsets may be particularly relevant in the practice of medicine, where individual expectations do not exist in isolation from one another.

Patients often enter the medical context with preexisting mindsets about health, disease, and treatments. When a physician shares information about disease and treatment with a patient, it is interpreted through the lens of the patient's mindsets, influencing his or her subsequent expectations.

The interactions that occur in the clinical context and the experience a patient has with his or her illness and treatment then shape existing mindsets, reinforcing or altering them. For example, a patient who has the mindset that their body is capable may preferentially attend to signals that their body is handling an illness well (Zion, Dweck, & Crum, 2018). They may expect their body to be able to manage an illness and the side effects of treatment. Critically, these expectations can often be self-fulfilling and an understanding and compassionate physician can help reinforce these adaptive mindsets (Howe, Goyer, & Crum, 2017).

4. SOCIAL AND CONTEXTUAL FACTORS INFORM PSYCHOLOGICAL PROCESSES

In Sections 2 and 3, the patient's internal psychological characteristics that contribute the placebo effect were discussed. Of course, these processes like implicit learning, conscious expectations, and mindsets do not exist in a vacuum; they are critically informed by the environment. Here, we break down the components of the social context and the treatment context that influence patient mindsets and contribute to the placebo effect in the practice of medicine (Fig. 2). First, we review some of the developmental and cultural factors that shape psychological processes outside the doctor's office. Next, we move into the social and contextual factors that influence psychological processes within the clinical context, including the patient–provider

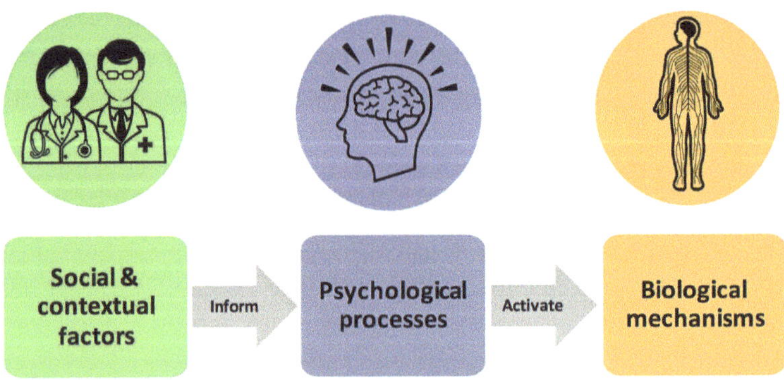

Fig. 2 The relationship between the components of the placebo effect. Social and contextual factors inform psychological processes, which in turn activate biological mechanisms.

relationship and social and observational learning. Finally, we review how the components of the treatment context—the characteristics of treatments and the physical environment—affect our psychological processes.

4.1 Developmental and Cultural Factors

The psychological processes that drive placebo effects may be set into motion long before a patient sets foot into a doctor's office. As children, our mindsets are shaped, in part, by our caregivers and our social environment (Gunderson et al., 2013; Mueller & Dweck, 1998). Mindsets can also be shaped less explicitly during these sensitive developmental periods through social influence and modeling (Bandura, 1977). Mindsets about health and illness may also develop early on from experiences with illness, visits to the pediatrician, and from observing family members and peers. As adults, these health mindsets are continually shaped through our interactions with the health care system and our positive and negative experiences with disease and treatment.

With age, our experiences shape our mindsets within cultural frameworks of norms and customs. Our mindsets are influenced by the culture in which we were raised, our social networks, religious customs, and the media (Markus & Kitayama, 2010). Culture, for instance, influences how pain is experienced. Individuals of Italian ancestry may focus on the immediacy of the pain and how it affects their current situation. American Jews and Protestants, conversely, tend to be more future oriented when experiencing pain, attending to the potential long-term implications of the experience (Zborowski, 1952). When managing one's health, individuals from western cultures in which individual agency and responsibility are often highly valued favor health promotion over illness prevention. Patients from East Asian cultures tend to take the opposite approach, favoring prevention over promotion and endorsing motivation for avoiding negative outcomes (Elliot, Chirkov, Kim, & Sheldon, 2001; Lockwood, Marshall, & Sadler, 2005).

The information we encounter can also have a particularly potent impact on our mindsets. For instance, the highly publicized claim citing a link between the MMR vaccination and autism—a link that has been thoroughly discredited by every major scientific organization—may have affected some patients' mindsets about the nature of vaccines (Godlee, Smith, & Marcovitch, 2011). Indeed, analyses years after the incident found that parents struggled to understand the true nature of the controversy or know which sources of information to trust (Hilton, Petticrew, & Hunt, 2007). This trust is important. We place our trust in figures of authority like physicians and base this trust on their degree of competence, compassion, reliability, and how they communicate information (Pearson & Raeke, 2000). We assign value to their claims, allowing these figures to shape mindsets in important ways—for better or for worse.

4.2 The Patient–Provider Relationship—Communication, Warmth, and Competence

The relationship between patient and provider is a critical factor in the quality of care a patient receives, but can also influence the beliefs, expectations, and mindsets patients have about health and disease. The patient–provider relationship shapes the way important medical information is communicated and this influences the mindsets a patient holds about health, illness, and treatments. The patient–provider relationship can also affect physiological health outcomes, both by motivating behavior and through its impact on the patient's internal psychological processes. The patient–provider relationship shapes the patient's internal psychological processes by both direct communication and the nonverbal cues that convey competence and warmth.

How information is framed and communicated can be a particularly strong influence on patients' mindsets. Imagine an emergency room physician meets with an incoming trauma patient. Telling the patient, "I am going to administer a dose of morphine, a safe but powerful pain killer that will alleviate your pain" activates a series of related beliefs and expectations that enhance the subjective and objective efficacy of that treatment. These verbal suggestions shift attention and motivation and affect brain regions associated with pain relief and reward (Benedetti, Amanzio, Vighetti, & Asteggiano, 2006). If the same doctor were to instead tell their patient that the morphine they were giving them was an addictive opioid that could cause severe side effects like respiratory depression, a very different network of beliefs would be activated (Kast & Loesch, 1961). Furthermore, positive interactions in which a diagnosis is clearly made and agreed upon by both the patient and the practitioner can speed recovery from illness (Bass et al., 1986). This effect also occurs in the opposite direction—patients of physicians who communicate poorly have a 19% higher risk of not adhering to medical advice and treatment regimens (Haskard-Zolnierek & DiMatteo, 2009).

It is not just the way treatments are described that can influence patients' mindsets. When doctors intentionally or inadvertently assign meaning to conditions or symptoms, it can shape how the patient experiences them. Cancer pain, for example, is often perceived as more unpleasant than post-operative pain. While cancer pain carries with it associations of sickness and death, postsurgical pain is often associated with a recovery processes (Ferrell, Dean, Grant, & Coluzzi, 1995; Smith, Gracely, & Safer, 1998). By assigning positive or negative meaning to pain, medical providers are activating preexisting mindsets. This can influence pain tolerance and modulate endogenous opioid and cannabinoid systems in the brain (Zubieta & Stohler, 2009).

Although there is never one correct way to communicate with a patient or an ideal model of the patient–provider relationship, certain characteristics appear to be universally important. Indeed, decades of social psychological research suggest that two qualities are of paramount importance: warmth and competence. When a patient meets a physician, he or she rapidly assesses the benevolence of the physician's intentions (warmth) and their ability to carry out these intentions (competence) (Fiske, Cuddy, & Glick, 2007). Warmth denotes a physician's understanding of the patient as a whole person, with a life, values, and goals outside of the health care context, while competence denotes a physician's understanding of medicine (i.e., the disease, prognosis, and treatment). Patient assessments of physician warmth and competence shape patient expectations about treatment, impact mindsets about illness, and modulate the magnitude of the placebo effect. In a recent study, an allergic reaction was induced in participants via a histamine skin prick. A placebo cream was administered with either positive (i.e., this cream will reduce your symptoms) or negative (i.e., this cream will exacerbate your symptoms) expectations under different conditions of provider warmth and competence. Expectations had a larger impact on the efficacy of the placebo cream when it was administered by a warm and competent provider and negated the effects when administered by a cold and incompetent provider (Howe et al., 2017). The social context in which the treatment was administered—seen here as the interaction with a provider who varied in warmth and competence—moderated the impact of patients' expectations on their allergic response.

The strength and quality of the patient–provider relationship is further shaped by empathy and trust. Physicians signal their warmth through their empathy, or ability to understand a patient's unique situation. Empathy is expressed explicitly through verbal information and implicitly through nonverbal cues like head and body position (Harrigan & Rosenthal, 1983). Displays of warmth and competence not only foster trusting patient–provider relationships, but they also help patients engage in adaptive psychological processes. For instance, physician empathy is significantly associated with reductions in patient anxiety and distress in addition to better clinical outcomes (Derksen, Bensing, & Lagro-Janssen, 2013). In a large retrospective study of over 20,000 diabetic patients, the patients of more empathetic physicians had significantly fewer metabolic complications (Del Canale et al., 2012). Practitioner characteristics such as empathy also affect biological markers of disease. In another study, patients who sought care for symptoms of the common cold were randomly assigned to either a standard interaction with a physician or an enhanced, empathetic interaction. Patients in the enhanced condition rated their physicians as more empathetic, reported lower severity of cold symptoms, and had a greater change in interleukin-8 (IL-8) and neutrophil counts 48 h after the interaction (Rakel et al., 2011).

4.3 Social Influence and Observational Learning

Medical treatments occur within a social environment that directly shapes their efficacy. Indeed, observational learning and social influence have long been suggested as potential mechanisms underlying certain placebo effects (Bootzin & Caspi, 2002). A patient who observes the behavior of another patient may modify their behaviors to more closely match those of the subject being observed (Zentall & Galef, 2013). Behavior modification, however, is not always needed to mediate the effect of observational learning on health outcomes. In an illuminating study, Colloca and Benedetti (2009) demonstrated that placebo effects can be experimentally induced through observational social learning (Colloca & Benedetti, 2009). In this study, participants who observed others undergoing an analgesic procedure experienced substantial placebo responses to the same paradigm. The placebo effect induced via social observation was comparable to those induced by conditioning and greater than those induced by verbal suggestion. The patient's degree of empathy was positively correlated with their placebo response, a finding that has implications for future research. The influence of social observation has important implications for certain integrative medical treatments and the design of novel interventions. For example, support groups or other forms of social interaction between patients may serve to encourage patients currently undergoing treatment, recruit some of these learning mechanisms to improve treatment efficacy, and complement the verbal and nonverbal expectations induced by a health care provider.

In addition to observational learning, placebo effects can be modulated by explicit social influence. After consuming bottled water that was labeled as caffeinated, participants exhibited increased alertness, increased motor function, and decreased cognitive interference. These effects were largest for subjects who heard a confederate report positive effects from the faux caffeinated water, suggesting an important role of social influence. Furthermore, these socially influenced subjects were more likely to purchase the product and endorse its effects to others (Crum, Phillips, Goyer, Akinola, & Higgins, 2016). These findings suggest that patients may be influenced by others who express confidence in specific treatments. While this social influence may come directly from other patients, it may also come from the media and advertising, which tend to portray new and innovative treatments with compelling success stories.

4.4 Treatment Type and Characteristics

The physical environment in which the treatment takes place and the specific characteristics of the treatment itself can activate certain beliefs, expectations, and mindsets, thereby influencing treatment outcomes. Many patients, for instance, exhibit a substantial but transient rise in blood pressure when it is measured by a physician in a medical setting. This so-called white coat syndrome is thought to result from a physiological stress response evoked by the symbolic status and authority that is represented by the physician's traditional white coat (Manios et al., 2008). Indeed, studies have demonstrated that both representations of social influence and authority can shape a patient's psychological processes and subsequently affect clinical outcomes (Den Hond, Celis, Vandenhoven, O'brien, & Staessen, 2003). However, if this blood pressure reading is used as the basis for a diagnosis of hypertension, it may lead to over medicating of patients who are simply exhibiting a physiological response to the meaning of an environmental cue.

Perceptual characteristics of the treatment itself have also been found to play a role in shaping both psychological processes and treatment outcomes. Seemingly inconsequential features, such as the color of a medication can impact perceptions of medication quality and consequently affect the actual efficacy of that treatment. A 1996 study indicated that patients relate the color of a drug to its underlying mechanism and efficacy (de Craen, Roos, Leonard de Vries, & Kleijnen, 1996). Shape and form also play a role according to a study in which patients perceived capsules to be stronger and more effective than tablets (Buckalew & Coffield, 1982).

Cost may serve as a mediator between the social valuation of an object and the impact that object has on the individual. Numerous studies have found that beliefs about the characteristics of consumer goods—unrelated to their objective characteristics—shape how they are perceived (Lee, Frederick, & Ariely, 2006). Knowing the price of a bottle of wine, for instance, influences how pleasant that wine is perceived to be and affects blood oxygen levels in the medial orbitofrontal cortex, a region of the brain that is involved in expectation and reward (Kringelbach, 2005; Plassmann et al., 2008). Price has also been found to moderate the efficacy of treatments in clinical populations. In a double-blind study, patients with Parkinson's disease were randomized to receive an injection of saline that was described as either an inexpensive or an expensive "novel injectable dopamine agonist" (Espay et al., 2015). Patients responded to both placebo treatments, but those who received the expensive placebo exhibited greater benefit. If given an expensive treatment first, patients exhibited a twofold increase in motor function over the cheaper placebo. These changes in motor function were associated with corresponding changes in activation of the left putamen, a major target for dopaminergic projections that govern motor activation in Parkinson's disease.

Similar effects have been found in patients' strong preferences for brand name rather than generic drugs. Although generic drugs and their brand name counterparts contain the same type and dose of medication, patients perceive generic drugs to be less effective and experience more side effects when taking them. Generic drugs are also viewed as less trustworthy, less powerful, and many patients do not feel they are appropriate for serious medical conditions (Figueiras et al., 2010; Himmel et al., 2005). In a related study, university students were given placebos and told they would be taking a new beta-blocker (an antihypertensive medication) to reduce performance anxiety. Subjects were randomized to remain taking the original "medication" or to switch to either a different brand or a generic condition. Those who remained on the same placebo treatment exhibited a greater reduction in blood pressure and anxiety compared to those who changed treatments. The switch to a generic beta-blocker yielded the lowest efficacy and the greatest number of adverse events (Faasse, Cundy, Gamble, & Petrie, 2013).

5. DISCUSSION

In this chapter, we have argued that the power of the placebo effect is not separate from but a critical component of medical treatment. In other words, the total effect of any treatment is the combined effect of the pharmaceutical agent and the psychosocial components that make up the placebo effect. These components include the psychological processes and social/contextual factors that drive neurobiological changes and influence subjective and objective treatment outcomes. They can powerfully shape the impact of an active treatment and should not be discounted in the practice of medicine where they can be harnessed to improve patient care.

The power of the placebo to boost active treatments is not novel to many clinicians who witness its effects in their own patients on a daily basis. However, for decades, the notion of the placebo effect was synonymous with deceiving patients by prescribing an inert treatment, a practice that is at odds with the core principles of the profession. So how can the placebo effect be ethically harnessed in the practice of medicine? First, we encourage a wider dissemination of knowledge about the nature of the placebo effect as a psychological process with disease-specific neurobiological effects that are shaped by the social environment and treatment context. This process then can be harnessed to improve patient care and treatment outcomes in a relatively simple and cost-efficient way.

Second, we suggest being aware of each patient's individual psychological characteristics and tendencies, with a specific focus on their mindsets and expectations. How does each patient think about themselves and the world? Recognizing patients who may hold maladaptive mindsets and observing how these mindsets shape their expectations and subsequent health, disease, and treatment is one way to harness the power of the placebo effect in the clinical encounter. Taking an active role in understanding how and why patients have the mindsets they have about health and healing could allow physicians to nondeceptively leverage the same forces that underlie placebo effects in the clinical encounter.

Finally, we encourage physicians to recognize that how they interact with their patients shapes the social environment and treatment context, which in turn influences patient health outcomes. Does the patient trust that the provider has his or her best intentions in mind? Does the patient feel like he or she is understood as a whole person, not just as a body with a disease? Does the provider exude characteristics of both warmth and competence? These variables are within the provider's control and can powerfully shape the ability to optimize treatment efficacy.

In understanding how the social environment and treatment context shape psychological processes, thereby affecting treatment outcomes, physicians today can harness the same power that Henry Beecher harnessed over 50 years ago—the power of the placebo phenomenon. This helps us move from a world in which the placebo effect is associated with deceptively administered sham treatments, to a world in which the placebo effect is recognized as the manifestation of the powerful social and psychological forces that scaffold all of medicine.

ACKNOWLEDGMENTS

We wish to acknowledge and thank the Stanford Mind & Body Lab, Kari Leibowitz, and Isaac Handley-Miner for their feedback and comments on the manuscript.

REFERENCES

Ader, R., & Cohen, N. (1982). Behaviorally conditioned immunosuppression and murine systemic lupus erythematosus. *Science, 215*(4539), 1534–1536.

Ader, R., & Cohen, N. (1993). Psychoneuroimmunology: Conditioning and stress. *Annual Review of Psychology, 44*, 53–85.

Amanzio, M., & Benedetti, F. (1999). Neuropharmacological dissection of placebo analgesia: Expectation-activated opioid systems versus conditioning-activated specific subsystems. *The Journal of Neuroscience: The Official Journal of the Society for Neuroscience, 19*(1), 484–494.

Bandura, A. (1977). *Social learning theory*. Englewood Cliffs, NJ: Prentice-Hall.

Bass, M. J., Buskc, C., Turner, L., Dickie, G., Pratt, G., & R.H. (1986). The physician's actions and the otucome of illnes in family practice. *The Journal of Family Practice, 23*, 43–47.

Beecher, H. K. (1955). The powerful placebo. *JAMA, 159*(17), 1602–1606.

Benedetti, F. (2008). Mechanisms of placebo and placebo-related effects across diseases and treatments. *Annual Review of Pharmacology and Toxicology, 48*, 33–60.

Benedetti, F., Amanzio, M., Baldi, S., Casadio, C., & Maggi, G. (1999). Inducing placebo respiratory depressant responses in humans via opioid receptors. *European Journal of Neuroscience, 11*(2), 625–631.

Benedetti, F., Amanzio, M., Vighetti, S., & Asteggiano, G. (2006). The biochemical and neuroendocrine bases of the hyperalgesic nocebo effect. *The Journal of Neuroscience, 26*(46), 12014–12022.

Benedetti, F., Carlino, E., & Pollo, A. (2011). How placebos change the patient's brain. *Neuropsychopharmacology, 36*(1), 339–354.

Benedetti, F., Maggi, G., Lopiano, L., Lanotte, M., Rainero, I., Vighetti, S., et al. (2003). Open versus hidden medical treatments: The patient's knowledge about a therapy affects the therapy outcome. *Prevention & Treatment, 6*(1).

Benedetti, F., Pollo, A., Lopiano, L., Lanotte, M., Vighetti, S., & Rainero, I. (2003). Conscious expectation and unconscious conditioning in analgesia, motor, and hormonal placebo/nocebo responses. *The Journal of Neuroscience: The Official Journal of the Society for Neuroscience, 23*(10), 4315–4323.

Bingel, U., Wanigasekera, V., Wiech, K., Ni Mhuircheartaigh, R., Lee, M. C., Ploner, M., et al. (2011). The effect of treatment expectation on drug efficacy: Imaging the analgesic benefit of the opioid remifentanil. *Science Translational Medicine*, *3*(70), 70ra14.

Bootzin, R. R., & Caspi, O. (2002). Explanatory mechanisms for placebo effects: Cognition, personality and social learning. In H. A. Guess, A. Kleinman, J. W. Kusek, & L. W. Engel (Eds.), *The science of the placebo: Toward an interdisciplinary research agenda* (pp. 108–132). London: BMJ Books.

Buckalew, L. W., & Coffield, K. E. (1982). An investigation of drug expectancy as a function of capsule color and size and preparation form. *Journal of Clinical Psychopharmacology*, *2*(4), 245–248.

Colloca, L., & Benedetti, F. (2009). Placebo analgesia induced by social observational learning. *Pain*, *144*(1), 28–34.

Colloca, L., Lopiano, L., Lanotte, M., & Benedetti, F. (2004). Overt versus covert treatment for pain, anxiety, and Parkinson's disease. *The Lancet. Neurology*, *3*(11), 679–684.

Crum, A. J., Corbin, W. R., Brownell, K. D., & Salovey, P. (2011). Mind over milkshakes: Mindsets, not just nutrients, determine ghrelin response. *Health Psychology*, *30*(4), 424–429.

Crum, A. J., & Langer, E. J. (2007). Mind-set matters exercise and the placebo effect. *Psychological Science*, *18*(2), 165–171.

Crum, A. J., Leibowitz, K. A., & Verghese, A. (2017). Making mindset matter. *BMJ (Clinical Research Ed.)*, *674*, j674.

Crum, A. J., Phillips, D. J., Goyer, J. P., Akinola, M., & Higgins, E. T. (2016). Transforming water: Social influence moderates psychological, physiological, and functional response to a placebo product. *PLoS One*, *11*(11), 1–18.

Crum, A. J., Salovey, P., & Achor, S. (2013). Rethinking stress: The role of mindsets in determining the stress response. *Journal of Personality and Social Psychology*, *104*(4), 716–733.

Crum, A., & Zuckerman, B. (2007). Changing mindsets to enhance treatment effectiveness. *JAMA*, *317*(20), 2063–2064.

de Craen, A. J., Roos, P. J., Leonard de Vries, A., & Kleijnen, J. (1996). Effect of colour of drugs: Systematic review of perceived effect of drugs and of their effectiveness. *BMJ (Clinical Research Ed.)*, *313*(7072), 1624–1626.

de la Fuente-Fernandez, R., & Stoessl, A. J. (2002). The placebo effect in Parkinson's disease. *Trends in Neurosciences*, *25*(6), 302–306.

Del Canale, S., Louis, D. Z., Maio, V., Wang, X., Rossi, G., Hojat, M., et al. (2012). The relationship between physician empathy and disease complications. *Academic Medicine*, *87*(9), 1243–1249.

Den Hond, E., Celis, H., Vandenhoven, G., O'brien, E., & Staessen, J. A. (2003). Determinants of white-coat syndrome assessed by ambulatory blood pressure or self-measured home blood pressure. *Blood Pressure Monitoring*, *8*(1), 37–40.

Derksen, F., Bensing, J., & Lagro-Janssen, A. (2013). Effectiveness of empathy in general practice: A systematic review. *British Journal of General Practice*, *63*(606), 76–84.

Elliot, A. J., Chirkov, V. I., Kim, Y., & Sheldon, K. M. (2001). A cross-cultural analysis of avoidance (relative to approach) personal goals. *Psychological Science*, *12*(6), 505–510.

Espay, A. J., Norris, M. M., Eliassen, J. C., Dwivedi, A., Smith, M. S., Banks, C., et al. (2015). Placebo effect of medication cost in Parkinson disease a randomized double-blind study. *Neurology*, *84*(8), 794–802.

Exton, M. S., Gierse, C., Meier, B., Mosen, M., Xie, Y., Frede, S., et al. (2002). Behaviorally conditioned immunosuppression in the rat is regulated via noradrenaline and β-adrenoceptors. *Journal of Neuroimmunology*, *131*(1–2), 21–30.

Faasse, K., Cundy, T., Gamble, G., & Petrie, K. J. (2013). The effect of an apparent change to a branded or generic medication on drug effectiveness and side effects. *Psychosomatic Medicine*, *75*(1), 90–96.

Ferrell, B. R., Dean, G. E., Grant, M., & Coluzzi, P. (1995). An institutional commitment to pain management. *Journal of Clinical Oncology, 13*(9), 2158–2165.

Figueiras, M. J., Cortes, M. A., Marcelino, D., Weinman, J., Joa, M., & Weinman, J. (2010). Lay views about medicines: The influence of the illness label for the use of generic versus brand. *Psychology and Health, 25*(9), 1121–1128.

Finniss, D. G., Kaptchuk, T. J., Miller, F., & Benedetti, F. (2010). Placebo effects: Biological, clinical and ethical advances. *Lancet, 375*(9715), 686–695.

Fiske, S. T., Cuddy, A. J. C., & Glick, P. (2007). Universal dimensions of social cognition: Warmth and competence. *Trends in Cognitive Sciences, 11*(2), 77–83.

Frensch, P. A., & Rünger, D. (2003). Implicit learning. *Current Directions in Psychological Science, 12*(1), 13–18.

Giang, D. W., Goodman, A. D., Schiffer, R. B., Mattson, D. H., Petrie, M., Cohen, N., et al. (1996). Conditioning of cyclophosphamide-induced leukopenia in humans. *The Journal of Neuropsychiatry and Clinical Neurosciences, 8*(2), 194–201.

Godlee, F., Smith, J., & Marcovitch, H. (2011). Wakefield's article linking MMR vaccine and autism was fraudulent. *BMJ (Clinical Research Ed.), 342*(7788), c7452.

Goebel, M. U., Meykadeh, N., Kou, W., Schedlowski, M., & Hengge, U. R. (2008). Behavioral conditioning of antihistamine effects in patients with allergic rhinitis. *Psychotherapy and Psychosomatics*, 77(4), 227–234.

Goebel, M. U., Trebst, A. E., Steiner, J., Xie, Y. F., Exton, M. S., Frede, S., et al. (2002). Behavioral conditioning of immunosuppression is possible in humans. *Official Publication of the Federation of American Societies for Experimental Biology, 16*(14), 1869–1873.

Gunderson, E. A., Gripshover, S. J., Romero, C., Dweck, C. S., Goldin-Meadow, S., & Levine, S. C. (2013). Parent praise to 1–3 year-olds predicts children's motivational frameworks 5 years later. *Child Development, 84*(5), 1526–1541.

Harrigan, J. A., & Rosenthal, R. (1983). Physicians' head and body positions as determinants of perceived rapport. *Journal of Applied Social Psychology, 13*(6), 496–509.

Haskard-Zolnierek, K. B., & DiMatteo, M. R. (2009). Physician communication and patient adherence to treatment: A meta-analysis. *Medical Care, 47*(8), 826–834.

Hilton, S., Petticrew, M., & Hunt, K. (2007). Parents' champions vs. vested interests: Who do parents believe about MMR? A qualitative study. *BMC Public Health*, 7(1), 42.

Himmel, W., Simmenroth-Nayda, A., Niebling, W., Ledig, T., Jansen, R. D.-D., Kochen, M. M., et al. (2005). What do primary care patients think about generic drugs? *International Journal of Clinical Pharmacology and Therapeutics, 43*(10), 472–479.

Howe, L. C., Goyer, J. P., & Crum, A. J. (2017). Harnessing the placebo effect: Exploring the influence of physician characteristics on placebo response. *Health Psychology, 36*(11), 1074–1082.

Kast, E. C. E., & Loesch, J. (1961). Influence of the doctor-patient relationship on drug action. *The Illinois Medical Journal, 119*, 390–393.

Kirsch, I. (1985). Response expectancy as a determinant of experience and behavior. *American Psychologist, 40*(11), 1189–1202.

Kringelbach, M. L. (2005). The human orbitofrontal cortex: Linking reward to hedonic experience. *Nature Reviews. Neuroscience, 6*(9), 691–702.

Lee, L., Frederick, S., & Ariely, D. (2006). Try it, you'll like it: The influence of expectation, consumption, and revelation on preferences for beer. *Psychological Science, 17*(12), 1054–1058.

Levine, J. D., & Gordon, N. C. (1984). Influence of the method of drug administration on analgesic response. *Nature, 312*(5996), 755–756.

Levine, J. D., Gordon, N. C., Bornstein, J. C., & Fields, H. L. (1979). Role of pain in placebo analgesia. *Proceedings of the National Academy of Sciences of the United States of America, 76*(7), 3528–3531.

Levine, J. D., Gordon, N. C., Smith, R., & Fields, H. L. (1981). Analgesic responses to morphine and placebo in individuals with postoperative pain. *Pain, 10*(3), 379–389.

Lockwood, P., Marshall, T. C., & Sadler, P. (2005). Promoting success or preventing failure: Cultural differences in motivation by positive and negative role models. *Personality and Social Psychology Bulletin*, *31*(3), 379–392.

Manios, E. D., Koroboki, E. A., Tsivgoulis, G. K., Spengos, K. M., Spiliopoulou, I. K., Brodie, F. G., et al. (2008). Factors influencing white-coat effect. *American Journal of Hypertension*, *21*(2), 153–158.

Markus, H. R., & Kitayama, S. (2010). Cultures and selves. *Perspectives on Psychological Science*, *5*(4), 420–430.

Montgomery, G. H., & Kirsch, I. (1997). Classical conditioning and the placebo effect. *Pain*, *72*(1–2), 107–113.

Mueller, C. M., & Dweck, C. S. (1998). Praise for intelligence can undermine children's motivation and performance. *Journal of Personality and Social Psychology*, *75*(1), 33–52.

Pearson, S. D., & Raeke, L. H. (2000). Patients' trust in physicians: Many theories, few measures, and little data. *Journal of General Internal Medicine*, *15*(7), 509–513.

Petrovic, P., Dietrich, T., Fransson, P., Andersson, J., Carlsson, K., & Ingvar, M. (2005). Placebo in emotional processing-induced expectations of anxiety relief activate a generalized modulatory network. *Neuron*, *46*(6), 957–969.

Plassmann, H., O'Doherty, J., Shiv, B., Rangel, A., Doherty, J. O., Shiv, B., et al. (2008). Marketing actions can modulate neural representations of experienced pleasantness. *Proceedings of the National Academy of Sciences of the United States of America*, *105*(3), 1050–1054.

Pollo, A., Torre, E., Lopiano, L., Rizzone, M., Lanotte, M., Cavanna, A., et al. (2002). Expectation modulates the response to subthalamic nucleus stimulation in Parkinsonian patients. *Neuroreport*, *13*(11), 1383–1386.

Pollo, A., Vighetti, S., Rainero, I., & Benedetti, F. (2003). Placebo analgesia and the heart. *Pain*, *102*, 125–133.

Price, D. D., Finniss, D. G., & Benedetti, F. (2008). A comprehensive review of the placebo effect: Recent advances and current thought. *Annual Review of Psychology*, *59*, 565–590.

Price, D. D., Milling, L. S., Kirsch, I., Duff, A., Montgomery, G. H., & Nicholls, S. S. (1999). An analysis of factors that contribute to the magnitude of placebo analgesia in an experimental paradigm. *Pain*, *83*(2), 147–156.

Rakel, D., Barrett, B., Zhang, Z., Hoeft, T., Chewning, B., Marchand, L., et al. (2011). Perception of empathy in the therapeutic encounter: Effects on the common cold. *Patient Education and Counseling*, *85*(3), 390–397.

Rescorla, R. A. (1988). Pavlovian conditioning: It's not what you think it is. *American Psychologist*, *43*(3), 151–160.

Seger, C. (1994). Implicit learning. *Psychological Bulletin*, *115*(2), 163–196.

Shetty, N., Friedman, J. H., Kieburtz, K., Marshall, F. J., & Oakes, D. (1999). The placebo response in Parkinson's disease. Parkinson Study Group. *Clinical Neuropharmacology*, *22*(4), 207–212.

Smith, W. B., Gracely, R. H., & Safer, M. A. (1998). The meaning of pain: Cancer patients' rating and recall of pain intensity and affect. *Pain*, *78*(2), 123–129.

Stewart Williams, S., & Podd, J. (2004). The placebo effect: Dissolving the expectancy versus conditioning debate. *Psychological Bulletin*, *130*, 324.

Voudouris, N. J., Peck, C. L., & Coleman, G. (1989). Conditioned response models of placebo phenomena: Further support. *Pain*, *38*(1), 109–116.

Voudouris, N. J., Peck, C. L., & Coleman, G. (1990). The role of conditioning and verbal expectancy in the placebo response. *Pain*, *43*(1), 121–128.

Wager, T. D., & Atlas, L. Y. (2015). The neuroscience of placebo effects: Connecting context, learning and health. *Nature Reviews. Neuroscience*, *16*(7), 403–418.

Walsh, B. T., Seidman, S. N., Sysko, R., & Gould, M. (2002). Placebo response in studies of major depression: Variable, substantial, and growing. *JAMA*, *287*(14), 1840–1847.

Wickramasekera, I. (1980). A conditioned response model of the placebo effect: Predictions from the model. *Biofeedback and Self-Regulation*, *5*(1), 5–18.
Zborowski, M. (1952). Cultural components in responses to pain. *Journal of Social Issues*, *8*(4), 16–30.
Zentall, T. R., & Galef, B. G., Jr. (2013). *Social learning: Psychological and biological perspectives*. Psychology Press.
Zion, S. R., Dweck, C. S., & Crum, A. J. (2018). In sickness and in health: Validation of a health mindset scale in healthy and chronically ill populations. In *Presented at the Society for Personality and Social Psychology, Atlanta, GA*.
Zubieta, J. K., & Stohler, C. S. (2009). Neurobiological mechanisms of placebo responses. *Annals of the New York Academy of Sciences*, *1156*(734), 198–210.

CHAPTER NINE

The Role of Social and Interpersonal Factors in Placebo Analgesia

Elizabeth A. Necka*[,1], Lauren Y. Atlas*[,†]

*Section on Affective Neuroscience and Pain, National Center for Complementary and Integrative Health, NIH, Bethesda, MD, United States

[†]National Institute on Drug Abuse, NIH, Bethesda, MD, United States

[1]Corresponding author: e-mail address: liz.necka@nih.gov

Contents

Abstract

Placebo effects are beneficial clinical outcomes that emerge as a result of nonspecific contextual factors, transmitted primarily by the treating physician and the social, physical, and behavioral cues he or she displays. The patient–provider therapeutic alliance is critical for determining placebo effects and health outcomes. In this chapter, we review the recent literature, suggesting that provider social characteristics modulate placebo and clinical outcomes. We highlight the importance of studying not only the provider but also the patient's *perception* of the provider, which is subject to the influence of the patient's psychosocial orientation, such as their psychosocial motivations and perceptions of their interpersonal relationships broadly. We argue that psychosocial orientation can exaggerate the influence of the patient–provider relationship on placebo effects and can directly affect the likelihood of placebo effects emerging by modulating the underlying biological systems that support them. Here, we examine patient loneliness, or perceived social isolation, as a case example for understanding how patients' psychosocial orientation may affect placebo effects across diseases. We propose psychosocial mechanisms by which loneliness might modulate placebo effects across medical outcomes, and focus in particular on how loneliness might specifically alter behaviorally

ISSN 0074-7742
https://doi.org/10.1016/bs.irn.2018.01.006

conditioned immune responses and placebo analgesia. Future studies should directly measure social factors to formally test the effects of social isolation on placebo effects and better elucidate the role of psychosocial and interpersonal factors in placebo effects and clinical outcomes.

Since the origins of modern medicine, the therapeutic alliance has been recognized as holding particular import for the treatment of disease. Hippocrates emphasized in the Book of Epidemics that "The art consists in three things – the disease, the patient, and the physician. The physician is the servant of the art, and the *patient must combat the disease along with the physician*" (emphasis added). The writings within the Hippocratic Corpus emphasize that disease cannot be reduced to its symptomology or physiopathology but is instead best understood through consideration of the patient as a whole individual. Hippocrates recognized the powerful effect that contextual factors could have on recovery from disease, what we would today refer to as placebo effects. Psychological or physiological changes which result from contextual factors, often transmitted by the treating physician and the social, physical, verbal, and behavioral cues he or she displays, are known as placebo effects (Benedetti, 2002; Di Blasi, Harkness, Ernst, Georgiou, & Kleijnen, 2001; Moerman, 2003). Placebo effects are the clinically relevant outcomes that cannot be attributed to any active ingredient of a treatment or mechanisms therein. A substantial body of theoretical and empirical work indicates that placebo effects arise from a number of psychological processes, including (1) learned associations (i.e., conditioning) between treatment context and healing, of which patients may or may not be consciously aware (Colloca & Benedetti, 2006; Colloca & Miller, 2011; Jensen, Kirsch, Odmalm, Kaptchuk, & Ingvar, 2015); (2) affective appraisals, or emotionally laden cognitive evaluations of cues or one's environment (Ashar, Chang, & Wager, 2017; Moerman & Jonas, 2002; Tracey, 2010; van der Meulen et al., 2017); and (3) cognitive beliefs or expectancies about treatments and treatment outcomes (Atlas & Wager, 2012; Büchel, Geuter, Sprenger, & Eippert, 2014; Kirsch, 1985, 1997).

In this review, we argue that an understudied but potentially significant factor in determining the placebo effect is variability in patients' orientations toward their social environments, including the patient–provider relationship as well as the patient's feelings about his or her broader social environment. Given that placebo effects emerge from the context surrounding treatment, the patient–provider relationship is a critical factor in predicting placebo outcomes (Dossett et al., 2015; Jensen & Kelley, 2016; Kaptchuk et al., 2008). As Hippocrates wrote, "the patient must combat the disease along with the physician." From this, we can infer that the psychosocial factors that shape the patients' reception of their physician and their broader interpersonal environment are also likely to influence responses to clinical interventions, including placebo effects. Here, we focus on the interpersonal aspects of placebo by integrating work from placebo, social psychology, and social neuroscience. We highlight the role of the provider in establishing explicit treatment expectations and examine the implicit expectations patients may derive from social interactions with their providers as a function of providers' verbal and behavioral cues. We also focus on how patients' relationships to their broader social environment can influence clinical outcomes and placebo effects. Because clinical interactions are necessarily dyadic in nature (that is, they include at least a patient and a medical professional), the social cues and behaviors that each interaction partner transmits and the lens through which each interaction partner interprets the others' cues mutually affect each other, with downstream effects on clinical outcomes and placebo effects (Fig. 1). As one such example, patients' loneliness—or perceived social isolation—affects both their own psychosocial motivations and interpersonal perceptions and behaviors, and the behaviors of others. Furthermore, loneliness has a demonstrable effect on health outcomes. Thus, loneliness is a relevant factor to consider for a more complete understanding of the placebo effect, both empirically and clinically. We suggest that patients' loneliness may modulate the impact of provider-generated expectations on placebo effects and directly impact the physiological processes which give rise to placebo effects and health outcomes.

1. PATIENT–PROVIDER INTERACTIONS AND PLACEBO EFFECTS

Characteristics of the physician critically influence the relation between patients' expectations and placebo effects, as revealed by studies that have measured or manipulated features of the medical provider (rather than focusing simply on the patient). In the clinic, a supportive patient–provider relationship can reduce symptom severity twice as much as receiving placebo treatment alone (Kaptchuk et al., 2008), and positive expectations delivered by a physician in a supportive and reassuring fashion reduce the use of active analgesic treatment for postoperative pain (Egbert, Battit, Welch, & Bartlett, 1964). Yet, just as Hippocrates reflected that a union

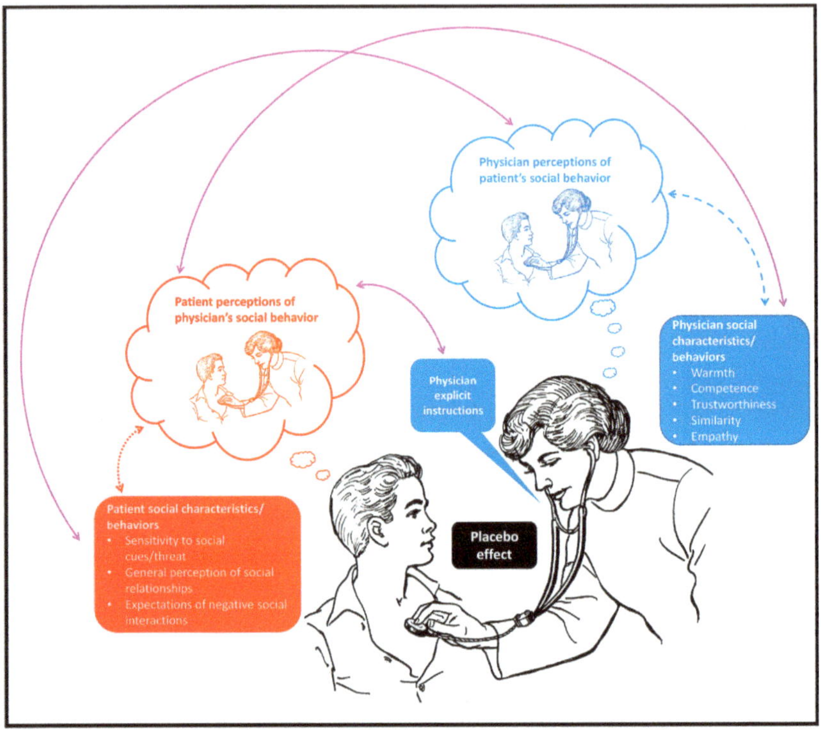

Fig. 1 A model of the patient–provider relationship in placebo. Recent literature demonstrates that both providers' social characteristics and patients' *perceptions* of their providers critically influence the relation between placebo expectations and effects (e.g., Howe, Goyer, & Crum, 2017; Losin, Anderson, & Wager, 2017). Here, we demonstrate a model of dynamic interpersonal interaction within the context of the therapeutic alliance. Both patient (*red*) and provider (*blue*) transmit social characteristics/cues (examples from the literature included), which are then interpreted by the other individual and can affect the other individuals' behavior in a dynamic process. *Purple solid lines* indicate linked processes between individuals, *blue dashed lines* represent linked processes within the provider, and *red dotted lines* indicate linked processes within the patient.

of patient and physician, side by side, is necessary to combat disease, positive clinician characteristics such as warmth, trustworthiness, and competence must operate in tandem with patients' expectations to induce placebo effects. If either the expectations or the manner in which they are conveyed (i.e., the social demeanor of the physician) is not suggestive of a positive outcome, the probability of inducing a placebo effect will be reduced. Indeed, recent work in the laboratory indicates that only in the presence of positive, certain expectations does physician positive affect increase patients' beliefs that treatment will reduce their pain (van Osch, van Dulmen, van Vliet, & Bensing, 2017). Similarly, physician warmth and competence only influence immune response to a placebo treatment in the presence of positive treatment expectations (Howe et al., 2017). Furthermore, just as the influence of provider characteristics on clinical outcomes is negligible in the absence of positive expectations, the influence of positive expectations on placebo effects may be attenuated or diminished completely when transmitted by a cold and incompetent provider (Howe et al., 2017). Taken together, these effects demonstrate that placebo effects emerge not only as a function of patients' expectations but also as a function of how those expectations are conveyed or shaped by the provider.

Yet physicians' words and behaviors cannot be considered in a vacuum. To more comprehensively understand the role of patient–provider interactions in placebo effects, we must consider not only the provider and their objective words and behaviors but also the patient's perspective, which critically determines the relationship between the providers' words and behaviors and the patient's own expectations. Indeed, how patients perceive physicians may be a more important determinant of their outcomes than the physicians' objective characteristics. For example, a recent study (Losin et al., 2017) examined the effect of clinician–patient similarity on pain using a minimal group paradigm, a common procedure used in social psychological experiments. Healthy volunteers were randomly assigned to be either a "clinician" (in which case, they administered noxious thermal stimulation) or a "patient" (in which case they received the thermal stimulation). Each "patient" received heat from two different "clinicians," one who supposedly shared their values and beliefs and was therefore more similar to them, and another who supposedly did not and was therefore dissimilar to them. Though in reality subjects were grouped together randomly (without attention to actual beliefs or values), all subjects believed the cover story that some "clinicians" and "patients" were more similar to each other than others. Yet, "patients" did not differentially experience pain as a function of whether their "clinician" was purportedly similar or dissimilar from them (i.e., the experimental manipulation). Instead, "patients'" *perceptions* of their "providers" trustworthiness and similarity to them influenced health outcomes (Losin et al., 2017). Those who felt that their provider was more trustworthy and more similar to them experienced less pain, suggesting that patients' social assessments are critical for determining clinical outcomes.

While these findings have yet to be replicated in a clinical context, similar findings have been obtained in other studies of paired social interaction. For instance, one study measured social influences on anxiety (Tedeschi, Weber, Prevost, Mischel, & Mobbs, 2015). Participants were told they would receive shocks when their study partner made mistakes on a study task that supposedly measured competence. The *perception* that one's partner was incompetent (based on visual appearance alone) was associated with greater anxiety about receiving pain. Note that perceptions of competence were independent of actual performance. These dynamics may be analogous to the patient–provider interaction, as a clinician's competence has direct consequences for the patient's well-being. In a clinical setting, perceptions that one's provider is incompetent may increase anxiety or uncertainty about potential analgesic effects of treatment. This in turn might impact placebo effects. For instance, both state anxiety (Morton, Watson, El-deredy, & Jones, 2009; Staats, Staats, & Hekmat, 2001) and uncertainty (Yoshida, Seymour, Koltzenburg, & Dolan, 2013) have been associated with increased pain in experimental settings and may inhibit the emergence of placebo analgesia (pain reduction). Increased anxiety may even lead to nocebo effects (increased pain; e.g., Benedetti, Amanzio, Vighetti, & Asteggiano, 2006). These links, though suggestive, are still preliminary and require more thorough investigation.

The findings above indicate that the interpersonal nature of the patient–provider interaction can influence clinical outcomes and placebo effects through the patient's perceptions of provider characteristics. However, this critical interpersonal relationship can also be influenced by patient characteristics that predate the dyadic nature of the interaction. Any factor that colors how an individual perceives his or her social conspecifics, integrates social information, or biases his or her social affective appraisals has the potential to influence placebo effects by altering the patient–provider relationship or how it is perceived. Potential factors are particularly worthy of investigation if there is evidence that they also directly modulate health outcomes, particularly if they modulate biological systems implicated in placebo effects. One candidate factor that meets both these criteria is individuals' loneliness, or perceived social isolation.

2. LONELINESS AS A POTENTIAL DETERMINANT OF PLACEBO EFFECTS

Perceived social isolation, i.e., loneliness, is a potent example of the influence psychosocial factors can have on health outcomes. Loneliness can be considered a psychosocial motivational state—just as hunger motivates eating behavior, loneliness motivates attempts at social reconnection (Qualter et al., 2015). Many people experience bouts of transitory loneliness from time to time and change their behavior to foster relationship development (Qualter, Brown, Munn, & Rotenberg, 2010; Vanhalst et al., 2015). However, for some individuals loneliness can develop into a chronic, dispositional trait, characterized by broad-ranging maladaptive physiological changes (Cacioppo, Cacioppo, Capitanio, & Cole, 2014; Hawkley & Cacioppo, 2003; Holt-Lunstad, Smith, Baker, Harris, & Stephenson, 2015), including increased pain (Jaremka, Andridge, et al., 2013; Jaremka, Fagundes, Glaser, et al., 2013; Wolf & Davis, 2014) and dysfunctional immune and inflammatory responses (Cole et al., 2007; Jaremka, Andridge, et al., 2013; Jaremka, Fagundes, Glaser, et al., 2013; Jaremka, Fagundes, Peng, et al., 2013; Kiecolt-Glaser et al., 1984). Loneliness is best predicted by a discrepancy in quality between ones' desired and actual relationships, not by the objective number of social contacts one has (Hawkley et al., 2008), suggesting that it is the individual's perception of his or her isolation that matters. Taken together, these findings strongly imply that loneliness-related changes in health and biological systems are manifestations of psychological processes in the brain, similar to the brain–body mechanisms that underlie placebo. Despite extensive research into the effects of chronic loneliness and social isolation on health outcomes, including increased risk of morbidity and mortality (Holt-Lunstad et al., 2015; Luo, Hawkley, Waite, & Cacioppo, 2012; Steptoe, Shankar, Demakakos, & Wardle, 2013), to our knowledge no work has examined whether chronic perceived social isolation modulates the placebo effect. Yet loneliness is characterized by (a) biases in social cognition that might impact the influence of social factors, including the patient–provider relationship, on clinical outcomes (Fig. 2) and (b) unique mind–body connections that might interact with the mechanisms of placebo. As described later, we believe that future work on clinical outcomes and placebo effects will benefit from formally considering the potential influence of loneliness and social isolation.

2.1 Loneliness and Social Dynamics in Placebo Effects and the Therapeutic Alliance

Individuals who feel socially isolated (i.e., lonely) expect that others will treat them poorly (Cacioppo & Hawkley, 2005; Jones, 1981; Lamster, Lincoln, Nittel, Rief, & Mehl, 2017), expect and perceive the quality of their social interactions to be negative (Duck, Pond, & Leatham, 1994; Hawkley,

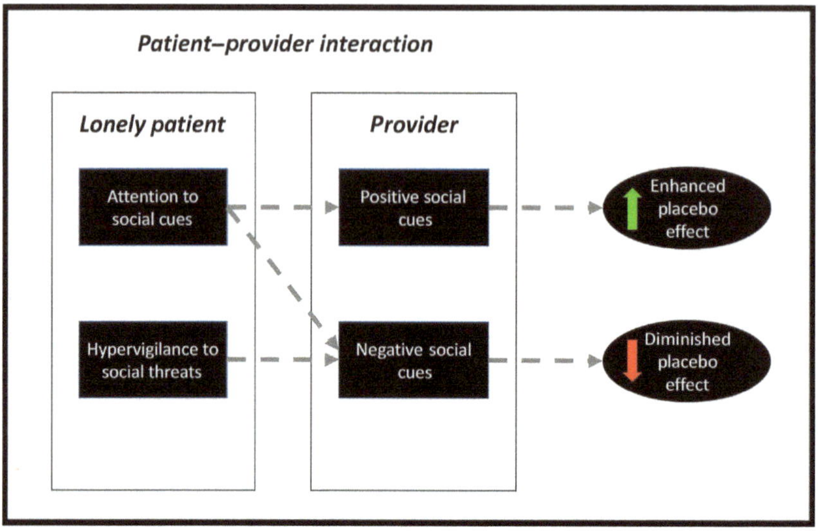

Fig. 2 A model of loneliness in placebo. Loneliness may affect placebo by altering the dynamics of the patient–provider relationship and its influence on placebo effects. *Rectangles* represent factors contributing to placebo effects, and *ovals* represent placebo outcomes. Loneliness increases attention to social cues, both positive and negative in nature (Bangee, Harris, Bridges, Rotenberg, & Qualter, 2014; Cacioppo, Norris, Decety, Monteleone, & Nusbaum, 2009; Gardner, Pickett, Jefferis, & Knowles, 2005; Lodder, Scholte, Goossens, Engels, & Verhagen, 2016; Qualter et al., 2013; Vanhalst et al., 2017), and can therefore be expected to increase the influence (both positive and negative) of the provider's social cues on placebo effects. However, hypervigilance to social threats in loneliness can also increase lonelier individuals' display of negative social behaviors, thereby increasing the likelihood that their interaction partners reciprocate with negative social cues (Rotenberg, Gruman, & Ariganello, 2002; Rotenberg & Kmill, 1992; Tsai & Reis, 2009). By wresting displays of negative social cues from providers, patient loneliness could decrease placebo effects.

Burleson, Berntson, & Cacioppo, 2003; Jones, 1981), and demonstrate heightened attention and sensitivity to social information (Gardner et al., 2005; Lodder et al., 2016), particularly negative social information and social threats (Bangee et al., 2014; Cacioppo et al., 2009; Qualter et al., 2013; Vanhalst et al., 2017; Fig. 2). In addition to being more attentive to such information, lonelier individuals' affective states are more strongly modulated by their own appraisals of their current social interactions than are less lonely individuals' (van Roekel et al., 2014). For example, when a lonelier individual perceives he is interacting with a highly negative social partner (i.e., someone who is perceived to be judgmental or threatening), he experiences greater negative affect and greater reductions in positive affect than his less lonely peer would, given the same perceptions of the social interaction partner, indicating that lonelier individuals are more responsive to their own affective appraisals. These biases in social cognition often impair the quality of otherwise salubrious social relationships and play a role in chronic maintenance of loneliness (Vanhalst et al., 2015).

In the context of placebo effects, lonelier individuals' hypersensitivity to potential social threats may drive them to be more attentive or responsive to social cues from others. A number of studies on social learning and empathy indicate that observing others' pain can influence one's own responses, placebo effects, and nocebo effects (Colloca & Benedetti, 2009; Debiec & Olsson, 2017; Koban, Jepma, Geuter, & Wager, 2017; Koban & Wager, 2016; Schenk, Krimmel, Colloca, & Sciences, 2017; Vögtle, Barke, & Kröner-Herwig, 2013). Lonely individuals may experience greater contagion or learn more quickly about pain or placebo effects due to heightened sensitivity, though to our knowledge no studies have measured the effects of perceived social isolation on social learning. Similarly, loneliness is likely to modulate the effect of the therapeutic alliance on placebo effects and clinical outcomes. Hypersensitivity to social cues among lonelier individuals might amplify the influence of positive provider characteristics such as warmth or empathy, which might in turn enhance the placebo effect (Fig. 2). Indeed, consistent with this hypothesis, preliminary work suggests that the effect of a supportive patient–provider interaction on symptomology in patients with irritable bowel syndrome is more positive in patients who report that they do not like to go out and meet people (a behavioral preference characteristic of the lonely social behavioral profile; Conboy et al., 2010). This work is suggestive of a link between loneliness and the effects of a positive patient-provider relationship on placebo, and demands further replication and extension. Taken together, these findings suggest that loneliness thus might enhance placebo effects by increasing attention to social interactions and enhancing the influence of the patient provider interaction or psychosocial context.

The hypothesis that loneliness would *enhance* placebo effects by modulating the effect of the patient–provider relationship on clinical outcomes relies on an assumption that the patient–provider relationship is positive in nature. Yet, in addition to increasing individuals' attention to social cues broadly, loneliness can also increase the amount of *negative* social cues one's interaction partner expresses. That is, patients' loneliness modulates both their own sensitivity to social cues and their interaction partners' social behaviors (Fig. 2). Lonely individuals behave differently from nonlonely individuals in social situations. In fact, lonely patients' altered behaviors might actually cause providers to respond in ways that *reduce* the likelihood of placebo effects and beneficial clinical outcomes. Specifically, lonelier individuals tend to have more negative self-perceptions in social interactions, consistent with others' assessments of them as being less friendly and less reasonable (Tsai & Reis, 2009). Evidence from the social psychological literature suggests that when individuals identify their interaction partners' loneliness, they may be less accepting, less sociable, and more negative when interacting with them (Rotenberg et al., 2002; Rotenberg & Kmill, 1992; Tsai & Reis, 2009). If physicians respond to their lonelier patients this way, their lonelier patients may actually experience more negative patient–provider interactions. As outlined earlier, we expect providers' interpersonal behaviors and characteristics to be stronger predictors of placebo in lonelier, relative to less lonely, patients. Thus, rather than loneliness enhancing placebo effects, the knock-on effect of loneliness on providers' demeanor (in reaction to patients' negative behaviors) might be particularly detrimental to the emergence of placebo effects in lonelier patients, potentially even extinguishing placebo effects or leading to nocebo effects. Critically, a model whereby patients' loneliness modulates the placebo effect by influencing both patients' and providers' behaviors relies upon the premise that the placebo effect cannot be determined by studying the patient or the provider independently of each other, but rather emerges as a function of the patient–provider therapeutic alliance (Figs. 1 and 2). As these hypotheses are built based on indirect associations, it will be important to test these links directly in the lab and in the clinic by studying these critical dyadic interactions.

2.2 Loneliness and the Biological Systems That Instantiate Placebo Effects

Beyond influencing the strength of the effect of social factors including the patient–provider interaction on placebo outcomes, loneliness may also affect placebo via its modulatory effect on biological systems implicated in placebo and clinical outcomes. As two examples, a large body of evidence links loneliness to dysregulated immune functioning and inflammatory responses (Cole et al., 2007; Jaremka, Andridge, et al., 2013; Jaremka, Fagundes, Glaser, et al., 2013; Jaremka, Fagundes, Peng, et al., 2013; Kiecolt-Glaser et al., 1984). Furthermore, lonelier individuals experience greater pain (Jaremka, Andridge, et al., 2013; Jaremka, Fagundes, Glaser, et al., 2013; Wolf & Davis, 2014). To the extent that lonelier individuals exhibit dysregulated physiological functioning, they may be less capable of mounting a response to placebos.

For instance, the immune system is susceptible to modulation by inert treatments in the context of behaviorally conditioned immune responses (Albring et al., 2012; Goebel et al., 2002; Wirth et al., 2011). Behaviorally conditioned immune responses depend upon both afferent signaling to the brain from the peripheral immune system during conditioning as well as efferent signaling from the brain to the immune system during treatment (Vits et al., 2011; Wendt, Albring, & Schedlowski, 2014). Because loneliness dysregulates proper immune functioning, one hypothesis might be that loneliness interrupts or obfuscates afferent signals to the brain during conditioning, thereby altering the degree to which patients exhibit these sorts of placebo-conditioned responses. Additionally, by altering beta-adrenergic signaling within the sympathetic nervous system (at least as indexed in the end organ of the heart; Cacioppo et al., 2002; Hawkley et al., 2003), loneliness could interrupt the efferent signals from the brain that engender placebo effects in the immune system during treatment. In other words, by clearly disrupting the biological system involved in engendering behaviorally conditioned immune responses, loneliness and its strong mind–body connections may also impact the likelihood of placebo effects emerging. A diminished or obsolete behaviorally conditioned immune response in lonelier individuals could also potentially explain in part their reduced responsivity to active medical treatments, such as their reduced antibody count in response to an influenza vaccine (Pressman et al., 2005). While clearly speculative, future studies of conditioned immunosuppression should directly measure social factors to formally test the effects of social isolation on immune responses to inert treatments, and the placebo component of responses to active treatments.

Placebo effects on pain (placebo analgesia) may also be influenced by patient loneliness. Placebo analgesia is thought to be mediated (at least in part) by the mu opioid system (Amanzio & Benedetti, 1999; Eippert et al., 2009; Levine, Gordon, & Fields, 1978; Wager, Scott, & Zubieta, 2007; Zubieta et al., 2005). Furthermore, past work demonstrates that experimental blockade of the opioid system enhances fear conditioning (Eippert, Bingel, Schoell, Yacubian, & Buchel, 2008) and social threat learning (Haaker, Yi, Petrovic, & Olsson, 2017), likely by blocking or modulating pain inhibitory pathways. The opioid system is also modulated by psychosocial factors associated with loneliness, including social rejection and acceptance (Hsu et al., 2013) and affective responses therein

(Bershad, Seiden, & de Wit, 2016). For example, individual differences in sensitivity to social rejection are associated with the mu opioid receptor gene (Way, Taylor, & Eisenberger, 2009). Specifically, carriers of the G allele on the SNP A118G of the OPRM1 mu opioid receptor gene demonstrate greater neural responsivity in the dorsal anterior cingulate cortex and the anterior insula to being socially rejected (Way et al., 2009). Carriers of this allele also have reduced availability of mu opioid receptors and exhibit reduced placebo effects (Peciña, Love, Stohler, Goldman, & Zubieta, 2014). Evidence that expectations of social rejection tend to be related to or even foster loneliness (Jackson, Fritch, Nagasaka, & Gunderson, 2002; London, Downey, Bonica, & Paltin, 2007) suggests a potential relationship between loneliness and dysregulation of the mu opioid system, but at present a link has yet to be clearly established. Though currently unsubstantiated, if the mu opioid system is indeed dysregulated among lonelier individuals, one hypothesis is that lonelier patients would have an impaired ability to regulate pain, leading to enhanced responses to learning (directly or through social observation) cues that predict aversive outcomes and diminished placebo analgesia. Future work on pain perception and relief and the psychological and neuromodulatory components therein should explore a potential role for social isolation.

2.3 How to Study Loneliness and Placebo

The study of loneliness in the clinic has the greatest potential to elucidate the effects of individual differences in social perceptions on placebo effects. Measuring loneliness before and after surgery or other medical treatment, succeeded by longitudinal follow-ups, will reveal the magnitude of the effect of loneliness on placebo effects. This will also permit researchers to determine whether the quality of the patient–provider relationship (as perceived by both the patient and the provider) modulates such effects, an inherently more complex question because it requires variability in the quality of patient–provider relationships (either as a product of natural variation or by explicit experimental manipulation). The gold standard for studying loneliness is currently self-report because loneliness arises from individuals' *perceptions* of their social isolation. As loneliness differs from *objective* social isolation, it cannot be measured by exclusively assessing quantity or frequency of social contacts. A number of well-validated scales exist for assessing loneliness, most of which avoid using the word "lonely" to minimize potential self-presentation demands biasing patients' reporting.

These scales include the UCLA Loneliness Scale (Russell, 1996; Russell, Peplau, & Cutrona, 1980) or its 3-item short-form version (Hughes, Waite, Hawkley, & Cacioppo, 2004), the De Jong Gierveld Loneliness Scale (De Jong Gierveld & Kamphuis, 1985; Gierveld & Van Tilburg, 2006), and for children or adolescents, the Louvain Loneliness Scale (Marcoen, Goossens, & Caes, 1987).

The laboratory's greatest strength in the study of loneliness and placebo is in utilizing classic social psychology paradigms that approximate the patient–provider interaction (e.g., using confederates, as in Howe et al., 2017, or assigning pairs of participants roles as "patients" or "providers," as in Losin et al., 2017). These paradigms permit researchers a high degree of experimental control over social factors relevant to placebo and can be paired with participants' self-reports of loneliness to test how perceived social isolation modulates placebo effects on behavior and brain responses. In the laboratory, while researchers may be constricted in their ability to experimentally manipulate loneliness, per se, there exist many paradigms which approximate the effects of loneliness on social cognition and behavior and can be used to study the cognitive or psychosocial mechanisms by which loneliness affects placebo. Examples include social exclusion manipulations (in which participants recall a past time when they were excluded or are led to expect a future alone; see Baumeister, Brewer, Tice, & Twenge, 2007) or hypnotic induction of loneliness (among highly hypnotizable participants; e.g., Cacioppo et al., 2006), both of which produce biases in social cognition that mirror those observed in lonelier individuals. Of course, both in the clinic and in the laboratory, it is important to study loneliness within the context of randomized controlled trials to rule out potential confounds and verify that there is a unique relation between loneliness and placebo effects.

3. WHAT LONELINESS AND THE STUDY OF INDIVIDUAL DIFFERENCES IN SOCIAL PERCEPTIONS CAN TELL US ABOUT PLACEBO

Psychosocial factors modulate endogenous physiological processes that impact clinical outcomes in placebo. Accumulating scientific literature echoes Hippocrates' ancient observation: the patient–provider therapeutic alliance is critical in determining placebo effects. In many cases, the patient's experience (that is, their perceptions) may be a more meaningful predictor of their likelihood to exhibit a placebo effect than more objective assessments of providers' social behaviors. Thus, a key source of variance in placebo effects and clinical outcomes may be patients' psychosocial orientation, or how they see their social world, including how socially isolated they perceive themselves to be. Robust effects of loneliness on health demonstrate that psychosocial perceptions exert powerful control on peripheral physiology, particularly in lonelier individuals. Loneliness, in particular, may thus be an important factor in understanding the effect of psychosocial context on clinical outcomes and placebo effects. Though still speculative, we identified several ways in which loneliness may differentially impact processes that contribute to placebo effects. Specifically, loneliness potentially upregulates the effect of positive social interactions on placebo effects while also downregulating the likelihood that positive social interactions occur, and loneliness dysregulates the biological systems that actualize the placebo effect via strong mind–body connections, potentially disrupting the emergence of placebo effects. Diverging hypotheses about the association between loneliness and placebo well-situate loneliness as a candidate factor for study to better understand the role of psychosocial and interpersonal factors in placebo effects and clinical outcomes.

ACKNOWLEDGMENT

This work was supported by the Intramural Research program of the NIH's National Center for Complementary and Integrative Health.

REFERENCES

Albring, A., Wendt, L., Benson, S., Witzke, O., Kribben, A., Engler, H., et al. (2012). Placebo effects on the immune response in humans: The role of learning and expectation. *PLoS One*, 7(11), e49477. https://doi.org/10.1371/journal.pone.0049477.

Amanzio, M., & Benedetti, F. (1999). Neuropharmacological dissection of placebo analgesia: Expectation-activated opioid systems versus conditioning-activated specific subsystems. *The Journal of Neuroscience*, *19*(1), 484–494. https://doi.org/10.1038/nrn3465.

Ashar, Y. K., Chang, L. J., & Wager, T. D. (2017). Brain mechanisms of the placebo effect: An affective appraisal account. *Annual Review of Clinical Psychology*, *13*, 73–98.

Atlas, L. Y., & Wager, T. D. (2012). How expectations shape pain. *Neuroscience Letters*, *520*(2), 140–148. https://doi.org/10.1016/j.neulet.2012.03.039.

Bangee, M., Harris, R. A., Bridges, N., Rotenberg, K. J., & Qualter, P. (2014). Loneliness and attention to social threat in young adults: Findings from an eye tracker study. *Personality and Individual Differences*, *63*, 16–23. https://doi.org/10.1016/j.paid.2014.01.039.

Baumeister, R. F., Brewer, L. E., Tice, D. M., & Twenge, J. M. (2007). Thwarting the need to belong: Understanding the interpersonal and inner effects of social exclusion. *Social and Personality Psychology Compass*, *1*(1), 506–520. https://doi.org/10.1111/j.1751-9004.2007.00020.x.

Benedetti, F. (2002). How the doctor's words affect the patient's brain. *Evaluation & the Health Professions*, *25*(4), 369–386. https://doi.org/10.1177/0163278702238051.

Benedetti, F., Amanzio, M., Vighetti, S., & Asteggiano, G. (2006). The biochemical and neuroendocrine bases of the hyperalgesic nocebo effect. *Journal of Neuroscience, 26*(46), 12014–12022. https://doi.org/10.1523/JNEUROSCI.2947-06.2006.

Bershad, A. K., Seiden, J. A., & de Wit, H. (2016). Effects of buprenorphine on responses to social stimuli in healthy adults. *Psychoneuroendocrinology, 63*, 43–49. https://doi.org/10.1016/j.psyneuen.2015.09.011.

Büchel, C., Geuter, S., Sprenger, C., & Eippert, F. (2014). Placebo analgesia: A predictive coding perspective. *Neuron, 81*(6), 1223–1239. https://doi.org/10.1016/j.neuron.2014.02.042.

Cacioppo, J. T., Cacioppo, S., Capitanio, J. P., & Cole, S. W. (2014). The neuroendocrinology of social isolation. *Annual Review of Psychology, 66*, 733–767. https://doi.org/10.1146/annurev-psych-010814-015240.

Cacioppo, J. T., & Hawkley, L. C. (2005). People thinking about people: The vicious cycle of being a social outcast in one's own mind. In K. D. Williams, J. P. Forgas, & W. von Hippel (Eds.), *Sydney symposium of social psychology series. The social outcast: Ostracism, social exclusion, rejection, and bullying* (pp. 91–108). New York, NY: Psychology Press.

Cacioppo, J. T., Hawkley, L. C., Crawford, L. E., Ernst, J. M., Burleson, M. H., Kowalewski, R. B., et al. (2002). Loneliness and health: Potential mechanisms. *Psychosomatic Medicine, 64*(3), 407–417. Retrieved from http://www.psychosomaticmedicine.org/content/64/3/407.full.

Cacioppo, J. T., Hawkley, L. C., Ernst, J. M., Burleson, M. H., Berntson, G. G., Nouriani, B., et al. (2006). Loneliness within a nomological net: An evolutionary perspective. *Journal of Research in Personality, 40*, 1054–1085.

Cacioppo, J. T., Norris, C. J., Decety, J., Monteleone, G., & Nusbaum, H. (2009). In the eye of the beholder: Individual differences in perceived social isolation predict regional brain activation to social stimuli. *Journal of Cognitive Neuroscience, 21*(1), 83–92. https://doi.org/10.1162/jocn.2009.21007.

Cole, S. W., Hawkley, L. C., Arevalo, J. M., Sung, C. Y., Rose, R. M., & Cacioppo, J. T. (2007). Social regulation of gene expression in human leukocytes. *Genome Biology, 8*(9), R189. https://doi.org/10.1186/gb-2007-8-9-r189.

Colloca, L., & Benedetti, F. (2006). How prior experience shapes placebo analgesia. *Pain, 124*(1–2), 126–133. https://doi.org/10.1016/j.pain.2006.04.005.

Colloca, L., & Benedetti, F. (2009). Placebo analgesia induced by social observational learning. *Pain, 144*(1–2), 28–34. https://doi.org/10.1016/j.pain.2009.01.033.

Colloca, L., & Miller, F. G. (2011). How placebo responses are formed: A learning perspective. *Philosophical Transactions of the Royal Society of London. Series B, Biological Sciences, 366*(1572), 1859–1869. https://doi.org/10.1098/rstb.2010.0398.

Conboy, L. A., Macklin, E., Kelley, J., Kokkotou, E., Lembo, A., & Kaptchuk, T. (2010). Which patients improve: Characteristics increasing sensitivity to a supportive patient-practitioner relationship. *Social Science and Medicine, 70*(3), 479–484. https://doi.org/10.1016/j.socscimed.2009.10.024.

Debiec, J., & Olsson, A. (2017). Social fear learning: From animal models to human function. *Trends in Cognitive Sciences, 21*(7), 546–555. https://doi.org/10.1016/j.tics.2017.04.010.

De Jong Gierveld, J., & Kamphuis, F. H. (1985). The development of a rasch type loneliness scale. *Applied Psychological Measurement, 9*, 289–299. https://doi.org/10.1177/014662168500900307.

Di Blasi, Z., Harkness, E., Ernst, E., Georgiou, A., & Kleijnen, J. (2001). Influence of context effects on health outcomes: A systematic review. *Lancet, 357*, 757–762.

Dossett, M. L., Mu, L., Davis, R. B., Bell, I. R., Lembo, A. J., Kaptchuk, T. J., et al. (2015). Patient-provider interactions affect symptoms in gastroesophageal reflux disease: A pilot randomized, double-blind, placebo-controlled trial. *PLoS One, 10*(9), 1–13. https://doi.org/10.1371/journal.pone.0136855.

Duck, S., Pond, K., & Leatham, G. (1994). Loneliness and the evaluation of relational events. *Journal of Social and Personal Relationships*, *11*(2), 253–276. https://doi.org/10.1177/0265407594112006.

Egbert, L. D., Battit, G. E., Welch, C. E., & Bartlett, M. K. (1964). Reduction of postoperative pain by encouragement and instruction of patients. *The New England Journal of Medicine*, *270*(16), 825–827.

Eippert, F., Bingel, U., Schoell, E., Yacubian, J., & Buchel, C. (2008). Blockade of endogenous opioid neurotransmission enhances acquisition of conditioned fear in humans. *Journal of Neuroscience*, *28*(21), 5465–5472. https://doi.org/10.1523/JNEUROSCI.5336-07.2008.

Eippert, F., Bingel, U., Schoell, E. D., Yacubian, J., Klinger, R., Lorenz, J., et al. (2009). Activation of the opioidergic descending pain control system underlies placebo analgesia. *Neuron*, *63*(4), 533–543. https://doi.org/10.1016/j.neuron.2009.07.014.

Gardner, W. L., Pickett, C. L., Jefferis, V., & Knowles, M. (2005). On the outside looking in: Loneliness and social monitoring. *Personality & Social Psychology Bulletin*, *31*(11), 1549–1560. https://doi.org/10.1177/0146167205277208.

Gierveld, J. D. J., & Van Tilburg, T. (2006). A 6-item scale for overall, emotional, and social loneliness confirmatory tests on survey data. *Research on Aging*, *28*, 582–598.

Goebel, M. U., Trebst, A. E., Steiner, J., Xie, Y. F., Exton, M. S., Frede, S., et al. (2002). Behavioral conditioning of immunosuppression is possible in humans. *The FASEB Journal*, *16*(14), 1869–1873. https://doi.org/10.1096/fj.02-0389com.

Haaker, J., Yi, J., Petrovic, P., & Olsson, A. (2017). Endogenous opioids regulate social threat learning in humans. *Nature Communications*, *8*, 15495. https://doi.org/10.1038/ncomms15495.

Hawkley, L. C., Burleson, M. H., Berntson, G. G., & Cacioppo, J. T. (2003). Loneliness in everyday life: Cardiovascular activity, psychosocial context, and health behaviors. *Journal of Personality and Social Psychology*, *85*(1), 105–120. https://doi.org/10.1037/0022-3514.85.1.105.

Hawkley, L. C., & Cacioppo, J. T. (2003). Loneliness and pathways to disease. *Brain, Behavior, and Immunity*, *17*, S98–S105.

Hawkley, L. C., Hughes, M. E., Waite, L. J., Masi, C. M., Thisted, R. A., & Cacioppo, J. T. (2008). From social structural factors to perceptions of relationship quality and loneliness: The Chicago health, aging, and social relations study. *The Journals of Gerontology Series B Psychological Sciences and Social Sciences*, *63*(6), S375–S384. Retrieved from http://www.pubmedcentral.nih.gov/articlerender.fcgi?artid=2769562&tool=pmcentrez&rendertype=abstract.

Holt-Lunstad, J., Smith, T. B., Baker, M., Harris, T., & Stephenson, D. (2015). Loneliness and social isolation as risk factors for mortality: A meta-analytic review. *Perspectives on Psychological Science*, *10*, 227–237. https://doi.org/10.1177/1745691614568352.

Howe, L. C., Goyer, J. P., & Crum, A. J. (2017). Harnessing the placebo effect: The influence of physician characteristics on placebo response. *Health Psychology*, *36*, 1074–1082.

Hsu, D. T., Sanford, B. J., Meyers, K. K., Love, T. M., Hazlett, K. E., Wang, H., et al. (2013). Response of the μ-opioid system to social rejection and acceptance. *Molecular Psychiatry*, *18*(11), 1211–1217. https://doi.org/10.1038/mp.2013.96.

Hughes, M. E., Waite, L. J., Hawkley, L. C., & Cacioppo, J. T. (2004). A short scale for measuring loneliness in large surveys: Results from two population-based studies. *Research on Aging*, *26*(6), 655–672. https://doi.org/10.1177/0164027504268574.

Jackson, T., Fritch, A., Nagasaka, T., & Gunderson, J. (2002). Towards explaining the association between shyness and loneliness: A path analysis with American college students. *Social Behavior and Personality: An International Journal*, *30*(3), 263–270.

Jaremka, L. M., Andridge, R. R., Fagundes, C. P., Alfano, C. M., Povoski, S. P., Lipari, A. M., et al. (2013). Pain, depression, and fatigue: Loneliness as a longitudinal risk factor. *Health Psychology*, *33*(9), 948–957. https://doi.org/10.1037/a0034012.

Jaremka, L. M., Fagundes, C. P., Glaser, R., Bennett, J. M., Malarkey, W. B., & Kiecolt-Glaser, J. K. (2013). Loneliness predicts pain, depression, and fatigue: Understanding the role of immune dysregulation. *Psychoneuroendocrinology*, *38*(8), 1310–1317. https://doi.org/10.1016/j.psyneuen.2012.11.016.

Jaremka, L. M., Fagundes, C. P., Peng, J., Bennett, J. M., Glaser, R., Malarkey, W. B., et al. (2013). Loneliness promotes inflammation during acute stress. *Psychological Science*, *24*(7), 1089–1097. https://doi.org/10.1177/0956797612464059.

Jensen, K., & Kelley, J. M. (2016). The therapeutic relationship in psychological and physical treatments, and their placebo controls. *Psychology of Conciousness: Theory, Research and Practice*, *3*(2), 132–145. https://doi.org/10.1037/cns0000057.

Jensen, K., Kirsch, I., Odmalm, S., Kaptchuk, T. J., & Ingvar, M. (2015). Classical conditioning of analgesic and hyperalgesic pain responses without conscious awareness. *Proceedings of the National Academy of Sciences of the United States of America*, *112*(25), 7863–7867. https://doi.org/10.1073/pnas.1504567112.

Jones, W. H. (1981). The persistence of loneliness: Self and other determinants. *Journal of Personality*, *49*(1), 27–48.

Kaptchuk, T. J., Kelley, J. M., Conboy, L. A., Davis, R. B., Kerr, C. E., Jacobson, E. E., et al. (2008). Components of placebo effect: Randomised controlled trial in patients with irritable bowel syndrome. *BMJ*, *336*(7651), 999–1003. https://doi.org/10.1136/bmj.39524.439618.25.

Kiecolt-Glaser, J. K., Garner, W., Speicher, C., Penn, G. M., Holliday, J., & Glaser, R. (1984). Psychosocial modifiers of immunocompetence in medical students. *Psychosomatic Medicine*, *46*(1), 7–14. Retrieved from http://www.ncbi.nlm.nih.gov/pubmed/6701256.

Kirsch, I. (1985). Response expectancy as a determinant of experience and behavior. *American Psychologist*, *40*(11), 1189–1202.

Kirsch, I. (1997). Response expectancy theory and application: A decennial review. *Applied & Preventive Psychology*, *6*, 69–79.

Koban, L., Jepma, M., Geuter, S., & Wager, T. D. (2017). What's in a word? How instructions, suggestions, and social information change pain and emotion. *Neuroscience and Biobehavioral Reviews*, *81*, 29–42. https://doi.org/10.1016/j.neubiorev.2017.02.014.

Koban, L., & Wager, T. D. (2016). Beyond conformity: Social influences on pain reports and physiology. *Emotion*, *16*(1), 24–32.

Lamster, F., Lincoln, T. M., Nittel, C. M., Rief, W., & Mehl, S. (2017). The lonely road to paranoia. A path-analytic investigation of loneliness and paranoia. *Comprehensive Psychiatry*, *74*, 35–43. https://doi.org/10.1016/j.comppsych.2016.12.007.

Levine, J., Gordon, N., & Fields, H. (1978). The mechanism of placebo analgesia. *The Lancet*, *312*(8091), 654–657.

Lodder, G. M. A., Scholte, R. H. J., Goossens, L., Engels, R. C. M. E., & Verhagen, M. (2016). Loneliness and the social monitoring system: Emotion recognition and eye gaze in a real life conversation. *British Journal of Psychology*, *107*(1), 135–153. https://doi.org/10.1111/bjop.12131.

London, B., Downey, G., Bonica, C., & Paltin, I. (2007). Social causes and consequences of rejection sensitivity. *Journal of Research on Adolescence*, *17*(3), 481–506. https://doi.org/10.1111/j.1532-7795.2007.00531.x.

Losin, E. A. R., Anderson, S. R., & Wager, T. D. (2017). Feelings of clinician-patient similarity and trust influence pain: Evidence from simulated clinical interactions. *The Journal of Pain*, *18*(7), 787–799. https://doi.org/10.1016/j.jpain.2017.02.428.

Luo, Y., Hawkley, L. C., Waite, L. J., & Cacioppo, J. T. (2012). Loneliness, health, and mortality in old age: A national longitudinal study. *Social Science & Medicine (1982)*, *74*(6), 907–914. https://doi.org/10.1016/j.socscimed.2011.11.028.

Marcoen, A., Goossens, L., & Caes, P. (1987). Loneliness in pre-through late adolescence: Exploring the contributions of a multidimensional approach. *Journal of Youth and Adolescence*, *16*(6), 561–577.

Moerman, D. (2003). Doctors and patients: The role of clinicians in the placebo effect. *Advances in Mind-Body Medicine*, *19*(1), 14–22. Retrieved from http://www.ncbi.nlm.nih.gov/pubmed/12825409.

Moerman, D. E., & Jonas, W. B. (2002). Deconstructing the placebo effect and finding the meaning response. *Annals of Internal Medicine*, *136*(6), 471–476.

Morton, D. L., Watson, A., El-deredy, W., & Jones, A. K. P. (2009). Reproducibility of placebo analgesia: Effect of dispositional optimism. *Pain*, *146*(1–2), 194–198. https://doi.org/10.1016/j.pain.2009.07.026.

Peciña, M., Love, T., Stohler, C. S., Goldman, D., & Zubieta, J.-K. (2014). Effects of the mu opioid receptor polymorphism (OPRM1 A118G) on pain regulation, placebo effects and associated personality trait measures. *Neuropsychopharmacology*, *1*, 1–9. https://doi.org/10.1038/npp.2014.272.

Pressman, S. D., Cohen, S., Miller, G. E., Barkin, A., Rabin, B. S., & Treanor, J. J. (2005). Loneliness, social network size, and immune response to influenza vaccination in college freshmen. *Health Psychology*, *24*(3), 297–306. https://doi.org/10.1037/0278-6133.24.3.297.

Qualter, P., Brown, S. L., Munn, P., & Rotenberg, K. J. (2010). Childhood loneliness as a predictor of adolescent depressive symptoms: An 8-year longitudinal study. *European Child & Adolescent Psychiatry*, *19*(6), 493–501. https://doi.org/10.1007/s00787-009-0059-y.

Qualter, P., Rotenberg, K., Barrett, L., Henzi, P., Barlow, A., Stylianou, M., et al. (2013). Investigating hypervigilance for social threat of lonely children. *Journal of Abnormal Child Psychology*, *41*(2), 325–338. https://doi.org/10.1007/s10802-012-9676-x.

Qualter, P., Vanhalst, J., Harris, R., Van Roekel, E., Lodder, G., Bangee, M., et al. (2015). Loneliness across the life span. *Perspectives on Psychological Science*, *10*, 250–264. https://doi.org/10.1177/1745691615568999.

Rotenberg, K. J., Gruman, J. A., & Ariganello, M. (2002). Behavioral confirmation of the loneliness stereotype. *Basic and Applied Social Psychology*, *24*(2), 81–89. https://doi.org/10.1207/S15324834BASP2402.

Rotenberg, K. J., & Kmill, J. (1992). Perception of lonely and non-lonely persons as a function of individual differences in loneliness. *Journal of Social and Personal Relationships*, *9*(2), 325–330. https://doi.org/10.1177/0265407592092009.

Russell, D. W. (1996). UCLA Loneliness Scale (version 3): Reliability, validity, and factor structure. *Journal of Personality Assessment*, *66*(1), 20–40. Retrieved from http://www.ncbi.nlm.nih.gov/pubmed/8576833.

Russell, D., Peplau, L. A., & Cutrona, C. E. (1980). The revised UCLA Loneliness Scale: Concurrent and discriminant validity evidence. *Journal of Personality and Social Psychology*, *39*(3), 472–480. Retrieved from http://www.ncbi.nlm.nih.gov/pubmed/7431205.

Schenk, L., Krimmel, S. R., & Colloca, L. (2017). Observe to get pain relief: Current evidence and potential mechanisms of socially-learned pain modulation. *Pain*, *158*(11), 2077–2081. https://doi.org/10.1097/j.pain.0000000000000943.

Staats, P. S., Staats, A., & Hekmat, H. (2001). The additive impact of anxiety and a placebo on pain. *Pain Medicine*, *2*(4), 267–279.

Steptoe, A., Shankar, A., Demakakos, P., & Wardle, J. (2013). Social isolation, loneliness, and all-cause mortality in older men and women. *Proceedings of the National Academy of Sciences of the United States of America*, *110*(15), 5797–5801. https://doi.org/10.1073/pnas.1219686110.

Tedeschi, E., Weber, J., Prevost, C., Mischel, W., & Mobbs, D. (2015). Inference of others' competence reduces anticipation of pain when under threat. *Journal of Cognitive Neuroscience*, *27*(10), 2071–2078. https://doi.org/10.1162/jocn.

Tracey, I. (2010). Getting the pain you expect: Mechanisms of placebo, nocebo and reappraisal effects in humans. *Nature Medicine*, *16*(11), 1277–1283. https://doi.org/10.1038/nm.2229.

Tsai, F., & Reis, H. T. (2009). Perceptions by and of lonely people in social networks. *Personal Relationships*, *16*, 221–238.

van der Meulen, M., Kamping, S., Anton, F., Meulen, M. V. D., Kamping, S., & Anton, F. (2017). The role of cognitive reappraisal in placebo analgesia: An fMRI study. *Social Cognitive and Affective Neuroscience*, *12*(7), 1128–1137. https://doi.org/10.1093/scan/nsx033.

Vanhalst, J., Gibb, B. E., Prinstein, M. J., Vanhalst, J., Gibb, B. E., & Lonely, M. J. P. (2017). Lonely adolescents exhibit heightened sensitivity for facial cues of emotion. *Cognition and Emotion*, *31*(2), 377–383. https://doi.org/10.1080/02699931.2015.1092420.

Vanhalst, J., Soenens, B., Luyckx, K., Petegem, S. V., Weeks, M. S., Asher, R., et al. (2015). Why do the lonely stay lonely? Chronically lonely adolescents' attributions and emotions in situations of social inclusion and exclusion. *Journal of Personality and Social Psychology*, *109*(5), 932–946.

van Osch, M., van Dulmen, S., van Vliet, L., & Bensing, J. (2017). Specifying the effects of physician's communication on patients' outcomes: A randomised controlled trial. *Patient Education and Counseling*, *100*, 1482–1489. https://doi.org/10.1016/j.pec.2017.03.009.

van Roekel, E., Goossens, L., Verhagen, M., Wouters, S., Engels, R. C. M. E., & Scholte, R. H. J. (2014). Loneliness, affect, and adolescents' appraisals of company: An experience sampling method study. *Journal of Research on Adolescence*, *24*(2), 350–363. https://doi.org/10.1111/jora.12061.

Vits, S., Cesko, E., Enck, P., Hillen, U., Schadendorf, D., & Schedlowski, M. (2011). Behavioural conditioning as the mediator of placebo responses in the immune system. *Philosophical Transactions of the Royal Society of London. Series B, Biological Sciences*, *366*(1572), 1799–1807. https://doi.org/10.1098/rstb.2010.0392.

Vögtle, E., Barke, A., & Kröner-Herwig, B. (2013). Nocebo hyperalgesia induced by social observational learning. *Pain*, *154*(8), 1427–1433. https://doi.org/10.1016/j.pain.2013.04.041.

Wager, T. D., Scott, D. J., & Zubieta, J.-K. (2007). Placebo effects on human mu-opioid activity during pain. *Proceedings of the National Academy of Sciences of the United States of America*, *104*(26), 11056–11061. https://doi.org/10.1073/pnas.0702413104.

Way, B. M., Taylor, S. E., & Eisenberger, N. I. (2009). Variation in the mu-opioid receptor gene (OPRM1) is associated with dispositional and neural sensitivity to social rejection. *Proceedings of the National Academy of Sciences of the United States of America*, *106*(35), 15079–15084. https://doi.org/10.1073/pnas.0812612106.

Wendt, L., Albring, A., & Schedlowski, M. (2014). Learned placebo responses in neuroendocrine and immune functions. In F. Benedetti, P. Enck, E. Frisaldi, & M. Schedlowski (Eds.), *Placebo* (pp. 159–181). Berlin, Heidelberg: Springer Berlin Heidelberg. https://doi.org/10.1007/978-3-662-44519-8_10.

Wirth, T., Ober, K., Prager, G., Vogelsang, M., Benson, S., Witzke, O., et al. (2011). Repeated recall of learned immunosuppression: Evidence from rats and men. *Brain, Behavior, and Immunity*, *25*, 1444–1451. https://doi.org/10.1016/j.bbi.2011.05.011.

Wolf, L. D., & Davis, M. C. (2014). Loneliness, daily pain, and perceptions of interpersonal events in adults with fibromyalgia. *Health Psychology*, *33*(9), 929–937. https://doi.org/10.1037/hea0000059.

Yoshida, W., Seymour, B., Koltzenburg, M., & Dolan, R. J. (2013). Uncertainty increases pain: Evidence for a novel mechanism of pain modulation involving the periaqueductal gray. *Journal of Neuroscience*, *33*(13), 5638–5646. https://doi.org/10.1523/JNEUROSCI.4984-12.2013.

Zubieta, J.-K., Bueller, J. A., Jackson, L. R., Scott, D. J., Yanjun, X., Koeppe, R. A., et al. (2005). Placebo effects mediated by endogenous opioid activity on mu-opioid receptors. *Journal of Neuroscience*, *25*(34), 7754–7762. https://doi.org/10.1523/JNEUROSCI.0439-05.2005.

CHAPTER TEN

What Is Minimally Required to Elicit Placebo Effects?

Karin B. Jensen[1]
Karolinska Institute, Stockholm, Sweden
[1]Corresponding author: e-mail address: karin.jensen@ki.se

Contents

Abstract

Placebo effects have often been attributed to cognitive processes described as "learning" and/or "expectancy," yet the role of conscious awareness is unclear. Furthermore, little is known about the placebo effects in patients with limited cognitive abilities, such as intellectual disability. Here, recent data on placebo mechanisms in patients with impaired cognitive function will be discussed, as well as experimental studies investigating how implicit cognitive processes may shape placebo effects. Together these studies comment on the minimum requirements in order to elicit placebo effects, both from the view of conscious awareness and from the perspective of Intelligence Quotient and basic brain function. Together with recent conceptualizations of placebo effects in terms of predictive coding, there is evidence to suggest that placebo effects are fundamental responses of the brain that have developed to promote survival.

International Review of Neurobiology, Volume 138
ISSN 0074-7742
https://doi.org/10.1016/bs.irn.2018.01.008

1. INTRODUCTION

Placebo effects are powerful examples of how psychological processes can affect disease symptoms. To date, placebo effects in humans have traditionally been attributed to higher-order cognitions, described as "expectancy effects" that require conscious access (Stewart-Williams & Podd, 2004) and prefrontal cortical control (Faria, Fredrikson, & Furmark, 2008). According to a predictive coding perspective on general brain function (Buchel, Geuter, Sprenger, & Eippert, 2014), the brain is constantly matching incoming sensory information with prior knowledge of the world. Placebo effects could thus occur as a result from "top-down" modulation, represented by a shift in balance between bottom-up sensory input and top-down prior beliefs (Buchel et al., 2014). As an example, studies of placebo analgesia show that expectations of pain relief, reflected in prefrontal brain areas, modulate nociceptive signals by activation of an opioid-dependent antinociceptive brain network (Eippert et al., 2009; Petrovic, Kalso, Petersson, & Ingvar, 2002). Yet, there are a number of reasons to assume that placebo effects can be elicited without higher-order cognitions and cortical control mechanisms (without dismissing the predictive coding perspective, as priors and top-down processes may be processed at different levels along the neural axis and also applies to animals (Gowaty & Hubbell, 2013)): (i) Placebo effects are seen in response to simple learning mechanisms such as conditioning (associative learning) without the formation of explicit expectancy (Jensen, Kirsch, Odmalm, Kaptchuk, & Ingvar, 2015). (ii) Placebo effects are reported in nonhuman animals with significantly different neocortical properties (Grochowicz et al., 1991; Munana, Zhang, & Patterson, 2010). (iii) Patients with severely impaired cognitive functions display placebo effects in clinical trials (Curie et al., 2015). (iv) Contextual information may reach the brain at different levels of awareness as our brain is hierarchically organized (Mesulam, 1998). Placebo conditioning can thus occur in response to stimuli that are not consciously perceived, because they are presented at low intensities or masked from conscious awareness (Bunce, Bernat, Wong, & Shevrin, 1999; Clark, Manns, & Squire, 2002; Esteves, Parra, Dimberg, & Ohman, 1994). (v) Furthermore, experimental data suggest that the placebo effect is not reduced when tested in presence of a high-demand cognitive task known to engage the same prefrontal brain areas seen in previous placebo studies (Buhle, Stevens, Friedman, & Wager, 2012). This suggests that access to prefrontal brain regions may not be a requirement to elicit placebo effects and open up for a new understanding of the underlying neural circuitry. (vi) It is possible that placebo responses occur as a response to internalized cultural expectations that do not rely on conscious awareness to shape behavior, for example, that a blue pill inherently signals tranquilizing and cool properties compared to a red pill in some cultures (Moerman & Jonas, 2002). Hence, there is evidence to suggest that contextual factors may affect placebo responses independently from verbal suggestions or learning procedures, e.g., through the name (Branthwaite & Cooper, 1981) or price of a treatment (Geuter, Eippert, Hindi Attar, & Buchel, 2013; Waber, Shiv, Carmon, & Ariely, 2008). One of the most prominent examples is the strong influence of patient–clinician relationships on placebo responses (Kaptchuk et al., 2008), where subtle social cues shape placebo responses, and clinicians who perform the same treatment protocol will differ markedly in effectiveness (Kelley et al., 2009).

In this chapter, the minimum requirements to elicit placebo effects will be addressed from several of these perspectives. Together with recent conceptualizations of placebo effects in terms of predictive coding, there is evidence to suggest that placebo effects are fundamental responses of the brain that have developed to promote survival.

2. PLACEBO AND BASIC BRAIN FUNCTION

2.1 Neuroimaging Techniques and the Functional Anatomy of Placebo Effects

The advent of neuroimaging techniques, such as Positron Emission Tomography (PET) and functional Magnetic Resonance Imaging (fMRI), has contributed to our current understanding of the neural correlates of placebo responses. Since the pioneering studies in the early 2000s (de la Fuente-Fernandez et al., 2001; Mayberg et al., 2002; Petrovic et al., 2002), results have revealed a number of brain regions implicated in placebo effects, both from experimental studies in healthy volunteers and studies in clinical groups. In brief, there has been strong emphasis on prefrontal areas, such as the ventrolateral and dorsolateral prefrontal cortices (vlPFC/dlPFC) (Petrovic et al., 2010), as well as areas of the mesolimbic reward system (Scott et al., 2009) (for a systematic review, see Faria et al., 2008). Early placebo neuroimaging studies focused on mapping the functional anatomy of placebo effects, and thereby provided objective correlates to subjective reports of reduced symptoms. Since then, placebo neuroimaging has evolved into more complex designs that aim for characterization of neural markers for placebo responders and include computational approaches to fMRI data (Hashmi et al., 2012). For example, prospective studies have suggested that the pattern of activity in some brain networks at baseline may predict subsequent placebo effects in a clinical trial (Pecina et al., 2015; Schmidt-Wilcke et al., 2014; Tetreault et al., 2016). More than self-reported treatment expectancies, it has been suggested that the accuracy of the brain's top-down predictions may be central for predicting placebo effects (Pecina & Zubieta, 2015).

In addition to neuroimaging studies, Transcranial Magnetic Stimulation (TMS) has been used to assess the role of the lateral prefrontal cortex in placebo analgesia (Klein et al., 2014). In the study by Krummenacher and colleagues, TMS was used to induce a temporary block of prefrontal function. In line with their hypothesis, the authors found that placebo analgesia was abolished when the dlPFC was experimentally blocked by means of inhibitory TMS (Krummenacher, Candia, Folkers, Schedlowski, & Schönbächler, 2010), corroborating the role of the prefrontal cortex in placebo analgesia.

2.2 Predictive Coding Framework: A Theoretical Approach to Brain Function

In line with data describing the basic brain mechanisms involved in placebo effects, an overarching theoretical framework based on Bayesian theory has emerged (Buchel et al., 2014). The theory posits that the brain is a prediction machine that is automatically matching incoming sensory data with an inner model of the world based on prior experience. In order to promote adaptation to our environment (and ultimately promote survival), we are dependent on successful perception and thus the "mismatch" between our inner model and the bottom-up sensory input should be minimized. In other words, successful perception requires the brain to minimize surprises (Clark, 2013). In the case of placebo effects, the predictive coding framework suggests that strong priors about a treatment outcome can shape sensory perception as a way to reconcile the mismatch between *what is expected* and *what is felt*. In this fine-tuned manner, our brains are constantly matching sensory input with our expectations, providing a foundation for prediction-driven learning that updates our inner models. As different situations require different predictive precision, the balance between bottom-up and top-down influences can be regulated by a volume control (Friston, 2009), which means that top-down influences have varying modulatory effect.

According to Clark (2013) this is good news, as we are at least not (entirely) slaves to our expectations. Yet, it should be added, the good news is that we are not slaves to the sensory signals that reach our sensory systems either.

While this is a description of how the brain adapts successfully to different contexts in general, it is also a way to describe the fundamental conditions for placebo effects. As the drive for adaptation has been essential in human brain development, the question of why placebo effects exist has been given an interpretation in evolutionary terms (Colloca & Miller, 2011; Humphrey, 2002). Importantly, the predictive coding framework is not unique for human cognition; it is a framework that encompasses nonhuman animals too. According to Gowaty and Hubbel all animals are Bayesian, as they are adaptively flexible individuals who "predict" ("visualize," "imagine") alternatives and make choices to serve fitness.

2.3 Experimental Tests of the Role of Prefrontal Cortex in Placebo

Before the first neuroimaging experiments of placebo effects, early PET studies of pain processing in the brain had been published, suggesting activation of the prefrontal cortex during pain anticipation (Ploghaus et al., 1999) and hypnotic suggestions (Rainville et al., 1999). These studies corroborated the notion that frontal cortices were likely to reflect top-down mechanisms involved in the reinterpretation of sensory experiences. In the study by Rainville et al. (1999) verbal suggestions were used to alter the painful experience, and the authors conclude that working memory processes (represented in the dlPFC) were likely necessary to keep the suggestions online and drive the top-down reinterpretation of the pain according to an inner model. In line with these early studies, theories emerged that emphasized executive function and cortical control in placebo effects. More specifically, theories postulate that executive processes are involved in the formation of expectations that in turn are necessary to recruit endogenous resources for symptom relief (Petrovic et al., 2010).

A decade after the first neuroimaging studies of placebo effects (de la Fuente-Fernandez et al., 2001; Mayberg et al., 2002; Petrovic et al., 2002) more recent studies have started to complement the role of higher order cognitive functions in placebo mechanisms. As a way to interrogate whether executive resources mediate placebo analgesia, Buhle et al. (2012) performed an experiment that combined placebo analgesia with a difficult cognitive task known to reduce pain (Buhle & Wager, 2010). Independent studies had shown that both the cognitive task and placebo analgesia would activate the same part of the prefrontal cortex, i.e., the dlPFC. The question was if a combination of the two would yield additive effects on pain reduction. Conversely, the logic of limited resources in the brain's modulatory systems would suggest that the combination would cause interference and thus reduce the analgesic effect. Buhle and colleagues found that executive demand did not reduce placebo analgesia; instead there was an additive effect when combining the two. Also, the placebo analgesia task did not affect working memory performance. In contrast to common belief, the results suggest two separate routes to pain control such that placebo analgesia is not dependent on executive functioning.

In order to reconcile with previous data, Buhle and Wager emphasize that evidence for the role of executive function in placebo analgesia has been indirect, e.g., studies demonstrating that neural regions believed to support executive function are also involved during placebo analgesia (Buhle & Wager, 2010). Also, it is possible that the need for measurable outcomes in experimental designs, such as explicit reports of expectations, has shaped the existing placebo paradigms and may have had an impact on the brain regions (and theories) found thus far. That way, the type of placebo tasks used in most neuroimaging experiments may have favored activation in dlPFC, even if this may not be a unique region for representing expectations of relief. This bridges over to the next section of this chapter, describing nonconscious activation of placebo effects, which is one way to address this methodological problem. What if study participants were not aware of the cues paired with expectations of pain relief?

3. PLACEBO AND CONSCIOUS AWARENESS

3.1 No Dichotomy Between Explicit Expectancy and Conditioning

There has been an often reported dichotomy between placebo effects shaped by verbal suggestions and those from conditioning. While verbal suggestion has been thought to form explicit expectations about the efficacy of a treatment, conditioning has been associated with implicit expectancy. Yet, empirical data have demonstrated that explicit and implicit representations overlap and conditioned placebo effects are thought to work through the creation of introspectable (consciously accessible) expectancies (Atlas & Wager, 2012; Kirsch, 2004). In other words, conditioning can be seen as a process that generates expectations about treatment outcomes. Even if conditioning is an automated learning mechanism per se, there will still be an explicit expectancy component as part of the response (Montgomery & Kirsch, 1997). We can imagine a placebo experiment where there is no verbal suggestion at all, just a conditioning procedure where an anesthetic substance (unconditioned stimulus) is paired with an inert topical cream (conditioned stimulus) and tested at a site where thermal pain is applied. In that scenario, the application of the cream will be consciously perceived by the patient. Also, the unconditioned behavioral response to the substance (reduced pain) can be consciously detectable. Thus, expectancy may be activated when attention is directed to it (Kirsch, 2004) even if there was no initial suggestion that the cream will reduce pain. Also, there is a possibility that there are inherent assumptions about certain treatment-related objects or behaviors that will affect the outcome without any conscious knowledge (de Craen, Roos, de Vries, & Kleijnen, 1996; Haug, 2011). With this said, it is clear that many instances include a mix of consciously accessible and nonconscious expectancies.

3.2 Nonconscious Processes Shape Behavior

There is general evidence suggesting that behavior can be shaped by nonconscious processes (Custers & Aarts, 2010; Dijksterhuis, Aarts, & Smith, 2005). In fact, fundamental aspects of human cognition are thought to operate outside of conscious awareness, for example, in reward processing (Bijleveld, Custers, & Aarts, 2010; Pessiglione et al., 2008), fear learning (Carlsson et al., 2004; Öhman, Carlsson, Lundqvist, & Ingvar, 2007), and social behavior (Doyen, Klein, Pichon, & Cleeremans, 2012; Strahan, Spencer, & Zanna, 2002). One way to study nonconscious influences on behavior is to use subliminal stimuli. Subliminal means that a stimulus is presented below (sub) the threshold (limen) for conscious recognition, yet the stimulus can still affect behavior as it has been registered at a basic level of perception (even if there is likely not a fixed threshold, for an overview of methods for subliminal validation, see Sand & Nilsson, 2016). For example, an experimental study that included phobic individuals showed that fearful reactions were triggered in response to subliminal images. Images of phobic stimuli were shown so quickly they could not be consciously perceived, yet they evoked increased arousal and negative valence ratings in phobic participants (Carlsson et al., 2004). In the case of fear reactions, it is thought that rapid responses may give a survival advantage (LeDoux, 1990). Hence, higher-order cognitive processes that include conscious awareness may be attenuated, whereas the performance of automatic responses is facilitated (Ohman, Carlsson, Lundqvist, & Ingvar, 2007). In favor of survival, there is evidence to suggest that predictive cues (signaling threat or reward) may be recognized subliminally and mediate nonconscious effects on human cognition and behavior.

3.3 Nonconscious Effects in Placebo and Nocebo Effects

Already in the late 1800s Peirce and Jastrow suggested that differences in somatosensory perception may be influenced by nonconscious mental processes "a difference which, though existing in sensation, could not be brought into consciousness by any effort of attention" (Peirce & Jastrow, 1885). Thereby, Peirce and Jastrow point to the fact that our perception of bodily symptoms is partly shaped by automatic processes that cannot be consciously detectable. It would thus be plausible that placebo effects could be modulated by nonconscious mental processes, as suggested in theoretical papers (Haug, 2011; Kaptchuk, 2011).

In a line of studies by Jensen and colleagues, conditioning was used to examine the role of conscious awareness in the *activation* and *acquisition* of conditioned placebo and nocebo effects (Jensen, Kaptchuk, Chen, et al., 2015; Jensen et al., 2012; Jensen, Kirsch, Odmalm, et al., 2015). In a first study (Jensen et al., 2012), it was tested if conditioned placebo and nocebo effects could be *activated* by means of subliminal cues. First, a conditioning paradigm that could induce robust placebo and nocebo effects was established, by pairing two clearly recognizable visual cues with either high- or low-intensity heat pain. The same visual cues were then paired with moderate heat during a test phase. In line with similar experiments results showed that moderate heat was rated as more painful when preceded by the high pain cue (nocebo-like effect), and less painful when preceded by the low pain cue (placebo-like effect), indicating successful conditioning of placebo- and nocebo-like effects. However, in a separate group of subjects, it was also assessed whether these conditioned responses could be activated by subliminal cues. Results revealed significant differences in pain ratings in response to the subliminal cues, confirming that conditioned placebo and nocebo effects can be activated by nonconscious cues. Even if there had been previous support for nonconscious activation of fear, this investigation translated the effect of nonconscious cues to placebo effects, a topic previously associated with formation of expectancies and higher-order cognitive functions. Also, the results moved the investigation of nonconscious effects to the clinical realm, by suggesting that health-related responses can be triggered by cues that are not consciously perceived, not only for pain, but also for other medical problems with demonstrated placebo effects.

As it had been demonstrated that placebo and nocebo effects could be activated nonconsciously, the question was if these effects could also be *acquired* by means of subliminal cues. Preliminary evidence suggests that fear can be acquired nonconsciously, as conditioned autonomic and motor changes occurred in response to subliminal threat cues (Bunce et al., 1999; Esteves et al., 1994; Raio, Carmel, Carrasco, & Phelps, 2012). In order to test if new conditioned placebo and nocebo pain responses could be formed outside conscious awareness, an experiment was performed that tested the full interaction between acquisition type (conscious/nonconscious) and activation type (conscious/nonconscious) (Jensen, Kirsch, Odmalm, et al., 2015). This study concluded that placebo and nocebo effects were indeed acquired in response to subliminal cues. In fact, similar placebo and nocebo effects were seen in response to conscious and nonconscious conditioning. Also, there were similar effects in response to conscious and nonconscious activation of placebo and nocebo effects (see Fig. 1). Hence, previous findings were extended by demonstrating that new learning of placebo and nocebo pain responses can take place in response to subliminal stimuli. In terms of brain function, this means that new associations can be acquired at a very basic level of information processing. Contrary to studies of fear learning in humans, where conditioning with subliminal stimuli demonstrated effects on autonomic and motor responses (Bunce et al., 1999; Esteves et al., 1994; Knight, Nguyen, & Bandettini, 2003; Raio et al., 2012), the study demonstrated how lower levels of the brain's hierarchical organization may affect learning in higher-order brain circuitry (as our primary outcome measured subjective pain reports). In a broader perspective, these results suggest that nonconscious stimuli may shape learning of complex cognitive processes (such as predictions about future pain relief) and not just affect rapid reactions to fear stimuli. Also, results validate that nonconscious learning may take place in the sensory domain (as our results were based on ratings of pain intensity) which indicates that similar learning is likely possible in other clinical contexts.

As previously mentioned, there is evidence for nonconscious processing of reward and fear responses. In a neuroimaging study of the neural correlates of monetary reward (Pessiglione et al., 2008), the authors found that subliminal reward cues were processed nonconsciously through direct activation of reward-related mesolimbic brain circuitry. Along the same lines,

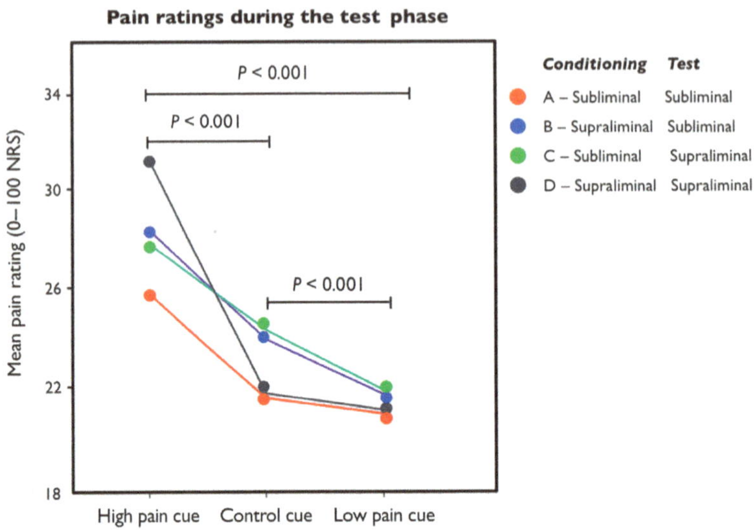

Fig. 1 Pain ratings during the test sequence of a placebo and nocebo experiment. In a pain experiment by Jensen et al. (2012), participants were conditioned with either supraliminal or subliminal visual cues. During a subsequent test sequence, the effect of conditioning was tested using either supraliminal or subliminal cues. All four combinations of supraliminal/subliminal conditioning and test sequence were tested. During conditioning, one cue was always paired with high thermal pain and one with low thermal pain. During the test sequence, the high cue, low cue, and a control cue were paired with a moderate temperature in order to test how predictive cues may change pain perception in a placebo-like (low cue) and nocebo-like (high cue) manner. Participants rated pain intensity on a 0–100 NRS. The plot represents pain ratings and interaction of cue type (High, Low, Control) by experimental group (A, B, C, and D). There was no significant difference in pain ratings between the experimental groups, which means that subliminal exposures led to the same effects as supraliminal exposures of the pain cues. *P* values reflect the pairwise comparisons between cue types across experimental groups (High–Control, Low–Control, High–Low).

neuroimaging studies have elucidated how fear cues can be conveyed via direct thalamic projections to the amygdala (LeDoux, 2000; Mesulam, 1998). This allows the brain to respond with rapid (but inflexible) processes when awareness is limited, and slower (highly flexible) processes when contextual awareness is available. As some instances of placebo effects involve reward-related processes, and nocebo responses are likely to involve fear-related processes in the brain, it is plausible that nonconscious placebo and nocebo effects would be manifested in such rapid fear and reward circuitry. In a neuroimaging experiment of conditioned placebo and nocebo pain responses, the neural pathways involved in nonconscious activation of such responses were determined (Jensen, Kaptchuk, Chen, et al., 2015). First of all, the neuroimaging results validated the previous findings by Jensen and colleagues, showing nonconscious activation of conditioned placebo and nocebo responses. The neuroimaging analyses revealed that nonconscious activation of *nocebo* (compared to conscious activation) was associated with activity in subcortical brain circuitry, previously implicated in processing of fear-relevant stimuli. Both the amygdala and hippocampus were activated during nonconscious nocebo trials, and neural activity in the amygdala correlated with participants' self-reported pain during nonconscious trials. This is in line with previous suggestions that so-called nonconscious brain processes interact with explicit ratings of pain or anxiety (Carlsson et al., 2004) and further the notion that nonconscious predictions may affect our decisions and goal pursuits (Custers & Aarts, 2010). In response to nonconscious *placebo* trials, there was more activity in the orbitofrontal cortex (Ofc), compared to conscious placebo trials. The Ofc is a cortical brain area but represents a phylogenetically old part of the prefrontal cortex (Thorpe, Rolls, & Maddison, 1983). The Ofc has previously been implicated in studies of placebo analgesia (Petrovic et al., 2010; Wager et al., 2004) and is thought to represent a reward-related component of pain relief rather than inhibition of the nociceptive input itself (Leknes, Lee, Berna, Andersson, & Tracey, 2011). In contrast to placebo-related activation of the dlPFC (and other parts of the lateral PFC), which has often been reported in placebo neuroimaging studies, it is possible that conscious awareness of the conditioned placebo cues is not necessary to activate the more archaic OFC (Jensen, Kaptchuk, Chen, et al., 2015). Taken together, these neuroimaging results validate that conditioned placebo and nocebo effects can be elicited outside of conscious awareness and suggest a hierarchical neural representation of conscious and nonconscious effects, respectively.

3.4 Priming

Priming is an implicit memory effect that has been used to investigate how nonconscious memory processes may affect behavior (Tulving & Schacter, 1990). The underlying assumption is that the exposure of one stimulus (the prime) can influence an individual's reaction to the following stimuli, and thereby influence one's behavior without understanding the connection to the prime (Tulving & Schacter, 1990). Even if priming has nonconscious influences on behavior, the prime does not need to be presented subliminally, as long as the connection between the prime and the associated behavior is not consciously known. For example, exposing a person to list of words that contain references to stereotypes of the elderly may nonconsciously trigger slower walking speed, as demonstrated in a well-cited (but poorly replicated) experiment on the automaticity of social behavior (Bargh, Chen, & Burrows, 1996). For more reading on the difficulties to replicate priming research in social psychology, see the "many-labs-study" (Klein et al., 2014) and more (Rohrer, Pashler, & Harris, 2015).

Priming research has suggested that nonconscious mental processes are important for understanding health-related behavior (Sheeran, Gollwitzer, & Bargh, 2013), and significant effects of priming have been reported for physical activity (Rebar, Loftus, & Hagger, 2015) and pain perception (Richter et al., 2014). Yet, until recently it was unknown if priming could affect placebo effects by increasing treatment expectations. In a study from 2016, Rosen, Kirsch, Kaptchuk, Ingvar, and Jensen (2017) performed a double-blind priming experiment where two types of primes were compared; one aimed at increasing positive expectations and one neutral control condition. A previously used priming paradigm, called the Scrambled Sentence Test (SST), was used to test the hypothesis that placebo analgesia could be enhanced by exposure to words that activate a positive mindset, compared to priming with neutral words. The hypothesis was rejected, as there was no evidence that implicit priming could enhance positive expectations and affect placebo effects. Instead, there was a high correlation between the participants' prior experience of pain relief and the placebo outcome. Priming is one of many ways to shape behavior, yet in the case of placebo analgesia it is likely that a person's prior experience of pain relief has a more powerful influence, which may compete with the subtle effects of priming. It is well known that learning is central for placebo effects (Benedetti et al., 2016; Colloca & Benedetti, 2006) and the experienced pain relief during an initial part of an experiment (acquisition phase) may determine the following placebo effects (Jensen et al., 2012; Jensen, Kirsch, et al., 2015). As there is significant influence of the patient–clinician relationship on placebo outcomes (Kelley, Kraft-Todd, Schapira, Kossowsky, & Riess, 2014), it is also possible that the influence of a trustworthy clinician may interfere with the creation of expectations through priming (Rosen et al., 2017). It is often emphasized how patients' expectations influence placebo effects; however, it is suggested that the expectations of the experimenter or clinician may also influence the outcome (Gracely, Dubner, Deeter, & Wolskee, 1985; Messer & Wampold, 2002). To date, priming studies have often been performed in a single-blinded manner, where the experimenter has been aware of the expected differences between two groups. Recent data demonstrate how the same priming procedure can yield opposite results depending on the beliefs of the experimenter (Doyen et al., 2012), highlighting the importance of double-blind designs in priming studies.

4. PLACEBO EFFECTS AMONG PATIENTS WITH INTELLECTUAL DISABILITY

4.1 Intellectual Disability

One way to challenge our current understanding of placebo mechanisms is to employ a "lesion model." Yet, instead of searching for individuals with specific brain lesions and study how it may affect placebo outcomes, patients with severe impairments in intellectual functioning have been studied in two recent meta-analyses (Curie et al., 2015; Jensen et al., 2017). Intellectual disability (ID) is a diagnosis that entails significant limitations in both intellectual functioning and in adaptive behavior (American Psychiatric Association, 2013). The intellectual impairments associated with ID include deficits in abstract thinking, reasoning, predicting the future, and learning from experience, i.e., the same mental abilities that are often implicated in placebo effects. Moreover, the overall Intelligence Quotient (IQ) will be lower than 70 in patients with ID, as measured with standardized intelligence testing. In the investigations by Jensen et al. (2017) and Curie et al. (2015) placebo effects were only studied in genetically determined ID (e.g., Down's, Fragile X, and Prader–Willi syndrome) in order to rule out the possibility that a normal brain development before ID onset would confound the results (i.e., ID acquired after birth). When data from placebo-controlled randomized controlled trials in ID were analyzed, there was evidence for significant placebo responses in the placebo arm of these trials (Curie et al., 2015). Hence, patients with severe impairment of the mental abilities associated with the formation of treatment expectations and placebo effects displayed placebo responses. The effect size was similar to that in studies including non-ID individuals, and placebo responses were seen both in response to subjective and objective outcomes, and in several behavioral and cognitive domains (Curie et al., 2015).

Yet, a placebo response that is defined by an improvement from baseline to study end (not including a natural history control group) will not be able to determine if the response was due to placebo mechanisms or influenced by factors such as regression to the mean or spontaneous remission (Kirsch, 2013) (even if the latter is unlikely in refractory conditions like ID). In response to this limitation, a second meta-analyses were performed (Jensen et al., 2017) where ID trials with different certainty of receiving genuine medication were compared. In open-label trials, the patients (and their doctors and families) know that there is a 100% chance of getting genuine medication, and thus the contribution of positive treatment expectations is likely to be high in that context. Conversely, placebo-controlled trials often have a 50% chance of receiving an active drug, which is associated with relatively lower expectations of improvement. The results confirmed that treatment expectations are likely to influence treatment outcomes in ID trials, as there were higher effect sizes in studies with 100% likelihood of receiving genuine medication, compared to both the drug and placebo arm of placebo-controlled trials. This result validates the findings by Curie et al. (2015) by demonstrating that patients with severe intellectual disabilities are susceptible to differences in treatment contexts and can display genuine placebo effects, which are not explained by factors such as spontaneous remission or regression to the mean (Jensen et al., 2017).

Even if the study by Curie et al. (2015) demonstrated robust placebo responses in patients with ID overall, there was a significant association between IQ and placebo effect size, suggesting that higher IQ correlated with higher placebo response (an effect that was not seen in response to genuine drugs). This may be an indication that placebo responses and effects do occur in patients with ID but only to a certain degree of cognitive impairment. Along those lines, the same meta-analysis suggests that placebo responses may occur before, but not after, the onset of dementia in patients with ID. This may be preliminary evidence to reconcile with a study where the placebo component of analgesic treatment (as investigated with the open–hidden administration paradigm) was lower among demented patients (Benedetti et al., 2006).

As there is evidence for placebo effects among individuals with ID, it is likely influenced by implicit social influences that are not dependent on reasoning, such as "placebo by proxy" effects (Grelotti & Kaptchuk, 2011). Placebo by proxy describes how a patient can be affected implicitly by treatment expectations among family members, and has been used to explain placebo effects in young children (Simmons et al., 2014; Weimer et al., 2013). In fact, the detection of expectations among the people that ensure one's physical and psychological integrity may be an excellent compensation for a lack of intellectual understanding of complex situations.

In that sense, placebo effects among individuals with ID could represent a "low road" adaptation, which is unrelated to higher-order cognitive functions, as compared to greater involvement of "high road" predictions and expectations described in healthy individuals.

5. FUTURE TOPICS

One decade after the first neuroimaging studies of the neural correlates to placebo effects there is still a strong emphasis on executive functions in theories of placebos. Yet, there are several reasons to believe that placebo effects may also be elicited through other neural pathways than those elucidated in phylogenetically recent parts of the human brain. In other words there are higher-order as well as lower-order systems that code the contextual information needed to elicit placebo effects.

The predictive coding perspective on human brain function has influenced cognitive neuroscience in recent years (Clark, 2013) and may direct placebo research in a new direction. As the adaptability and evolution of the human brain are promoted by the predictive coding framework, it is possible that future studies will provide translation between animal and human studies of placebo effects. Also, the redundancy of conscious awareness in placebo effects and focus on lower-order neural pathways may improve translational research since human consciousness and animal cognition are an inherent paradox in behavioral science.

REFERENCES

American Psychiatric Association. (2013). *Diagnostic and statistical manual of mental disorders* (5th ed.). . Arlington, VA: American Psychiatric Association.

Atlas, L. Y., & Wager, T. D. (2012). How expectations shape pain. *Neuroscience Letters, 520*(2), 140–148.

Bargh, J. A., Chen, M., & Burrows, L. (1996). Automaticity of social behavior: Direct effects of trait construct and stereotype-activation on action. *Journal of Personality and Social Psychology, 71*(2), 230–244.

Benedetti, F., Arduino, C., Costa, S., Vighetti, S., Tarenzi, L., Rainero, I., et al. (2006). Loss of expectation-related mechanisms in Alzheimer's disease makes analgesic therapies less effective. *Pain, 121*(1–2), 133–144

Benedetti, F., Frisaldi, E., Carlino, E., Giudetti, L., Pampallona, A., Zibetti, M., et al. (2016) Teaching neurons to respond to placebos. *The Journal of Physiology, 594*, 5647–5660.

Bijleveld, E., Custers, R., & Aarts, H. (2010). Unconscious reward cues increase invested effort, but do not change speed-accuracy tradeoffs. *Cognition, 115*(2), 330–335.

Branthwaite, A., & Cooper, P. (1981). Analgesic effects of branding in treatment of headaches. *British Medical Journal (Clinical Research Ed.), 282*(6276), 1576–1578.

Buchel, C., Geuter, S., Sprenger, C., & Eippert, F. (2014). Placebo analgesia: A predictive coding perspective. *Neuron, 81*(6), 1223–1239.

Buhle, J. T., Stevens, B. L., Friedman, J. J., & Wager, T. D. (2012). Distraction and placebo: Two separate routes to pain control. *Psychological Science, 23*(3), 246–253.

Buhle, J., & Wager, T. D. (2010). Performance-dependent inhibition of pain by an executive working memory task. *Pain, 149*(1), 19–26.

Bunce, S. C., Bernat, E., Wong, P. S., & Shevrin, H. (1999). Further evidence for unconscious learning: Preliminary support for the conditioning of facial EMG to subliminal stimuli. *Journal of Psychiatric Research, 33*, 341–347.

Carlsson, K., Petersson, K. M., Lundqvist, D., Karlsson, A., Ingvar, M., & Öhman, A. (2004). Fear and the amygdala: Manipulation of awareness generates differential cerebral responses to phobic and fear-relevant (but nonfeared) stimuli. *Emotion, 4*(4), 340–353.

Clark, A. (2013). Whatever next? Predictive brains, situated agents, and the future of cognitive science. *The Behavioral and Brain Sciences, 36*(3), 181–204.

Clark, R. E., Manns, J. R., & Squire, L. R. (2002). Classical conditioning, awareness, and brain systems. *Trends in Cognitive Sciences, 6*(12), 524–530.

Colloca, L., & Benedetti, F. (2006). How prior experience shapes placebo analgesia. *Pain, 124*(1–2), 126–133.

Colloca, L., & Miller, F. G. (2011). How placebo responses are formed: A learning perspective. *Philosophical Transactions of the Royal Society of London. Series B, Biological Sciences, 366*, 1859–1869.

Curie, A., Yang, K., Kirsch, I., Gollub, R., des Portes, V., Kaptchuk, T., et al. (2015). Placebo responses in genetically determined intellectual disability: A meta-analysis. *PLoS One, 10*(7), e0133316.

Custers, R., & Aarts, H. (2010). The unconscious will: How the pursuit of goals operates outside of conscious awareness. *Science, 329*, 47–50.

de Craen, A. J., Roos, P. J., de Vries, A. L., & Kleijnen, J. (1996). Effect of colour of drugs: Systematic review of perceived effect of drugs and of their effectiveness. *BMJ, 313*(7072), 1624–1626.

de la Fuente-Fernandez, R., Ruth, T., Sossi, V., Schulzer, M., Calne, D., & Stoessl, A. (2001). Expectation and dopamine release: Mechanism of the placebo effect in Parkinson's disease. *Science, 293*, 1164–1166.

Dijksterhuis, A., Aarts, H., & Smith, P. K. (2005). *The power of the subliminal: On subliminal persuation and other potential applications. The new unconscious—Social cognition and social neuroscience.* New York: Oxford University Press.

Doyen, S., Klein, O., Pichon, C. L., & Cleeremans, A. (2012). Behavioral priming: It's all in the mind, but whose mind? *PLoS One*, 7(1), e29081.

Eippert, F., Bingel, U., Schoell, E. D., Yacubian, J., Klinger, R., Lorenz, J., et al. (2009). Activation of the opioidergic descending pain control system underlies placebo analgesia. *Neuron, 63*(4), 533–543.

Esteves, F., Parra, C., Dimberg, U., & Ohman, A. (1994). Nonconscious associative learning: Pavlovian conditioning of skin conductance responses to masked fear-relevant facial stimuli. *Psychophysiology, 31*(4), 375–385.

Faria, V., Fredrikson, M., & Furmark, T. (2008). Imaging the placebo response: A neurofunctional review. *European Neuropsychopharmacology, 18*(7), 473–485.

Friston, K. (2009). The free-energy principle: A rough guide to the brain? *Trends in Cognitive Sciences, 13*(7), 293–301.

Geuter, S., Eippert, F., Hindi Attar, C., & Buchel, C. (2013). Cortical and subcortical responses to high and low effective placebo treatments. *NeuroImage, 67*, 227–236.

Gowaty, P. A., & Hubbell, S. P. (2013). Bayesian animals sense ecological constraints to predict fitness and organize individually flexible reproductive decisions. *The Behavioral and Brain Sciences, 36*(3), 215–216.

Gracely, R., Dubner, R., Deeter, W., & Wolskee, P. (1985). Clinicians' expectations influence placebo analgesia. *Lancet, 1*(8419), 43.

Grelotti, D. J., & Kaptchuk, T. J. (2011). Placebo by proxy. *BMJ, 343*, d4345.

Grochowicz, P. M., Schedlowski, M., Husband, A. J., King, M. G., Hibberd, A. D., & Bowen, K. M. (1991). Behavioral conditioning prolongs heart allograft survival in rats. *Brain, Behavior, and Immunity, 5*(4), 349–356.

Hashmi, J. A., Baria, A. T., Baliki, M. N., Huang, L., Schnitzer, T. J., & Apkarian, A. V. (2012). Brain networks predicting placebo analgesia in a clinical trial for chronic back pain. *Pain, 153*(12), 2393–2402.

Haug, M. (2011). Explaining the placebo effect: Aliefs, beliefs, and conditioning. *Philosophical Psychology, 1*(20), 1–20.

Humphrey, N. (2002). *The mind made flesh: Essays from the frontiers of evolution and psychology*. chapter 19 (pp. 255–285). Oxford University Press.

Jensen, K. B., Kaptchuk, T. J., Chen, X., Kirsch, I., Ingvar, M., Gollub, R. L., et al. (2015). A neural mechanism for nonconscious activation of conditioned placebo and nocebo responses. *Cerebral Cortex, 25*(10), 3903–3910.

Jensen, K. B., Kaptchuk, T. J., Kirsch, I., Raicek, J., Lindstrom, K. M., Berna, C., et al. (2012). Nonconscious activation of placebo and nocebo pain responses. *Proceedings of the National Academy of Sciences of the United States of America, 109*(39), 15959–15964.

Jensen, K. B., Kirsch, I., Odmalm, S., Kaptchuk, T. J., & Ingvar, M. (2015). Classical conditioning of analgesic and hyperalgesic pain responses without conscious awareness. *Proceedings of the National Academy of Sciences of the United States of America, 112*(25), 7863–7867.

Jensen, K. B., Kirsch, I., Ponten, M., Rosen, A., Yang, K., Gollub, R. L., et al. (2017). Certainty of genuine treatment increases drug responses among intellectually disabled patients. *Neurology, 88*(20), 1912–1918.

Kaptchuk, T. (2011). Placebo studies and ritual theory: A comparative analysis of Navajo, acupuncture and biomedical healing. *Philosophical Transactions of the Royal Society B, 366*, 1849–1858.

Kaptchuk, T. J., Kelley, J. M., Conboy, L. A., Davis, R. B., Kerr, C. E., Jacobson, E. E., et al. (2008). Components of placebo effect: Randomised controlled trial in patients with irritable bowel syndrome. *BMJ, 336*(7651), 999–1003.

Kelley, J. M., Kraft-Todd, G., Schapira, L., Kossowsky, J., & Riess, H. (2014). The influence of the patient-clinician relationship on healthcare outcomes: A systematic review and meta-analysis of randomized controlled trials. *PLoS One, 9*(4), e94207.

Kelley, J. M., Lembo, A. J., Ablon, J. S., Villanueva, J. J., Conboy, L. A., Levy, R., et al. (2009). Patient and practitioner influences on the placebo effect in irritable bowel syndrome. *Psychosomatic Medicine, 71*, 789–797. https://doi.org/10.1097/PSY.0b013e3181acee12.

Kirsch, I. (2004). Conditioning, expectancy, and the placebo effect: Comment on Stewart-Williams and Podd (2004). *Psychological Bulletin, 130*(2), 341–343. [discussion 344–345].

Kirsch, I. (2013). The placebo effect revisited: Lessons learned to date. *Complementary Therapies in Medicine, 21*(2), 102–104.

Klein, R. A., Ratliff, K. A., Vianello, M., Adams, R. B., Jr., Bahník, S., Bernstein, M. J., et al. (2014). Investigating variation in replicability: A "many labs" replication project. *Social Psychology, 45*, 142–152.

Knight, D. C., Nguyen, H. T., & Bandettini, P. A. (2003). Expression of conditional fear with and without awareness. *Proceedings of the National Academy of Sciences of the United States of America, 100*(25), 15280–15283.

Krummenacher, P., Candia, V., Folkers, G., Schedlowski, M., & Schönbächler, G. (2010). Prefrontal cortex modulates placebo analgesia. *Pain, 148*(3), 368–374.

LeDoux, J. (1990). *The emotional brain*. New York: Simon & Schuster.

LeDoux, J. E. (2000). Emotion circuits in the brain. *Annual Review of Neuroscience, 23*, 155–184.

Leknes, S., Lee, M., Berna, C., Andersson, J., & Tracey, I. (2011). Relief as a reward: Hedonic and neural responses to safety from pain. *PLoS One*, *6*(4), e17870.
Mayberg, H. S., Silva, J. A., Brannan, S. K., Tekell, J. L., Mahurin, R. K., McGinnis, S., et al. (2002). The functional neuroanatomy of the placebo effect. *The American Journal of Psychiatry*, *159*(5), 728–737.
Messer, S. B., & Wampold, B. E. (2002). Let's face facts: Common factors are more potent than specific therapy ingredients. *Clinical Psychologist*, *9*, 21–25.
Mesulam, M. M. (1998). From sensation to cognition. *Brain: A Journal of Neurology*, *121*(Pt. 6), 1013–1052.
Moerman, D. E., & Jonas, W. B. (2002). Deconstructing the placebo effect and finding the meaning response. *Annals of Internal Medicine*, *136*(6), 471–476.
Montgomery, G. H., & Kirsch, I. (1997). Classical conditioning and the placebo effect. *Pain*, *72*(1–2), 107–113.
Munana, K. R., Zhang, D., & Patterson, E. E. (2010). Placebo effect in canine epilepsy trials. *Journal of Veterinary Internal Medicine*, *24*(1), 166–170.
Ohman, A., Carlsson, K., Lundqvist, D., & Ingvar, M. (2007). On the unconscious subcortical origin of human fear. *Physiology & Behavior*, *92*(1–2), 180–185.
Pecina, M., Bohnert, A. S., Sikora, M., Avery, E. T., Langenecker, S. A., Mickey, B. J., et al. (2015). Association between placebo-activated neural systems and antidepressant responses: Neurochemistry of placebo effects in major depression. *JAMA Psychiatry*, *72*(11), 1087–1094.
Pecina, M., & Zubieta, J. K. (2015). Molecular mechanisms of placebo responses in humans. *Molecular Psychiatry*, *20*(4), 416–423.
Peirce, C., & Jastrow, J. (1885). On small differences in sensation. *Memoirs of the National Academy of Sciences*, *3*, 73–83.
Pessiglione, M., Petrovic, P., Daunizeau, J., Palminteri, S., Dolan, R., & Frith, C. (2008). Subliminal instrumental conditioning demonstrated in the human brain. *Neuron*, *59*(4), 561–567.
Petrovic, P., Kalso, E., Petersson, K. M., Andersson, J., Fransson, P., & Ingvar, M. (2010). A prefrontal non-opioid mechanism in placebo analgesia. *Pain*, *150*(1), 59–65.
Petrovic, P., Kalso, E., Petersson, K. M., & Ingvar, M. (2002). Placebo and opioid analgesia—Imaging a shared neuronal network. *Science*, *295*(5560), 1737–1740.
Ploghaus, A., Tracey, I., Gati, J. S., Clare, S., Menon, R. S., Matthews, P. M., et al. (1999). Dissociating pain from its anticipation in the human brain. *Science*, *284*(5422), 1979–1981.
Rainville, P., Hofbauer, R. K., Paus, T., Duncan, G. H., Bushnell, M. C., & Price, D. D. (1999). Cerebral mechanisms of hypnotic induction and suggestion. *Journal of Cognitive Neuroscience*, *11*(1), 110–125.
Raio, C. M., Carmel, D., Carrasco, M., & Phelps, E. A. (2012). Nonconscious fear is quickly acquired but swiftly forgotten. *Current Biology*, *22*(12), R477–479.
Rebar, A. L., Loftus, A. M., & Hagger, M. S. (2015). Cognitive control and the nonconscious regulation of health behavior. *Frontiers in Human Neuroscience*, *9*, 122.
Richter, M., Schroeter, C., Puensch, T., Straube, T., Hecht, H., Ritter, A., et al. (2014). Pain-related and negative semantic priming enhances perceived pain intensity. *Pain Research & Management*, *19*(2), 69–74.
Rohrer, D., Pashler, H., & Harris, C. R. (2015). Do subtle reminders of money change people's political views? *Journal of Experimental Psychology. General*, *144*(4), e73–85.
Rosen, A., Yi, J., Kirsch, I., Kaptchuk, T. J., Ingvar, M., & Jensen, K. B. (2017). Effects of subtle cognitive manipulations on placebo analgesia—An implicit priming study. *European Journal of Pain*, *21*, 594–604.
Sand, A., & Nilsson, M. E. (2016). Subliminal or not? Comparing null-hypothesis and Bayesian methods for testing subliminal priming. *Consciousness and Cognition*, *44*, 29–40.

Schmidt-Wilcke, T., Ichesco, E., Hampson, J. P., Kairys, A., Peltier, S., Harte, S., et al. (2014). Resting state connectivity correlates with drug and placebo response in fibromyalgia patients. *NeuroImage. Clinical, 6*, 252–261.

Scott, D. J., Stohler, C. S., Egnatuk, C. M., Wang, H., Koeppe, R. A., & Zubieta, J.-K. (2009). Individual differences in reward responding explain placebo-induced expectations and effects. *Neuron, 55*, 325–336.

Sheeran, P., Gollwitzer, P. M., & Bargh, J. A. (2013). Nonconscious processes and health. *Health Psychology, 32*(5), 460–473.

Simmons, K., Ortiz, R., Kossowsky, J., Krummenacher, P., Grillon, C., Pine, D., et al. (2014). Pain and placebo in pediatrics: A comprehensive review of laboratory and clinical findings. *Pain, 155*(11), 2229–2235.

Stewart-Williams, S., & Podd, J. (2004). The placebo effect: Dissolving the expectancy versus conditioning debate. *Psychological Bulletin, 130*(2), 324–340.

Strahan, E. J., Spencer, S. J., & Zanna, M. P. (2002). Subliminal priming and persuasion: Striking while the iron is hot. *Journal of Experimental Social Psychology, 38*(1), 556–558.

Tetreault, P., Mansour, A., Vachon-Presseau, E., Schnitzer, T. J., Apkarian, A. V., & Baliki, M. N. (2016). Brain connectivity predicts placebo response across chronic pain clinical trials. *PLoS Biology, 14*(10), e1002570.

Thorpe, S. J., Rolls, E. T., & Maddison, S. (1983). The orbitofrontal cortex: Neuronal activity in the behaving monkey. *Experimental Brain Research, 49*(1), 93–115.

Tulving, E., & Schacter, D. L. (1990). Priming and human memory systems. *Science, 247*(4940), 301–306.

Waber, R. L., Shiv, B., Carmon, Z., & Ariely, D. (2008). Commercial features of placebo and therapeutic efficacy. *JAMA, 299*(9), 1016–1017.

Wager, T. D., Rilling, J. K., Smith, E. E., Sokolik, A., Casey, K. L., Davidson, R. J., et al. (2004). Placebo-induced changes in fMRI in the anticipation and experience of pain. *Science, 303*(5661), 1162–1167.

Weimer, K., Gulewitsch, M. D., Schlarb, A. A., Schwille-Kiuntke, J., Klosterhalfen, S., & Enck, P. (2013). Placebo effects in children: A review. *Pediatric Research, 74*(1), 96–102.

FURTHER READING

Petrovic, P., & Ingvar, M. (2002). Imaging cognitive modulation of pain processing. *Pain, 95*(1–2), 1–5.

CHAPTER ELEVEN

Critical Life Functions: Can Placebo Replace Oxygen?

Fabrizio Benedetti[*,†,1], **Diletta Barbiani**[*,†], **Eleonora Camerone**[‡]
[*]University of Turin Medical School, Turin, Italy
[†]Plateau Rosà Laboratories, Plateau Rosà, Italy/Switzerland
[‡]Institute of Psychiatry, Psychology & Neuroscience, King's College, London, United Kingdom
[1]Corresponding author: e-mail address: fabrizio.benedetti@unito.it

Contents

Abstract

A crucial question in placebo research is related to which conditions and physiological functions are affected by placebos. Here we present evidence that critical life functions, like ventilation, oxygenation, circulation, and perfusion, can be sensitive to placebo treatments in some circumstances. Indeed, we have investigated the role of placebo effects at an altitude of 3500 m, where oxygen pressure is 64% compared to the sea level. In these extreme conditions, hypoxia triggers several compensatory responses, such as hyperventilation, increased cardiac output, and increased brain perfusion. A conditioned placebo procedure was found to mimic the effects of oxygen on these compensatory responses, and these effects are still present at altitudes as high as 4500 and 5500 m, where oxygen pressure is only 57% and 50%, respectively, compared to the sea level. Thus, placebo effects also take place for those functions that are critical for life and whereby oxygen is the key element.

International Review of Neurobiology, Volume 138
ISSN 0074-7742
https://doi.org/10.1016/bs.irn.2018.01.009

1. INTRODUCTION

One of the most important challenges of placebo research is to understand where, when, and how placebos work. "Where" means in which medical conditions and therapeutic interventions, "When" means in which circumstance, e.g., in the hospital or at home, and "How" refers to the psychobiological underpinnings. Possibly, we want to also understand "why" placebo effects exist at all, which refers to their emergence and significance during the course of evolution. As to the first question "where," so far the most studied conditions have been pain, motor disorders, especially Parkinson's disease, behavioral conditioning of the immune and endocrine system, some psychiatric conditions such as depression and anxiety, as well as physical performance in sport (Benedetti, 2014).

Overall, what has emerged over the past years is that, not surprisingly, the magnitude of placebo effects varies across different conditions, from small to large. In general, although no detailed study has been done so far, it would not be surprising to find no placebo effect for infectious diseases, whereby a placebo (fake) antibiotic is unlikely to kill bacteria. Likewise, it is unlikely that a placebo (fake) anesthetic gas would induce general anesthesia. Therefore, there are surely a range of conditions from no placebo effect at all to large effects, and even in conditions, where placebo effects are very likely to occur, there still is a huge variability in their magnitude.

As to the question "how," it is now clear that placebos activate mechanisms similar to those activated by drugs. For example, there is compelling experimental evidence that placebos may modulate the same receptors that are modulated by narcotics and cannabinoids (Benedetti, Amanzio, Rosato, & Blanchard, 2011; Petrovic, Kalso, Petersson, & Ingvar, 2002). Similarly, placebos can activate dopamine receptors, which are the very same activated by dopaminergic anti-Parkinson drugs (De la Fuente-Fernández et al., 2002, 2001), and can modulate the cyclooxygenase (COX) pathway in the same way as nonsteroid antiinflammatory drugs (Benedetti & Dogue, 2015; Benedetti, Durando, Giudetti, Pampallona, & Vighetti, 2015; Benedetti, Durando, & Vighetti, 2014).

On the basis of these considerations, physiological functions that are critical for life, such as ventilation, are particularly interesting for several reasons. In fact, they give rise to challenging and provocative questions such as: Can we breathe without oxygen, by merely replacing oxygen with a placebo? The answer is straightforward: No. However, we can reframe this question in the following way: Can oxygen be partially replaced by a placebo? Now the answer is: Yes. Indeed, the aim of this chapter is to present experimental evidence that this is really possible, namely, that oxygen can be partially replaced by a placebo (fake oxygen) with little changes, if any, of the overall functioning of the body.

2. CRITICAL LIFE FUNCTIONS: WHAT THEY ARE AND HOW THEY CAN BE STUDIED

Ventilation, oxygenation, circulation, and perfusion are considered the four most important critical life functions. Oxygen is at the core of these functions. It goes from atmospheric air to the lungs, and it is exchanged with carbon dioxide in the lungs and spreads to the blood stream, to then reach all the peripheral tissues and cells, e.g., the brain and neurons (Fig. 1). When

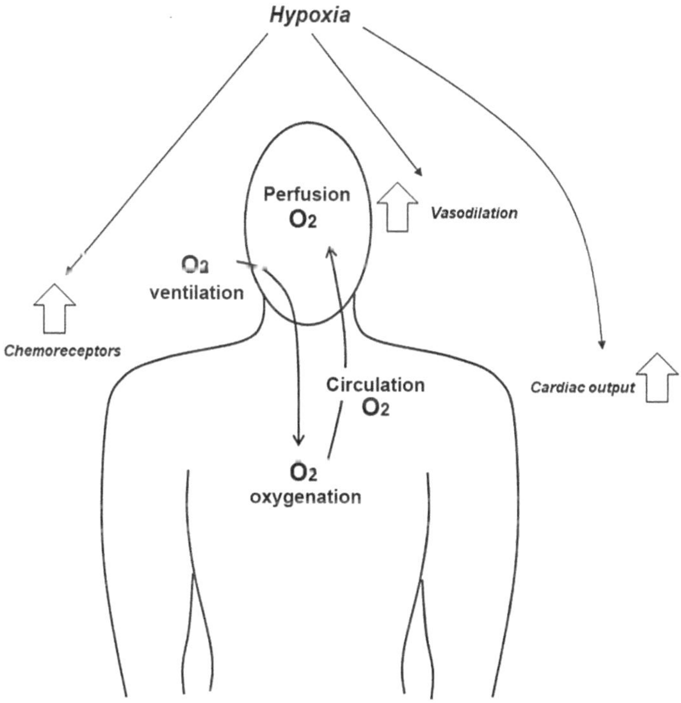

Fig. 1 The four critical life functions (ventilation, oxygenation, circulation, and perfusion) and the effects of hypoxia. In particular, hypoxia triggers several compensatory responses, for example, hyperventilation through the chemoreceptors, increased cardiac output through increased heart rate, and increased brain perfusion through vasodilation.

little oxygen is available in the air, a condition called hypoxia, several body functions counterbalance the oxygen shortage by triggering at least three fundamental compensatory responses: increase in ventilation (hyperventilation) through the activation of chemoreceptors, increase in circulation through increased cardiac output, e.g., heart rate increase, and increase in perfusion through vasodilation, e.g., cerebral vasodilation, whereby prostaglandins (PG) such as PGE2 have been found to be involved (Benedetti & Dogue, 2015) (Fig. 1).

A crucial question is to understand whether placebo effects have any role in these functions, that is, whether placebo effects occur for any physiological function in which oxygen is involved. In other words, we return to the previous question: Can we replace oxygen with a placebo? Or else, can a placebo procedure mimic the effects of oxygen in conditions where the latter runs low? It goes without saying that this poses several ethical problems in the clinical setting, because we would need to study patients in hypoxic conditions, namely, with low blood oxygen saturation (SO_2). For example, replacing even a tiny amount of oxygen with a placebo in a patient with either respiratory failure or admitted to an intensive care unit would certainly be unethical, and investigating the effects of placebo treatments on SO_2 or impaired ventilation would be unfeasible at the bedside. Therefore, what is needed in order to study placebo effects in critical life functions, such as ventilation and oxygenation, is a model that is both ethical and amenable to scientific investigation.

3. THE HIGH ALTITUDE MODEL OF OXYGEN SHORTAGE

A natural model of hypoxia is represented by high altitude: as altitude increases, oxygen decreases. What is important is not so much oxygen concentration but rather oxygen partial pressure. As altitude increases, oxygen pressure decreases. For example, oxygen pressure at sea level is 159 mmHg, whereas it drops to 102 mmHg at 3500 m. Mountain medicine indicates three altitude regions that reflect the lowered amount of oxygen in the atmosphere: high altitude (1500–3500 m), very high altitude (3500–5500 m), and extreme altitude (above 5500 m) (International Society for Mountain Medicine, 2017). On the basis of the way our body responds to oxygen shortage, high altitude can be subdivided into four zones of adaptation or acclimatization (Schmidt & Thews, 1989) (Fig. 2), although several other subdivisions do exist. The zone from the sea level to about 1800–2000 m is called "indifferent," as there are no appreciable effects on the body, at least in

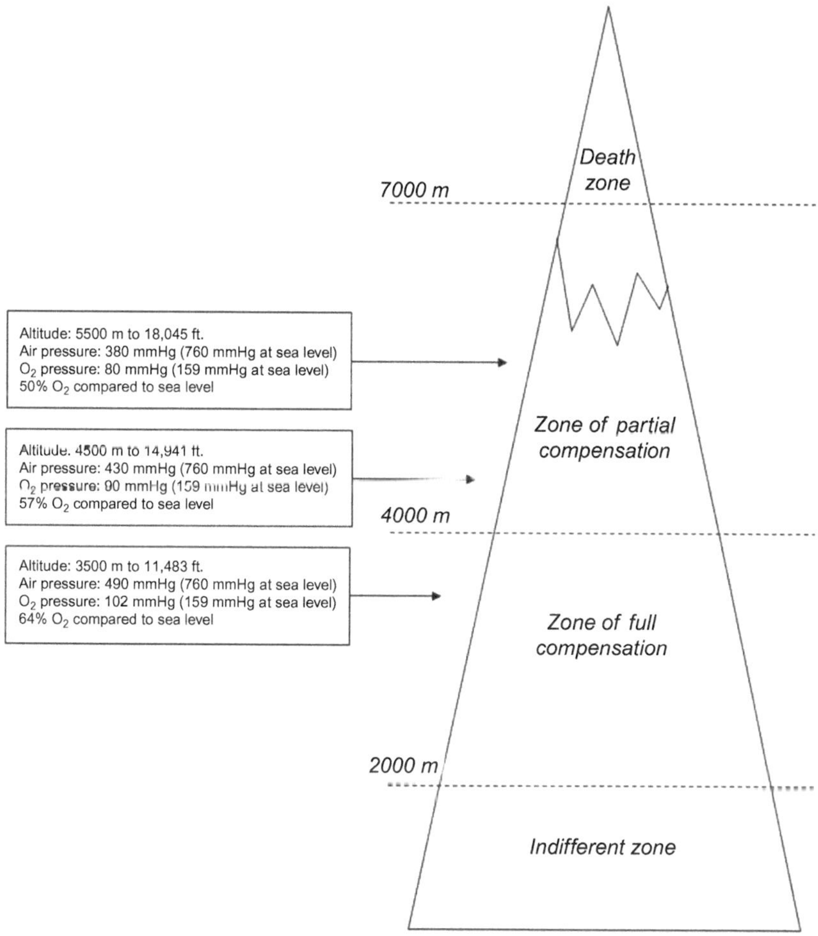

Fig. 2 General subdivision of different altitude zones, and the altitudes where we have data about placebo effects.

most people. The zone from 2000 m to about 4000 m is called "full compensation zone," as many physiological responses compensate the low oxygen pressure, e.g., hyperventilation, increased heart rate, and cerebral vasodilation. The zone from 4000 m to about 7000 m is called "partial compensation zone," as many physiological responses are not capable of compensating for oxygen shortage. Here, many severe symptoms can develop, such as pulmonary edema and cerebral edema, which can result in severe permanent damage and even death. The zone above about 7000 m is called "death zone": most people die at this altitude, as the amount of oxygen is insufficient to sustain human life and the body is incapable of acclimatizing. The concept of the *death zone* (originally the *lethal zone*) was first conceived in 1953 by Edouard Wyss-Dunant, a Swiss doctor, in an article about acclimatization (Wyss-Dunant, 1953). Although this is a very general subdivision and description (Fig. 2), we believe it gives a general idea of the high altitude model of hypoxia.

Most of the studies on the placebo effects at high altitude have been performed in the zone of full compensation, where oxygen availability is 64% compared to the sea level (Fig. 2). The reason for that is simple: we need a degree of hypoxia in which several physiological compensatory responses develop, yet they are not threatening for life. This is made possible by one of the locations where we work, namely, the Center for Hypoxia at the Plateau Rosa Laboratories, located at an altitude of 3500 m in the Matterhorn area at the border between Italy and Switzerland. Only a couple of studies have been performed in the zone of partial compensation, at altitudes of 4500 and 5500 m, where oxygen availability is 57% and 50%, respectively, compared to the sea level (Benedetti, 2017) (Fig. 2). The studies at 4500 m have been performed at the Regina Margherita Hut on the Monte Rosa in Italy, whereas the studies at 5500 m have been done at the Simon Bolivar and Cristobal Colon twin peaks in Colombia and at the Denali (former McKinley) mountain in Alaska. The main problem with these very high altitudes is the small number of subjects who can participate in the study, which is justified by the inherent difficulty of bringing healthy volunteers to high altitude. For example, only six subjects were tested at the Regina Margherita Hut, five at the Denali, and four at the Colombian twin peaks. Although these findings will require further investigation with larger groups, they nonetheless have provided some important information on the placebo response in severe hypoxic conditions, as discussed later.

The clinical condition triggered by hypobaric (low oxygen pressure) hypoxia at high altitude is known as acute mountain sickness (AMS) (Imray, Wright, Subudhi, & Roach, 2010; Wilson, Newman, & Imray, 2009), which is usually diagnosed by means of the Lake Louise Score questionnaire (Sutton, Coates, Houston, & Oelz, 1992). This is aimed at detecting several symptoms, such as headache, nausea/vomiting, dizziness, insomnia, and neurological symptoms, which emphasize the complex nature of this hypoxia-related clinical syndrome. Headache is the cardinal symptom and indeed the most common symptom even at relatively low altitudes. Headache is also one of the most studied symptoms of AMS, although the details of its pathophysiology are partly unknown. As far as we know, there are at least two pathways triggering high altitude headache. The first is represented by the acute effects of hypoxia on PG synthesis through the COX enzyme, with the formation of PGD2, PGF2, PGE2, PGI2 (prostacyclin), and thromboxane (TX) A2 (Benedetti et al., 2014; Richalet et al., 1991). One of the most important effects of these eicosanoids, particularly PGE2, is represented by vasodilation, which is thought to be the principal factor inducing acute hypoxia headache (Busse, Fosterman, Matsuda, & Pohl, 1984; Davis et al., 2004; Fredericks, Liu, Rusch, & Lombard, 1994; Messina, Sun, Koller, Wolin, & Kaley, 1992; Ray, Abbas, Coney, & Marshall, 2002), although the direct stimulation of nociceptive afferents cannot be ruled out (Kawabata, 2011). However, a second pathway that is important for high altitude headache is represented by the hypoxia-related hyperventilation that, in turn, induces the excessive elimination of carbon dioxide (CO_2) with the consequent increase in blood pH (alkalosis) (West, 2004). To support the important role of alkalosis in AMS and high altitude headache is the therapeutic effect of blood pH reduction by means of acetazolamide (Leaf & Goldfarb, 2007). Besides the reduction of alkalosis with acetazolamide, there are at least two other effective treatments for high altitude headache: oxygen inhalation (Bartsch, Baumgartner, Waber, Maggiorini, & Oelz, 1990; Benedetti et al., 2015) and oral COX inhibitors, such as aspirin (Benedetti et al., 2014; Burtscher, Likar, Nachbauer, & Philadelphy, 1998). Whereas the former restores blood oxygen saturation (SO_2), thus decreasing hyperventilation and alkalosis, the latter inhibits the hypoxia-activated COX and PG synthesis, thus reducing cerebral vasodilation (Benedetti & Dogue, 2015).

It can be seen from this very brief description of the effects of high altitude on the body that physiological compensatory responses to hypobaric hypoxia can be approached by focusing on different parameters, ranging from ventilation to heart rate and from PG-mediated effects to AMS. The following section is a general overview of the effects of placebos on these physiological responses to hypobaric hypoxia at high altitude. They represent an excellent model to understand how placebos may influence critical life functions at the core of which oxygen represents the crucial element.

4. PLACEBO EFFECTS AT HIGH ALTITUDE: CAN OXYGEN BE REPLACED BY PLACEBO?

As already stated in Section 1, the question "Can we breathe without oxygen by replacing it with a placebo?" can be reframed with the question "Can oxygen be partially replaced with a placebo?". In fact, whereas the quite obvious answer to the first question is "no," the answer to the second question is "yes." Indeed, there is today experimental evidence that, by using the high altitude model of hypoxia, placebos can mimic the effects of oxygen for oxygen reductions up to 50% compared to the sea level. As anticipated in the previous section, most of the evidence comes from studies at 3500 m (64% oxygen compared to the sea level), 4500 m (57% oxygen), and 5500 m (50% oxygen) (Benedetti, 2017), as well as from high altitude simulations in the lab at the sea level (Torres-Peralta et al., 2016). These studies show that placebos can affect critical life functions like ventilation, circulation, perfusion, as well as overall performance, but not oxygenation.

4.1 Oxygenation

In no case a change in SO_2 was found after placebo administration (see the studies described below), which indicates that placebos cannot in any way affect blood oxygenation. This is not surprising after all, as blood oxygenation is due to the direct diffusion of oxygen from the lungs to the blood stream, without any involvement of the central nervous system. Indeed, it is unlikely that a placebo can change the way oxygen diffuses from air to blood. As it will be discussed in detail later, this is one of the most interesting findings of all these studies, because it shows that placebos may affect many body functions without any change in oxygen content of the body.

4.2 Ventilation

The first evidence that ventilation can be affected by placebos comes from early studies with narcotics (Benedetti, Amanzio, Baldi, Casadio, & Maggi, 1999). In fact, placebo respiratory depression can be induced after repeated administrations of the partial opioid agonist buprenorphine. The placebo respiratory depressant effect that results from the buprenorphine conditioning was found to be completely blocked by a dose of 10 mg of naloxone, indicating that it was mediated by endogenous opioids. Therefore, placebos act, via the activation of opioid receptors, not only on pain mechanisms (Amanzio & Benedetti, 1999) but on the respiratory centers as well, thus mimicking a typical side effect of narcotics. The procedure used by Benedetti et al. (1999) did not produce any expectation of respiratory depression and, similarly, the subjects did not notice any sign of respiratory discomfort; thus the placebo respiratory depression cannot be explained on the basis of cognitive or motivational mechanisms. Rather, it appears to be a sequence effect due to learning, thus suggesting a conditioning mechanism mediated by endogenous opioids.

A similar hypoventilation effect can be elicited by a placebo after oxygen preconditioning. In fact, the typical compensatory hyperventilation of high altitude can be inhibited by administering oxygen and this can be mimicked by placebo. It can be seen in Fig. 3 that, whereas placebo given for the first time does not produce any effect on minute ventilation (V_{min}), expressed as L/min, a placebo after prior oxygen exposure for *n* times induces substantial changes of V_{min}. This effect is present with 64% oxygen in our laboratories at 3500 m (Benedetti & Dogue, 2015), as well as with 57% oxygen at the Regina Margherita Hut and 50% oxygen at the Denali and Colombian twin peaks (Benedetti, 2017). Two important points need to be emphasized for ventilation. First, as for narcotic preconditioning (Benedetti et al., 1999), oxygen preconditioning is crucial for a placebo respiratory effect to occur (Benedetti & Dogue, 2015). Therefore, the placebo respiratory effect is basically a conditioned hypoventilation response. Second, this effect on ventilation can be seen with an oxygen reduction up to 50%.

The effects on ventilation by placebo are quite important, as they are not limited to ventilation itself, but extend to blood pH. In fact, hyperventilation at high altitude is accompanied by an increase in blood pH (alkalosis). Oxygen decreases both hyperventilation and blood pH. Likewise, placebo after oxygen preconditioning can produce the very same effect, namely, a decrease in both hyperventilation and blood pH (Fig. 3). This effect is present even at 5500 m, where oxygen is 50% compared to the sea level. Therefore, placebos may affect both ventilation and its consequences on blood pH, thus suggesting that other blood changes can take place during conditioned placebo effects. This is certainly a challenge for future placebo research.

It is crucial to note in Fig. 3 that, not surprisingly, SO_2 increases after oxygen administration. Conversely, no change in SO_2 occurs after placebo administration, which indicates that hypoventilation and pH responses are not due to SO_2 per se, but rather to a SO_2-independent learning mechanism.

4.3 Circulation

Conditioned bradycardia has long been known and has been found to involve the endogenous opioid systems, as pharmacological manipulation with opioid antagonists, like naloxone, impairs learning of classically conditioned heart rate changes (Harris & Fitzgerald, 1989; Hernandez & Powell, 1983; Hernandez, Powell, & Gibbs, 1990; Hernandez & Watson, 1997).

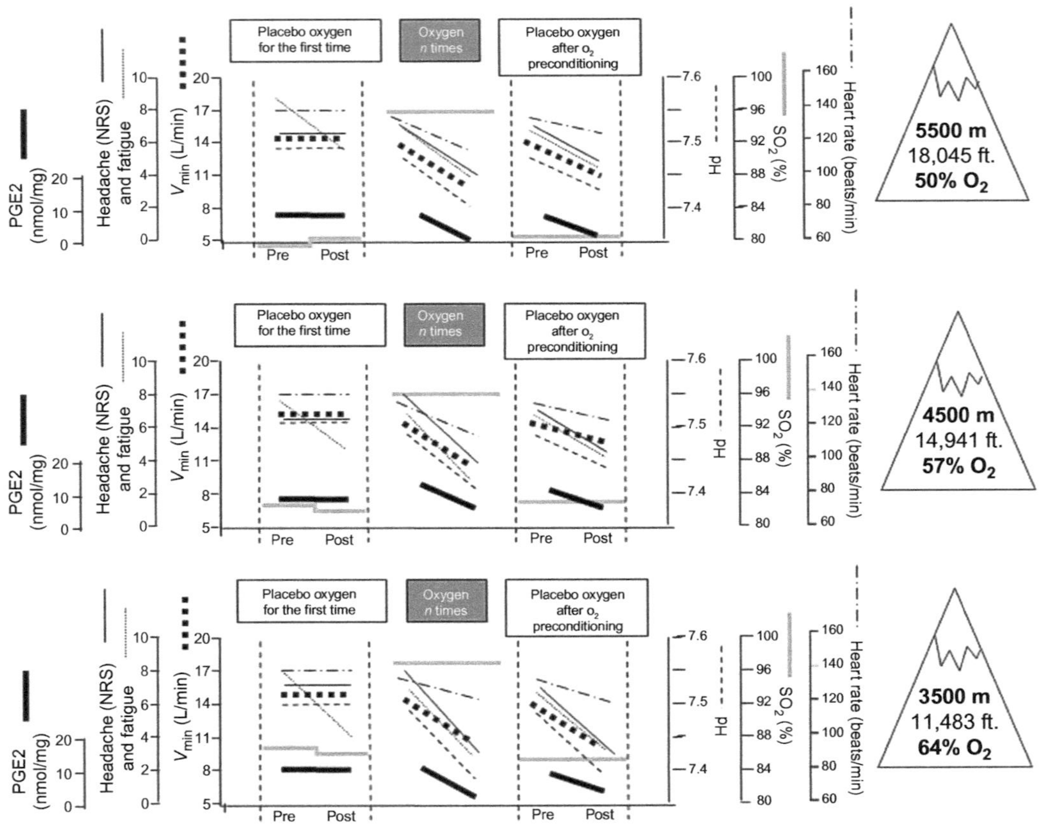

Fig. 3 The effects of placebo given for the first time, of oxygen, and of placebo after oxygen preconditioning on PGE2, headache, fatigue, minute ventilation, blood pH, SO_2, and heart rate at different altitudes, from 3500 m, where oxygen is 64% compared to the sea level to 4500 m, where oxygen is 57% and to 5500 m, where oxygen is 50%.

The first study on placebo and conditioned bradycardia was done in subjects who underwent induction of experimental pain, along with the suggestion that it was a painkiller (Pollo, Vighetti, Rainero, & Benedetti, 2003). Besides the assessment of the analgesic effect, both heart rate and heart rate variability were measured. It was found that the opioid antagonist naloxone completely antagonized both placebo analgesia and the concomitant reduced heart rate, whereas the beta-blocker propranolol antagonized the placebo heart rate reduction but not placebo analgesia. By contrast, both placebo effects were present during muscarinic blockade with atropine, indicating no involvement of the parasympathetic system.

As for ventilation at high altitude, a similar bradycardic effect can be elicited by a placebo after oxygen preconditioning. In fact, the typical compensatory tachycardia of high altitude can be inhibited by administering oxygen and this effect can be mimicked by the placebo. In Fig. 3, it can be seen that, whereas a placebo given for the first time does not produce any effect on heart rate, expressed as beats/min, a placebo given after prior oxygen exposure for *n* times induces substantial changes in heart rate. This effect is present with 64%, 57%, and 50% oxygen compared to the sea level. Therefore, as for ventilation, oxygen preconditioning is crucial for a placebo bradycardic response to occur (Benedetti & Dogue, 2015). We can thus consider the placebo bradycardic effect as a conditioned bradycardic response. Most interestingly, this effect on heart activity can be seen with an oxygen reduction up to 50%.

Again, note in Fig. 3 that whereas SO_2 increases after oxygen administration, no change in SO_2 occurs after placebo administration, which indicates that bradycardia is not due to SO_2 per se, but rather to a SO_2-independent learning mechanism.

4.4 Perfusion

The experimental evidence that placebos may affect cerebral perfusion through vasodilation is indirect, as no direct measurement of cerebral blood flow has been performed so far. Evidence comes from the assessment of some biomarkers that are known to be involved in cerebral vasodilation. In particular, PGE2 is known to induce vasodilation, a typical compensatory response to high altitude hypoxia, and is thought to be the principal factor inducing acute hypoxia headache (Busse et al., 1984; Davis et al., 2004; Fredericks et al., 1994; Messina et al., 1992; Ray et al., 2002). Therefore, headache is the clinical expression of cerebral vasodilation, which represents an interesting parameter to assess in order to understand the effects of PGE2 on cerebral blood vessels.

As for ventilation and heart rate at high altitude, a similar reduction in PGE2 can be elicited by a placebo after oxygen preconditioning. The typical PGE2 increase at high altitude can be blocked by administering oxygen and this effect can be mimicked by the placebo. Fig. 3 shows that whereas a placebo given for the first time does not produce any effect on PGE2, a placebo given after prior oxygen exposure for *n* times induces a substantial decrease in PGE2. Again, this effect is present with 64%, 57%, and 50% oxygen compared to the sea level. Therefore, as for ventilation and heart rate, oxygen preconditioning is crucial for a placebo PGE2 reduction to occur (Benedetti & Dogue, 2015). Thus, this placebo effect is basically a conditioned PGE2 response. It is worth noting that this effect on heart activity can be seen with an oxygen reduction up to 50%.

In the same way as PGE2 can be considered an indirect measure of cerebral vasodilation, headache can be considered the clinical expression of cerebral vasodilation. It can be seen in Fig. 3 that placebos can reduce headache only after oxygen preconditioning, thus supporting the findings for PGE2. It is important to note that two types of headache have been tested: at rest and after exercise. Only postexercise headache has been found to be affected by placebos, whereas no effect was observed for headache at rest.

Once more, it can be seen in Fig. 3 that whereas SO_2 increases after oxygen administration, no change in SO_2 occurs after placebo administration, which indicates that PGE2 modulation and postexercise headache are not due to SO_2 per se, but rather to a SO_2-independent learning mechanism.

4.5 Performance

Overall performance in hypoxic conditions is reduced because of the changes in ventilation, heart activity, and perfusion of different organs, including brain and muscles. However, what is emerging today is that SO_2 is not completely responsible for overall performance, at least within some limits, central nervous system mechanisms, including complex psychological factors, playing an important role. These factors are related to the so-called central command or central governor, a mechanism implying that several physiological parameters, like heart rate, arterial blood pressure, pulmonary ventilation, and muscle performance, could be altered by manipulating the subject's perception of exercise (Krogh & Lindhard, 1913). Placebos have been proposed to impact on this central governor of fatigue. The output of this center would continuously regulate exercise performance to avoid reaching maximal physiological capacity. This would provide protection against damage on one hand, and constant availability of a reserve capacity on the other (Hampson, St Clair Gibson, Lambert, & Noakes, 2001; Lambert, St Clair Gibson, & Noakes, 2005). By altering expectations, placebos could then represent a psychological means to signal to the central governor to release the brake, allowing an increase in performance. Muscle fatigue has also been found to be affected by a central governor (Lambert et al., 2005; St Clair Gibson et al., 2003, 2006). In many studies, athletes are usually asked to perform at their limit, and placebos apparently act by pushing this limit forward (Piedimonte, Benedetti, & Carlino, 2015). To support this perspective, many studies suggest that fatigue and overall performance are of central rather than peripheral origin. It is interesting to note that the concept of a central governor has been applied to health as well, thus extending from fatigue and physical performance to the healing environment (Humphrey & Skoyles, 2012). However, for the sake of this chapter, we focus our attention only on those studies related to hypoxia.

Indeed, as it can be seen in Fig. 3, fatigue is the only parameter that does not require oxygen preconditioning in order to be sensitive to placebos (Benedetti, 2017). In other words, differently from the above-mentioned conditions, a placebo given for the first time, along with positive verbal suggestions of fatigue reduction and performance improvement, is sufficient to reduce fatigue. The effect of placebo alone, without any oxygen preconditioning, is so powerful that it is also present with an oxygen reduction of 50% compared to the sea level. It is crucial to note that this effect on fatigue is independent of SO_2, thus supporting the central command model of perceived fatigue and performance.

The perceived fatigue has a powerful influence on overall performance. Barbiani and Benedetti (2017) showed that the time to complete 3000 steps with a stepper worsened from sea level to high altitude, and that performance returned to that of sea level following administration of both oxygen and placebo (Fig. 4). Importantly, a placebo was given without any oxygen preconditioning, and this "first-time" placebo administration was sufficient to reduce perceived fatigue and increase performance. Therefore, the increase in SO_2 after oxygen delivery was not completely responsible for the decrease in time to complete 3000 steps, as performance improved also with a placebo and without any SO_2 increase.

In a simulation of high altitude hypoxia in the lab at the sea level (Torres-Peralta et al., 2016), volunteers exercised in severe acute hypoxia (oxygen = 73 mmHg). Upon exhaustion, subjects were asked to continue

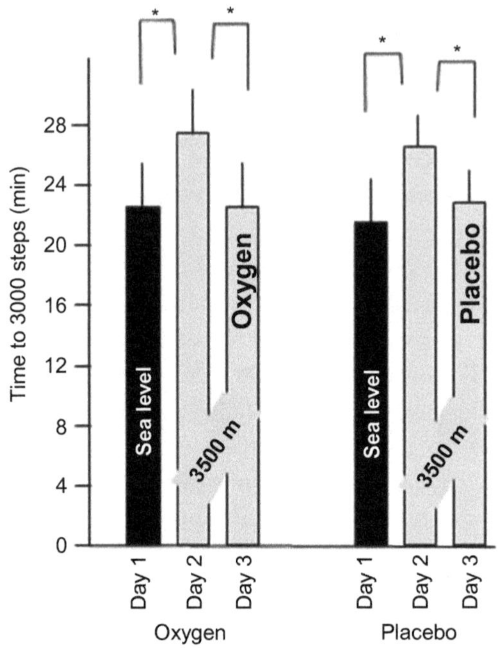

Fig. 4 Here performance is measured as the time needed to complete 3000 steps on a stepper. It can be seen that in the subjects of the oxygen group time increases (performance worsens) from the sea level to an altitude of 3500 m, but oxygen administration restores sea-level performance. In the placebo group, a placebo produced the same effects as oxygen. *Asterisks* indicate significant differences.

exercising, while the breathing gas mixture was swiftly changed to a placebo (oxygen = 73 mmHg) or to 82, 92, 99, and 142 mmHg, and the exercise continued until a new exhaustion. At the second exhaustion, the breathing gas was changed to room air (normoxia) and the exercise continued until the final exhaustion. The switch to a placebo or higher oxygen pressures allowed for the continuation of exercise in all instances. The fact that at exhaustion in severe hypoxia the exercise was continued with the placebo–gas mixture demonstrates that a central and psychological mechanism is of crucial importance for performance.

5. NOCEBO EFFECTS AT HIGH ALTITUDE

No specific study of the nocebo effect at high altitude has been done in relation to oxygenation and ventilation. However, within this context it is worth mentioning the effects of nocebo and negative expectations of some physiological and biochemical parameters that have been measured at high altitude. For example, it has been shown that the cyclooxygenase–prostaglandins (COX–PG) pathway can be modulated by nocebos (Benedetti et al., 2014). In fact, negative expectations about headache pain, the crucial element of the nocebo effect, lead to the enhancement of the COX—PG pathway which, in turn, induces pain worsening. It is interesting to note that PGE2 and headache are particularly sensitive to nocebo, which suggests that nocebos may affect cerebral vasodilation, as already discussed for placebos (see above). Thus, this could be considered as indirect evidence that nocebos, namely, negative expectations, may influence cerebral perfusion. Further research is certainly needed to support this view.

In addition, it is interesting to note that the very same study (Benedetti et al., 2014) showed that placebo administration to headache sufferers at high altitude inhibited the nocebo-related component of pain and PG synthesis, which indicates that the COX pathway can be modulated by both nocebos and placebos. In particular, the same authors ran two clinical trials of aspirin vs placebo for the control of high altitude headache. The first trial was performed in controls, whereas the second trial was done in individuals exposed to social learning by propagation of negative information. No placebo response was noted in the first trial: aspirin was effective in reducing both pain and prostaglandin synthesis in the controls, whereas placebo was completely inefficacious. Conversely, high placebo responses were observed in the second trial: both aspirin and placebo reduced pain in the socially affected individuals. The difference in placebo effect between the two groups is attributable to the different baseline values of headache pain and prostaglandins—e.g., PGE2 induced by the dissemination of negative information. A placebo effect was observed only for the "socially infected" individuals, as the placebo acted only on the nocebo component of the prostaglandin and pain increase. Overall, this study suggests that placebos could be effective only when a preexisting nocebo component is present.

6. CONCLUSIONS AND FUTURE CHALLENGES

A summary of our current knowledge about the effects of placebos in hypoxic conditions is shown in Fig. 5. In this figure, both critical life functions, such as ventilation and heart rate, and other parameters, such as headache and performance, are shown in order to present the general framework of different placebo responses when oxygen is as low as 50% compared to the sea level. It can be seen that the reduction of all parameters by oxygen is

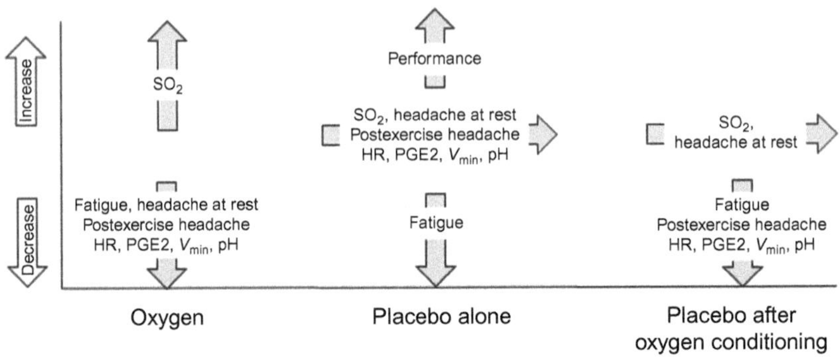

Fig. 5 Summary of the effects of oxygen, placebo alone, and placebo after oxygen conditioning for all the parameters described in this chapter.

accompanied by an increase of SO_2. By contrast, no SO_2 increase is induced by the placebo, yet this can influence some parameters in different ways. Fatigue and overall performance are very sensitive to placebo alone, that is, a placebo given for the first time along with verbal suggestions of reduced fatigue and improved performance. Instead, verbal suggestions alone have no effect on all the other parameters. However, if the placebo is given after prior oxygen conditioning, a response occurs for ventilation, blood pH, heart activity, PGE2, and postexercise headache as well, thus indicating that learning plays a crucial role in these responses. The fact that these placebo effects take place without any change in SO_2 suggests that psychobiological mechanisms are sometimes and in some circumstances more important than or as important as the oxygen content of the body.

The future challenge is threefold. First, we need to understand how these physiological placebo effects occur and how they can bypass the oxygen shortage in the body. Second, we need to identify the limits of these placebo effects, for oxygen reductions beyond 50%. Third, we need to analyze other physiological and biochemical parameters in order to understand which body functions are affected by placebos.

REFERENCES

Amanzio, M., & Benedetti, F. (1999). Neuropharmacological dissection of placebo analgesia: Expectation-activated opioid systems versus conditioning-activated specific subsystems. *Journal of Neuroscience*, *19*(1), 484–494.

Barbiani, D., & Benedetti, F. (2017). In *Differential effects of placebo oxygen on fatigue and pain in hypoxic conditions at high altitude. Proceedings of the first SIPS conference on placebo studies, Leiden, the Netherlands.*

Bartsch, P., Baumgartner, R. W., Waber, U., Maggiorini, M., & Oelz, O. (1990). Comparison of carbondioxide-enriched, oxygen-enriched, and normal air in treatment of acute mountain sickness. *The Lancet, 336*, 772–775.

Benedetti, F. (2014). *Placebo effects* (2nd ed.). Oxford: Oxford University Press.

Benedetti, F. (2017). In *Placebo effects at great heights. Proceedings of the first SIPS conference on placebo studies, Leiden, the Netherlands.*

Benedetti, F., Amanzio, M., Baldi, S., Casadio, C., & Maggi, G. (1999). Inducing placebo respiratory depressant responses in humans via opioid receptors. *European Journal of Neuroscience, 11*(2), 625–631.

Benedetti, F., Amanzio, M., Rosato, R., & Blanchard, C. (2011). Nonopioid placebo analgesia is mediated by CB1 cannabinoid receptors. *Nature Medicine, 17*(10), 1228–1230.

Benedetti, F., & Dogue, S. (2015). Different placebos, different mechanisms, different outcomes: Lessons for clinical trials. *PLoS One, 10*(11). e0140967.

Benedetti, F., Durando, J., Giudetti, L., Pampallona, A., & Vighetti, S. (2015). High altitude headache: The effects of real versus sham oxygen administration. *Pain, 156*, 2326–2336.

Benedetti, F., Durando, J., & Vighetti, S. (2014). Nocebo and placebo modulation of hypobaric hypoxia headache involves the cyclooxygenase-prostaglandins pathway. *Pain, 155*(5), 921–928.

Burtscher, M., Likar, R., Nachbauer, W., & Philadelphy, M. (1998). Aspirin for prophylaxis against headache at high altitudes: Randomised, double blind, placebo controlled trial. *British Medical Journal, 316*, 1057–1058.

Busse, R., Fosterman, U., Matsuda, H., & Pohl, U. (1984). The role of prostaglandins in the endothelium-mediated vasodilatory response to hypoxia. *Pflügers Archiv, 401*, 77–83.

Davis, R. J., Murdoch, C. E., Ali, M., Purbrick, S., Ravid, R., Baxter, G. S., et al. (2004). EP4 prostanoid receptor-mediated vasodilation of human middle cerebral arteries. *British Journal of Pharmacology, 141*, 580–585.

De la Fuente-Fernández, R., Phillips, A. G., Zamburlini, M., Sossi, V., Calne, D. B., Ruth, T. J., et al. (2002). Dopamine release in human ventral striatum and expectation of reward. *Behavioural Brain Research, 136*(2), 359–363.

De la Fuente-Fernández, R., Ruth, T. J., Sossi, V., Schulzer, M., Calne, D. B., & Stoessl, A. J. (2001). Expectation and dopamine release: Mechanism of the placebo effect in Parkinson's disease. *Science, 293*(5532), 1164–1166.

Fredericks, K. T., Liu, Y., Rusch, N. J., & Lombard, J. H. (1994). Role of endothelium and arterial K+ channels in mediating hypoxic dilation of middle cerebral arteries. *American Journal of Physiology, 267*, H580–H586.

Hampson, D. B., St Clair Gibson, A., Lambert, M., & Noakes, T. D. (2001). The influence of sensory cues on the perception of exertion during exercise and central regulation of exercise performance. *Sports Medicine, 31*, 935–952.

Harris, G. C., & Fitzgerald, R. D. (1989). Impaired learning of classically conditioned bradycardia in rats following fourth ventricle administration of D-Ala2-methionine-enkephalinamide. *Behavioral Neuroscience, 103*, 77–83.

Hernandez, L. L., & Powell, D. A. (1983). Naloxone induces multiple effects on aversive Pavlovian conditioning in rabbits. *Behavioral Neuroscience, 97*, 478–491.

Hernandez, L. L., Powell, D. A., & Gibbs, C. M. (1990). Amygdaloid central nucleus neuronal activity accompanying Pavlovian cardiac conditioning. Effects of naloxone. *Behavioural Brain Research, 41*, 71–79.

Hernandez, L. L., & Watson, K. L. (1997). Opioid modulation of attention-related responses: Delta receptors modulate habituation and conditioned bradycardia. *Psychopharmacology, 131*, 140–147.

Humphrey, N., & Skoyles, J. (2012). The evolutionary psychology of healing: A human success story. *Current Biology, 22*(17), R695–R698.

Imray, C., Wright, A., Subudhi, A., & Roach, R. (2010). Acute mountain sickness: Pathophysiology, prevention and treatment. *Progress in Cardiovascular Diseases*, *52*, 467–484.

International Society for Mountain Medicine. (2017). http://www.ismm.org/index.php/normal-acclimatization.html.

Kawabata, A. (2011). Prostaglandin E2 and pain—An update. *Biological and Pharmaceutical Bulletin*, *34*, 1170–1173.

Krogh, A., & Lindhard, J. (1913). The regulation of respiration and circulation during the initial stages of muscular work. *Journal of Physiology*, *47*, 112–136.

Lambert, E. V., St Clair Gibson, A., & Noakes, T. D. (2005). Complex systems model of fatigue: Integrative homoeostatic control of peripheral physiological systems during exercise in humans. *British Journal of Sports Medicine*, *39*, 52–62.

Leaf, D. E., & Goldfarb, D. S. (2007). Mechanisms of action of acetazolamide in the prophylaxis and treatment of acute mountain sickness. *Journal of Applied Physiology*, *102*(1), 313–322.

Messina, E. J., Sun, D., Koller, A., Wolin, M. S., & Kaley, G. (1992). Role of endothelium-derived prostaglandins in hypoxia elicited arteriolar dilation in rat skeletal muscle. *Circulation Research*, *71*, 790–796.

Petrovic, P., Kalso, E., Petersson, K. M., & Ingvar, M. (2002). Placebo and opioid analgesia: Imaging a shared neuronal network. *Science*, *295*(5560), 1737–1740.

Piedimonte, A., Benedetti, F., & Carlino, E. (2015). Placebo-induced decrease in fatigue: Evidence for a central action on the preparatory phase of movement. *European Journal of Neuroscience*, *41*(4), 492–497.

Pollo, A., Vighetti, S., Rainero, I., & Benedetti, F. (2003). Placebo analgesia and the heart. *Pain*, *102*, 125–133.

Ray, C. J., Abbas, M. R., Coney, A. M., & Marshall, J. M. (2002). Interactions of adenosine, prostaglandins and nitric oxide in hypoxia-induced vasodilatation: In vivo and in nitro studies. *Journal of Physiology*, *544*, 195–209.

Richalet, J-P., Hornych, A., Rathat, C., Aumont, J., Larmignat, P., & Rémy, P. (1991). Plasma prostaglandins, leukotrienes and thromboxane in acute high altitude hypoxia. *Respiration Physiology*, *85*, 205–215.

Schmidt, R. F., & Thews, G. (Eds.), (1989). *Human physiology*. Heidelberg: Springer.

St Clair Gibson, A., Baden, D. A., Lambert, M. I., Lambert, E. V., Harley, Y. X., Hampson, D., et al. (2003). The conscious perception of the sensation of fatigue. *Sports Medicine*, *33*(3), 167–176.

St Clair Gibson, A., Lambert, E. V., Rauch, L. H. G., Tucker, R., Baden, D. A., Foster, C., et al. (2006). The role of information processing between the brain and peripheral physiological systems in pacing and perception of effort. *Sports Medicine*, *36*(8), 705–722.

Sutton, J. R., Coates, G., Houston, C. S., & Oelz, O. (1992). The Lake Louise consensus on the definition and quantification of altitude illness. In J. R. Sutton, G. Coates, & C. S. Houston (Eds.), *Hypoxia and mountain medicine* (pp. 327–330). Burlington, Vermont: Queen City Printers.

Torres-Peralta, R., Losa-Reyna, J., Morales-Alamo, D., González-Izal, M., Pérez-Suárez, I., Ponce-González, J. G., et al. (2016). Increased PIO2 at exhaustion in hypoxia enhances muscle activation and swiftly relieves fatigue: A placebo or a PIO2 dependent effect? *Frontiers in Physiology*, *7*, 333.

West, J. B. (2004). The physiologic basis of high-altitude diseases. *Annals of Internal Medicine*, *141*, 789–800.

Wilson, M. H., Newman, S., & Imray, C. S. (2009). The cerebral effects of ascent to high altitudes. *The Lancet Neurology*, *8*, 175–191.

Wyss-Dunant, E. (1953). Acclimatisation. In M. Kurz (Ed.), *The mountain world* (pp. 110–117). New York: Harper & Brothers.

CHAPTER TWELVE

Placebos Without Deception: Outcomes, Mechanisms, and Ethics

Luana Colloca[*,†,1], **Jeremy Howick**[‡,§]
[*]School of Nursing, University of Maryland, Baltimore, MD, United States
[†]School of Medicine, University of Maryland, Baltimore, MD, United States
[‡]University of Oxford, Oxford, United Kingdom
[§]Faculty of Philosophy, University of Oxford, Oxford, United Kingdom
[1]Corresponding author: e-mail address: colloca@umaryland.edu

Contents

Abstract

Scientific research indicates that open-label and dose-extending placebos (that patients know are placebos) can elicit behavioral, biological, and clinical outcome changes. In this chapter, we present the state-of-the-art evidence and ethical considerations about open-label and dose-extending placebos, discussing the perspective of giving placebos with a rational, as dose extension of active drugs, or expectancy boosters. Previous comprehensive reviews of placebo use have considered how to harness placebo effects in medicine and the need to focus on elements of the clinical encounter as well as

International Review of Neurobiology, Volume 138
ISSN 0074-7742
https://doi.org/10.1016/bs.irn.2018.01.005

patient–clinician relations. Here, we illustrate the similarities and differences between standard (deceptive) placebos, open-label placebos and dose-extending placebos. We conclude that placebos without deception would override ethical barriers to their clinical use. This paves the way to future large-scale, pragmatic randomized trials that investigate the potential of ethical open-label and dose-extending placebos to improve patients' outcomes, and reduce side effects.

1. INTRODUCTION

Three factors are of major importance in the suffering of badly wounded men [during the Second World War]: pain; mental distress; and thirst. Therapy has been almost entirely directed to pain, and this usually limited to the administration of morphine in large dosage.

Henry Knowles Beecher, American anesthetist and medical ethicist

Surveys from around the world consistently find that healthcare practitioners prescribe placebos quite often (Colloca, Enck, & DeGrazia, 2016; Fassler, Meissner, Schneider, & Linde, 2010; Howick et al., 2013). Placebo use, however, is criticized as being unethical for two reasons. First, placebos are supposedly ineffective (or less effective than "real" treatments), so the ethical requirement of beneficence (and "relative" nonmaleficence) renders their use unethical. Second, they allegedly require deception for their use, violating patient autonomy. Here, we take it as given that at least for some conditions, placebos have effects (see Howick, 2017; Howick et al., 2013 for discussion). The recent research on open-label placebos suggests that the second objection—namely the claim that placebos require unethical deception—is also invalid. If placebos can have effects even when patients are told they are placebos, then placebos do *not* require deception and ethical objections to placebo use lose their force.

2. DO PLACEBOS REQUIRE DECEPTION? THE MYSTERIOUS CASE OF OPEN-LABEL PLACEBOS

A handful of studies have shown that long-term placebo effects can also be elicited under open-label conditions, in which patients are *explicitly* informed that they will receive a placebo (Blease, Colloca, & Kaptchuk, 2016; Charlesworth et al., 2017). This is counterintuitive since placebos supposedly work because people believe they do, but (presumably) *knowing* a treatment is a mere sugar pill makes it difficult to believe they will work. In spite of the lack of intuitive appeal, numerous studies have demonstrated that deception may not be needed to elicit placebo effects and have also demonstrated potential effectiveness at improving significant clinical outcomes in patients with irritable bowel syndrome (IBS) (Kaptchuk et al., 2010), chronic low back pain (Carvalho et al., 2016), depression (Park & Covi, 1965), attention-deficit hyperactivity disorder (ADHD) (Sandler, Glesne, & Geller, 2008), rhinitis (Schaefer, Harke, & Denke, 2016), and cancer-related fatigue (Hoenemeyer, Kaptchuk, Mehta, & Fontaine, 2018).

3. CLINICAL OPEN-LABEL PLACEBO TRIALS

Park and Covi (1965) were the first to attempt we are aware of to test the apparent paradoxical effect of open-label placebos in patients suffering from a range of anxiety symptoms (Park & Covi, 1965). The patients were told that they would have received sugar pills but that they would have perceived benefits in terms of symptom relief. Despite the small number of enrolled patients (14), significant symptom improvement was reported at 1 week of taking open-labeled placebos. In pain medicine, other studies have recently shed light on the potential efficacy of open-labeled placebos without deception in patients suffering from IBS (Kaptchuk et al., 2010) and low back pain (Carvalho et al., 2016).

In a more recent trial conducted by Kaptchuk et al., 80 patients diagnosed with IBS were randomized to receive either open-label placebo pills or no treatment (Kaptchuk et al., 2010). The open-label placebo was presented as follows:

> *The provider clearly explained that the placebo pill was an inactive (i.e., "inert") substance like a sugar pill that contained no medication and then explained in an approximately 15 minute a priori script the following "four discussion points:" 1) the placebo effect is powerful, 2) the body can automatically respond to taking placebo pills like Pavlov's dogs who salivated when they heard a bell, 3) a positive attitude helps but is not necessary, and 4) taking the pills faithfully is critical (Kaptchuk et al., 2010).*

Investigators then measured the effect of the treatment on the IBS Global Improvement Scale (IBS-GIS, stated primary outcome). Open-label placebo produced significantly higher mean global improvement scores (IBS-GIS) at both 11-day midpoint and 21-day endpoint.

In Carvalho et al.'s (2016) study, 83 patients with at least 3 months of chronic lower back pain were randomized to receive two open-label placebo tablets, taken twice daily, or treatment as usual, for 3 weeks (Carvalho et al., 2016). Patients were told that the placebo pill was an inactive substance, like a flour pill, that contained no active medication in it. Patients were also taught about placebo effects using four "discussion points." These were: (1) the placebo effect can be powerful, (2) the body automatically can respond to taking placebo pills like Pavlov dogs that salivated when they heard a bell, (3) a positive attitude can be helpful but is not necessary, and (4) taking the pills faithfully for 21 days is critical. All participants were also shown a video clip (1 min and 25 s) of a television news report, in which participants in an OLP trial of IBS were interviewed (excerpted from http://www.nbcnews.com/video/nightly-news/40787382#40787382). Primary outcomes were mean weekly retrospective pain assessments (0–10) and the Roland–Morris Disability Questionnaire (RMDQ) assessed at 3 weeks. The open-label placebo demonstrated a statistically significant benefit over treatment as usual (TAU).

Gathering all these studies together, a recent meta-analysis found that open-label placebos can lead to positive therapeutic effects when compared to no-treatment. The clinical conditions were IBS, depression, allergic rhinitis, back pain, and ADHD (Charlesworth et al., 2017). However, the meta-analysis involves only five trials that were small, had different control groups (TAU vs waiting-list group), often included positive suggestions alongside the open-label placebos, one involved elements of partial conditioning (ADHD amphetamine treatment and placebos) and were rated as having a moderate risk of bias. Importantly, all these trials are characterized by the lack of blinding that can be achieved by comparing open placebos vs hidden placebos and the comparison with the best available treatment to estimate the relevance of open placebo potential effectiveness. Recently, open label placebos have been tested in 74 cancer survivors in a 21-day assessor blinded, randomized-controlled trial that compared an open-label placebo to TAU for fatigue (Hoenemeyer et al., 2018). Two placebo pills taken twice induced a 29% improvement in fatigue severity, and a 39% improvement in fatigue-disrupted quality of life. Open label placebos were tested in 74 cancer survivors ($N=74$) in a 21-day assessor blinded, randomized-controlled trial that compared an open-label placebo to TAU for fatigue (Hoenemeyer et al., 2018). Two placebo pills taken twice induced a 29% improvement in fatigue severity, and a 39% improvement in fatigue-disrupted quality of life (Table 1). These and the provocative studies on dose-extending placebos (see Section 6) may open up new research avenues with a focus on translational and mechanistic approaches.

Table 1 Clinical Studies With Open Label Placebos

Study	Country	Condition	No. Participants	Control Treatment	Intervention Timing	Primary Outcome Measure	Outcome Estimation (Pooled Effect Size, and 95% CI)
Carvalho et al. (2016)	Portugal	Chronic low back pain	83	Treatment as usual	21 days	Roland–Morris Disability Questionnaire	0.75 [0.31, 1.20]
Hoenemeyer et al. (2018)	United States	Cancer related fatigue	73	Treatment as usual	21 days	Fatigue Symptom Inventory (FSI-14)	0.32 [−0.14, 0.79]
Kaptchuk et al. (2010)	United States	IBS	80	No treatment	21 days	IBS Global improvement	0.78 [0.32, 1.24]
Kelley, Kaptchuk, Cusin, Lipkin, and Fava (2012)	United States	Major Depressive Disorder	20	Waitlist	14 days	17-item Hamilton Scale for Depression	0.51 [−0.38, 1.41]
Schaefer et al. (2016)	Germany	Allergic Rhinitis	25	No treatment	14 days	Allergy symptoms	1.12 [0.26, 1,97]

4. LABORATORY (OPEN-LABEL) PLACEBO STUDIES

Placebo (conditioned) hypoalgesic effects persisted after revealing that the cream used to reduce the experimental heat pain was merely Vaseline (Schafer, Colloca, & Wager, 2015). Participants were told that the study aimed "to compare the analgesic effects of a topical cream with an active analgesic component (placebo cream) to a topical cream with no active ingredients (control cream)." An initial calibration phase, either long or short conditioning, and a test phase were performed. Placebo analgesia was tested before and after participants were told the treatment was a vaseline cream. Placebo analgesia was defined as the difference in pain reports between placebo and control stimulations at identical heat temperatures. Schafer and colleagues demonstrated that experiencing multiple conditioning sessions leads to robust placebo analgesia that persist even when the true nature of the placebo treatment is convincingly revealed to study participants (Schafer et al., 2015). Additionally, this study provided evidence that experienced placebo analgesia can be uncorrelated with expected analgesia. Conditioned placebo analgesia can be mediated by processes not accessible by reportable expectancy, and that there is a transition from expectancy-mediated processes to more involuntary analgesic processes as those initial expectancies are reinforced through repeated experience. These results parallel emerging evidence from other studies, suggesting that placebo analgesia might sometimes occur in the absence of belief (Charlesworth et al., 2017). Therefore, informing study participants about the realm of the placebo phenomenon and its mechanisms presents no negative reactions or negative consequences and may help engage potential factors that favor outcome improvements.

Open-label placebos have also been explored in healthy participants in a standardized experimental heat pain modulation paradigm and research has shown that they are as effective as deceptive placebos when accompanied by a rationale (Locher et al., 2017). Locher et al. (2017) explored the effectiveness of open-label placebo given with a rationale (Locher et al., 2017) as compared to open-label placebo without a rationale, deceptive placebo, and no-intervention in 160 participants who were randomly assigned to the experimental groups. All groups received an application of a placebo cream except the no-intervention group. Baseline and posttreatment measurements of pain tolerance, pain intensity, and pain unpleasantness ratings were assessed as primary outcomes. Those who received the placebo given with a rationale and the deceptive placebo compared with those who received the placebo treatment without a rationale reported significantly less pain intensity and unpleasantness ratings. These changes in pain experience were independent from the individual level of pain tolerance. A rationale given along with the placebo was as efficacious as the deceptive placebo (Locher et al., 2017) in an experimental setting.

Locher et al. findings are in contrast with other results obtained in a clinical context. Placebo analgesic effects appeared to be larger when full deceptive disclosures are given to patients with postoperative pain (Colloca, 2017; Pollo et al., 2001). The authors compared clinical pain outcomes in acute pain patients who received saline solution and were told nothing about the analgesic effect (natural history) of a basal infusion in the postoperative setting; were told that the treatment could have been either a potent painkiller or a placebo (similar to a double-blind clinical trial); or were told that the basal infusion was a potent painkiller (full-deceptive administration) (Pollo et al., 2001). Operationally, the placebo effect was defined as the change in the amount of requested doses of buprenorphine over the 3-day postoperative pain treatment when a continuous saline infusion was added to the active buprenorphine treatment. Overall the buprenorphine requests decreased by 20.8% in the double-blind group, and by 33.8% in the deceptive group in comparison to the natural history, respectively, leading to a significant reduction of the buprenorphine opioid intake (Pollo et al., 2001). A genuine placebo-induced analgesic effect was observed with the more transparent disclosures although the deceptive group elicited the larger reduction of buprenorphine requests.

5. DOSE-EXTENDING (OPEN-LABEL) PLACEBOS

Placebos have been given as "dose extenders" to prone the brain–body systems to create conditioned responses (CR) that are similar to the effects of the treatments (US) when deliberate conditioning stimuli (CSs) are paired with the US.

For example, Goebel et al. administered cyclosporine A (2.5 mg/kg, US) along with a green-colored, strawberry-flavored milk drink (CS) in healthy participants to test the hypothesis that cyclosporine-like effects can be detected when placebos are given with the CS in place of the cyclosporine (Goebel et al., 2002). Placebos administered with the flavored drink significantly suppressed immune functions in terms of interleukin-2 (IL-2) and interferon gamma (IFN-gamma) mRNA expression, in vitro release of

IL-2 and IFN-gamma, as well as lymphocyte proliferation, suggesting that placebos can act as "dose extenders" of the cyclosporine action (Goebel et al., 2002). The duration of such a conditioned effect (e.g., suppression of T-cell function) extinguished after 14 unreinforced exposures to the CS drink. However, administering subtherapeutic dosages of cyclosporine A (0.25 mg/kg) along with the CS drink prevented the extinction of the conditioned immunosuppression (Albring et al., 2014). The intrinsic action of dose-extending placebos is illustrated in human research that demonstrated that pharmacological conditioning is effective in extending the response to morphine (Amanzio & Benedetti, 1999; Benedetti, Pollo, & Colloca, 2007; Guo, Wang, & Luo, 2010). Robust analgesic responses were documented when the administration of morphine for two consecutive days was replaced by a placebo on the third day (Amanzio & Benedetti, 1999). Importantly, different schedules of pharmacological conditioning worked in eliciting morphine-mimic effects, at least in the range of days and weeks (Benedetti et al., 2007). These observations suggest that a pharmacological conditioning procedure creates a learned response that can be reevoked. Similar results have been found in mice using pharmacological opioid and nonopioid conditioning (Guo et al., 2010).

Clinically speaking, a recent study demonstrated the effectiveness of dose-extending placebo in patients with psoriasis treated with corticosteroids (Ader et al., 2010). Patients were treated under a partial schedule of pharmacological (corticosteroid) reinforcement in which a full dose was given 25%–50% of the time and substituted by placebos the other times as compared to a dose control group, in which patients received the full dose 25%–50% of the time but not placebos, and a group receiving the full dose of active corticosteroids (100%). The frequency of relapse under partial reinforcement was lower (26.7%) than in the control group (61.5%) and clinically comparable to the reduction in symptoms induced by a full dose of corticosteroids (22.2%). Thus, dose-extending placebos given with the partial schedule of pharmacotherapeutic reinforcement (Ader et al., 2010) with corticosteroids given one quarter or half as frequently as currently prescribed along with dose-extending placebos were sufficient to mitigate psoriasis relapses.

Importantly, a study in children with ADHD (Sandler & Bodfish, 2008) used a methodological twist in which open placebos and partial reinforcement were merged. In fact, placebo use was described to both parents and children transparently. Following a model for preauthorized placebo use, patients and parents were explicitly informed that placebos (e.g., lactose or talc pills) will be given to extend medication effects (amphetamines). Children were assigned to three arms. Those in arm 1 received a placebo pill paired with a 50%-reduced dose of amphetamine. The same reduction of treatment was performed in arm 2 but without a controlled conditioned cue (control group). Children in arm 3 received a full dose of amphetamine treatment. Pairing a conditioned stimulus (CS) with amphetamines produced placebo-conditioned responses that allowed children with ADHD to be treated effectively with a lower dose of stimulant medication. Moreover, Perlis and colleagues randomized patients with chronic insomnia to distinct regimes of 10mg zolpidem including nightly treatment with 10 or 5mg, intermittent treatment with 10mg, or partial reinforcement treatment with placebos and 10mg for 12 weeks (Perlis et al., 2015). The partial reinforcement group maintained treatment response that were similar to the full treatment groups and better than the outcomes observed in those patients assigned to the intermittent treatment who exhibited poorer sleep quality. These pioneering clinical trials could potentially merge open-label placebos, authorized deception, and (evidence-based) rationale for using placebos, clinically.

6. POTENTIAL MECHANISMS

6.1 Pharmacological Memory

The mechanisms by which open-label placebos given without a formal conditioning are complex and remain to be confirmed (see Fig. 1 and 2). It is possible that sugar pills labeled as placebos work because they retrieve a pharmacological memory, therefore acting as a conditioned cue that elicits previously learned responses in line with the learning theories including classical and nonclassical forms of conditioning (Colloca & Miller, 2011a, 2011c).

In addition to (subconscious) conditioning, conscious expectancy could play a role in how open-label placebos work. Often, the open-label placebos were delivered in addition to the TAUs and importantly with explicit positive suggestions (Carvalho et al., 2016; Hoenemeyer et al., 2018; Kaptchuk et al., 2010; Schaefer et al., 2016) proving a rationale (e.g., "Pavlovian conditioning") and instilling some hope of improvement. The only study that lacked any positive framing and instruction sets had the smallest effect size (Kelley et al., 2012). It is known that the expectation of pain relief has been found to modulate the central regulation of pain through, in particular, the dopamine reward system and the endogenous opioid system (Price, Finniss, & Benedetti, 2008). There is also a growing body of evidence that in addition to what practitioners say, the way in which they deliver these

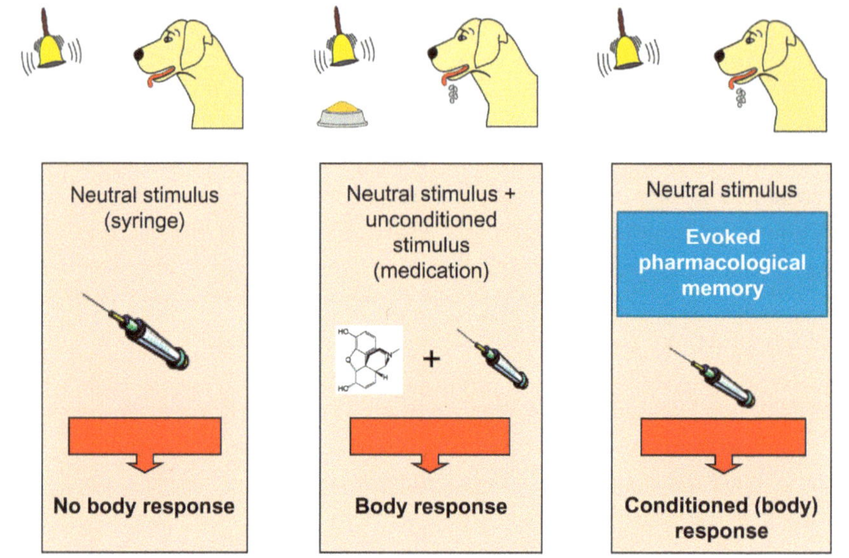

Fig. 1 Pharmacological conditioning and evoked drug memory.

messages (for example, with more or less empathy) can also affect health outcomes (Annoni & Miller, 2016; Caspi & Bootzin, 2002; Friedman, Sedler, Myers, & Benson, 1997).

6.2 Partial Reinforcement and Classical Conditioning

Dose-extending placebos rely on the ability of a nonhuman and human being to acquire a physiological reflex via associative learning processes (e.g., classical conditioning and partial reinforcement learning paradigms) (Ader, 1989, 1990; Colloca & Miller, 2011a). Classical conditioning experiments demonstrated that dogs would salivate (CR) in response to a bell (CS) that had previously been paired with the administration of food (unconditioned stimulus, US), Fig. 1. For dogs conditioned in this way, a ringing bell implied food, causing such automatic physiological responses as salivation (Pavlov, 1927). Similarly, visual, tactile, and gustatory cues can be associated with the US of active medication, through repeated pairing, to elicit responses (Colloca & Miller, 2011a; Enck, Bingel, Schedlowski, & Rief, 2013). As described earlier, these learning mechanisms can account for responses elicited using dose-extending placebos. Although the CS–US pairing mechanisms can explain most of the conditioned responses described in Section 5, further studies are needed to understand how adaptive responses that compensate for the primary drug effect can develop. Opposite

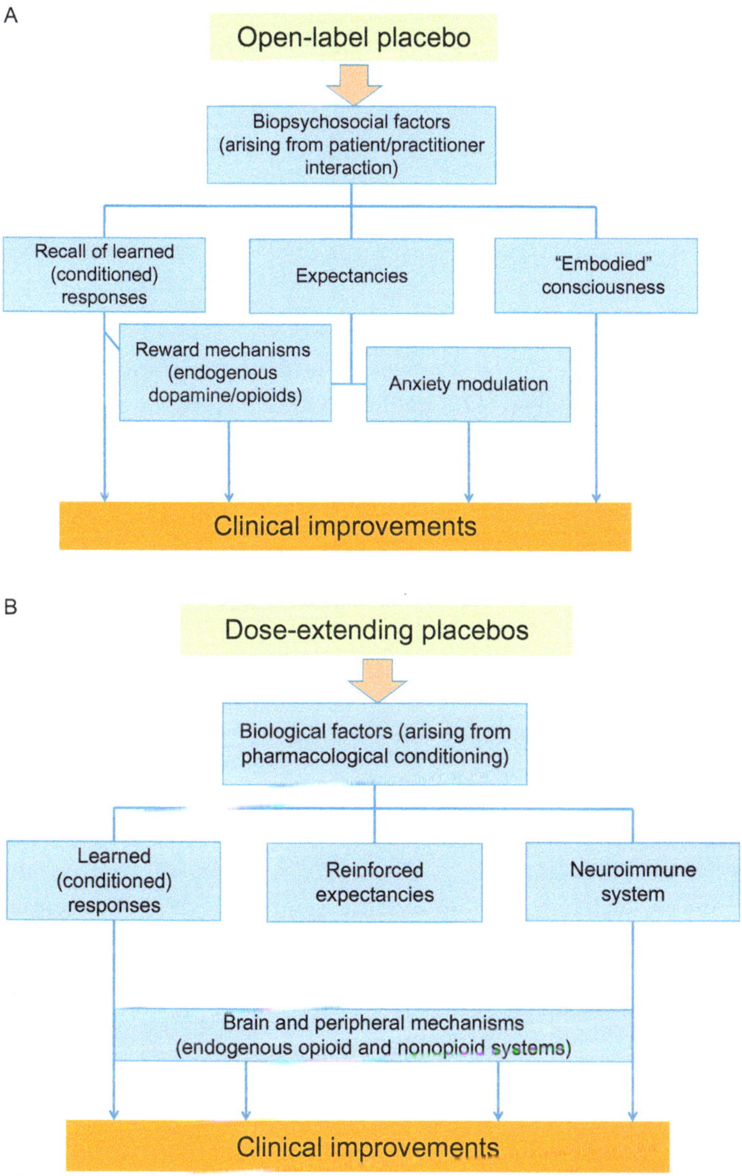

Fig. 2 Potential mechanisms for dose-extending and open-label placebos.

conditioned responses (e.g., tachycardia) can occur when tolerance, a decreased response to a drug within the course of administrations, develops (Siegel, Baptista, Kim, McDonald, & Weise-Kelly, 2000). Dogs treated with epinephrine every few days presented *tachycardic* responses but when epinephrine was replaced by placebo, *bradycardic* response was observed (Subkov & Zilov, 1937). Despite potential limitations, dose-extending placebos work by means of learning effects and can enhance treatment outcomes with transparent use of placebos give as adjunct treatments Fig. 2A. In the real-world setting of health care, these putative placebo mechanisms are likely to operate in unison. Therefore, it is reasonable to assume that these proposed mechanisms are combined, to differing degrees depending on the individual and their disease, to demonstrate effectiveness.

6.3 Embodied Cognition

"Embodied cognition" is a relatively new theory that beginning to help explain how open-label placebos might work but is currently at the speculative stage. According to this theory (Shapiro, 2014) our physical interaction with the world influences or even determines our cognitions (Kemmerer, Miller, Macpherson, Huber, & Tranel, 2013). For example, the sound of the dentist's drill might trigger a specific bodily sensation (Thompson, Ritenbaugh, & Nichter, 2009). Hence, sensory signals could evoke different reactions including those involved in positive and negative healing experiences (Fuchs & Schlimme, 2009). Embodied cognition is related to conditioning because it operates at a subconscious level and is automatic. However, it also differs in important respects. For one, it does not require a specific conditioning procedure (such as the learned pairing of a bell ringing with food). Relatedly, the cognitions arise directly from bodily experiences that are not mediated by the brain Fig. 2A. Some healthcare settings in which open-label placebos are delivered could induce the body to react in a way that subsequently leads to cognitions, which, in turn, induce the brain to produce endogenous substances such as analgesic endorphins. Further work is warranted to investigate the role of embodied condition in explaining how open-label placebos work.

In the real-world setting of health care, these putative placebo mechanisms are likely to operate in unison. Therefore, it is reasonable to assume that these proposed mechanisms are combined, to differing degrees depending on the individual and their disease, to demonstrate efficacy.

7. CLINICAL IMPLICATIONS

Whether placebos can be prescribed to achieve similar or better outcomes compared with usual medical care, whether and how physicians may recommend treatments that lack any specific efficacy remains controversial (Colloca, 2014; Comaroff, 1976; Henriksen & Hansen, 2004).

A number of studies reported that placebos are indeed used by clinicians across different countries (Fassler, Meissner, Schneider, & Linde, 2010; Kermen, Hickner, Brody, & Hasham, 2010; Louhiala, 2012; Meissner, Hofner, Fassler, & Linde, 2012; Nizan, Barash, Valinsky, Lichter, & Manulis, 1997; Tilburt, Emanuel, Kaptchuk, Curlin, & Miller, 2008) including United States (Kermen et al., 2010; Sherman & Hickner, 2008; Tilburt et al., 2008), Canada (Harris & Raz, 2012; Raz et al., 2011), Germany (Linde et al., 2013; Meissner, 2005), Switzerland (Fassler, Gnadinger, Rosemann, & Biller-Andorno, 2009), Denmark (Hrobjartsson & Norup, 2003), United Kingdom (Howick et al., 2013), Israel (Nitzan & Lichtenberg, 2004), India (Shah, Panchal, Vyas, & Patel, 2009), Saudi Arabia (Hassan, Fauzi, & Hasan, 2011), and New Zealand (Holt & Gilbey, 2009). A systematic review of 22 studies from 12 different countries reported that between 17% and 80% of interviewed clinicians administered sugar pills or saline injections during their careers (Fassler et al., 2010). Still physicians feel that there is a lack of harm and even a potential benefit associated with placebo use but that deception is essential to elicit placebo effects (Bishop et al., 2014). Interestingly, patients feel that potential benefit outweighs the importance of transparency in use. In the United States patients viewed deceptive placebo use acceptable (70%), and approximately 79% would prefer transparency over deception (Hull et al., 2013).

Other factors are also likely to play a role, including biopsychosocial forces arising from contact with a healthcare practitioner (Holt-Lunstad, Smith, & Layton, 2010), and lowered patient anxiety due to a positive expectation of recovery (Benedetti, Carlino, & Pollo, 2011; Darragh et al., 2016).

7.1 Patients' Perspective

Information regarding the views of patients, especially surrounding the concept of deception, would help inform the clinical use of placebos (Bishop et al., 2014; Cohen & Shapiro, 2013; Gold & Lichtenberg, 2014; Hull et al., 2013; Justman, 2013). Ortiz and colleagues performed a qualitative analysis of part of a US national survey to uncover underlying patient attitudes about the use of placebo in the face of deception or transparency. A total of 853 participants participated in a telephone survey (Hull et al., 2013). Adults seen in an outpatient clinic for a chronic health problem at least once in the prior 6 months were invited to participate in the survey. Respondents were women (61%) and men (39%), with an average age of 45 years. 58% were white and 42% were nonwhite. 44% had at least an undergraduate college education. Given the large size of this survey, relative frequencies of patients' attitudes and how demographic characteristics (e.-g., sex, age, race, and level education) influence such attitudes were explored. Lack of harm and potential benefit are the most common themes to justify acceptability of placebo use. Of the minority of respondents who judged it never acceptable for doctors to recommend placebo treatments, the majority referred to the doctors' obligation to do further clinical tests. The demographic characteristics that emerged as relevant were the level of education and age. Those participants with higher education mentioned potential benefit as a reason for using placebos clinically. Older age was associated with likelihood to identify overall physician–patient relationships, as opposed to treatments as relevant factors for optimal care.

Moreover, participants were asked their opinions about disclosing the use of a placebo. The majority of participants thought that physicians should not lie to patients when actively asked by the patient, a view that was based on the *patient's right to know*, the value of *honesty*, and the chance to harness the *power of the mind*. Only a minority of participants felt the patient should not be informed of the use of placebos for reasons related to *potential harm*, *obligation to do more*, and potential *lack of benefit* in being told about the use of a placebo.

8. ETHICAL IMPLICATIONS

8.1 Authorized Deception and Placebo Use

Despite the common belief that placebo research has to involve elements of deception consisting of deliberately communicating misleading information about the goal of the research study and the nature of experimental procedures, theoretical and experimental research is advocating the possibility of obtaining authorization to use placebos (Miller & Kaptchuk, 2008; Miller, Wendler, & Swartzman, 2005; O'Neil & Miller, 2009). Importantly several authors have created elegant normative work as well as empirical results on the use of deception in placebo research and pain. For example, Martin and Katz (2010) tested the inclusion of authorized deception in the informed consent process by randomly assigning participants to an authorized deception group or a deception group without authorized deception. Interestingly, the authors found that authorized deception did not influence the size of placebo-induced placebo analgesia, recruitment, and retention of participants. Martin and Katz found that informing participants about the nature of the placebo manipulation does not cause distress and lack of trust in research (Martin & Katz, 2010). More recently, Corsi and Colloca (2017) and Colloca, Pine, Ernst, Miller, and Grillon (2016) published findings that have been obtained with a preauthorization to use deceptive information. Namely the study participants were told that the research would have involved deception during the informed consent process with a verbal and written section that describes deception in the consent form as follows:

> *"Use of Deception - At some point during the study, we will provide you with misleading information. After the study is completed we will give you a written explanation on how the information was not true and why. We will also answer any questions that you have about the procedure and the use of any misleading information."*

Participants had been informed about the deceptive component of this study (why and when the reduced dose was used) and had the opportunity to withdraw the data after they are done participating. The so-called authorized deception approach did not impact the possibility to observe robust placebo and nocebo effects in behavioral and pharmacological research with healthy participants. Importantly, the same approach had been recently used in fibromyalgic pain patients. Perceptions about participation in an authorized deception study were examined in fibromyalgia patients and healthy controls. The majority of participants expressed little or no concern about the deception, still trusted the scientific process, and found the debriefing procedure helpful and worthwhile (personal communication).

8.2 Ethics of Open Label Placebos

Since open-label placebos do *not* require deception, their existence and effectiveness undermine the ethical objections to placebo use. The blooming research in this area warrants an investigation and perhaps revision of ethical standards surrounding placebo use. There are, however, a few notes of caution that must be issued about the ethics of open-label placebos.

First, the arguments that placebos can be ethical since they do not require deception apply to *open-label* placebos and not deceptive placebos. (Recall that deceptive placebos may be ethical but the objection that they violate autonomy is hard to eradicate altogether.) Second, up to half of patients in clinical trials do not recall or understand what they consented to (Tam et al., 2015). If not, then perhaps patients who consent to taking an "open-label placebo" might actually believe it is a real treatment. In two open-label placebo trials investigating this possibility, patients did indeed appear to understand that they were taking placebos (Carvalho et al.,

2016; Kaptchuk et al., 2010). Third, the positive suggestions often delivered alongside open-label placebos could involve an element of deception, depending on their wording. This is, strictly speaking, a distinct issue and we leave a discussion of the ethics of therapeutic communication to another study (Annoni & Miller, 2016). In short, the research demonstrating the effects and mechanisms of open-label placebos demands a reanalysis of ethical strictures on placebo use, and further discussion on how open-label placebos might be implemented is warranted.

8.3 Ethics of Dose-Extending Placebos

An area where dose-extending placebo use could be particularly interesting and ethical is in comparison with standard full regimens of medication. Dose-extending placebos—wherever effective—have several benefits. First, extending the effects of a medication through the use of dose-extending placebos—rather than using only medication for a treatment of equal duration—may reduce the side effects associated with the medicine, a speculation that has been confirmed in some studies in which side effects were monitored (Sandler & Bodfish, 2008; Sandler et al., 2008). However, side effects (nocebo effects) may respond to conditioning and learning mechanisms (Colloca & Miller, 2011b), so there is a risk of conditioned side effects instead of conditioning drug efficacy when dose-extending placebos are paired with active treatments. Second, dose-extending placebos may decrease physiological or psychological dependence to medication and habit-forming behaviors toward medication. Third, using dose-extending placebos for part of the therapeutic strategy rather than using medication for the entire treatment will lower costs by reducing the total intake of the required medication. Last but not least, dose-extending use of placebos provided with patients' education will raise awareness about the body's capacity for self-healing. Finally, by including the primary therapy alongside the dose-extending placebo, the benefits of conditioning could boost the placebo effect.

An obvious area where placebo use could improve patient outcomes is the treatment of pain, where opioid overuse has become a crisis (Belcher, Ferré, Martinez, & Colloca, 2017; Colloca, 2017). Using open-label placebos, perhaps as dose extenders, could reduce the harms caused by the opioid epidemics (Colloca, Enck, & DeGrazia, 2016). Although animal studies can be used to inform clinical studies, further human studies are needed to determine in which diseases and conditions the use of dose-extending placebos can be effective and safe. Methodologically, the ideal study protocol should include three arms: (1) an arm with a partial schedule of pharmacological reinforcement in which the full dose is given 25%–50% of the time and substituted by placebos at other times; (2) a control arm in which the full dose is given 25%–50% of the time and no placebos are administrated; and (3) a comparator arm in which full dose of medication. Whenever feasible, this study design would help circumscribe changes in the efficacy outcome measures due to spontaneous remission, regression to the mean, and natural history of the disease.

Some factors stand in the way of using dose-extending placebos. These include the irreversibility of a disease, clinical contraindication to introduce treatment reductions, and the pharmacokinetic properties of the agent (e.g., US–CS pairings). Safety, optimization, and feasibility studies will help obtain a meaningful investigation of dose-extending placebos.

9. CONCLUSIONS

In conclusion, the recent flourishing of open-label and dose-extending placebo research shows that placebo effects do not necessarily require deception to produce their effects bypassing at least some of the conventional ethical barriers to their clinical use. Future large scale, pragmatic randomized trials should investigate the potential of open-label and dose-extending placebos to improve outcomes and reduce side effects. A parallel body of evidence is needed to inform us about the mechanisms underpinning how open-label and dose-extending open placebos work. Open label placebo research demands a reanalysis of ethical barriers to clinical placebo use.

ACKNOWLEDGMENTS

This work was supported by the National Institute of Dental and Craniofacial Research (NIDCR, R01DE025946, L.C.) and MPowering the State Grant (L.C.).

REFERENCES

Ader, R. (1989). Conditioned immune responses and pharmacotherapy. *Arthritis Care and Research*, *2*(3), S58–64.

Ader, R. (1990). Immune-derived modulation of behavior. *Annals of the New York Academy of Sciences*, *594*, 280–288.

Ader, R., Mercurio, M. G., Walton, J., James, D., Davis, M., Ojha, V., et al. (2010). Conditioned pharmacotherapeutic effects: A preliminary study. *Psychosomatic Medicine*, *72*(2), 192–197. https://doi.org/10.1097/PSY.0b013e3181cbd38b.

Albring, A., Wendt, L., Benson, S., Nissen, S., Yavuz, Z., Engler, H., et al. (2014). Preserving learned immunosuppressive placebo response: Perspectives for clinical application. *Clinical Pharmacology and Therapeutics*, *96*(2), 247–255. https://doi.org/10.1038/clpt.2014.75.

Amanzio, M., & Benedetti, F. (1999). Neuropharmacological dissection of placebo analgesia: Expectation-activated opioid systems versus conditioning-activated specific subsystems. *The Journal of Neuroscience*, *19*(1), 484–494.

Annoni, M., & Miller, F. G. (2016). Placebo effects and the ethics of therapeutic communication: A pragmatic perspective. *Kennedy Institute of Ethics Journal*, *26*(1), 79–103. https://doi.org/10.1353/ken.2016.0004.

Belcher, A. M., Ferré, S., Martinez, P. E., & Colloca, L. (2017). Role of placebo effects in pain and neuropsychiatric disorders. *Progress in Neuro-Psychopharmacology & Biological Psychiatry*, https://doi.org/10.1016/j.pnpbp.2017.06.003. PubMed PMID: 28595945; PubMed Central PMCID: PMC5722709 (Review; Epub ahead of print).

Benedetti, F., Carlino, E., & Pollo, A. (2011). How placebos change the patient's brain. *Neuropsychopharmacology*, *36*(1), 339–354. https://doi.org/10.1038/npp.2010.81.

Benedetti, F., Pollo, A., & Colloca, L. (2007). Opioid-mediated placebo responses boost pain endurance and physical performance: Is it doping in sport competitions? *The Journal of Neuroscience*, *27*(44), 11934–11939. https://doi.org/10.1523/JNEUROSCI.3330-07.2007.

Bishop, F. L., Howick, J., Heneghan, C., Stevens, S., Hobbs, F. D., & Lewith, G. (2014). Placebo use in the UK: A qualitative study exploring GPs' views on placebo effects in clinical practice. *Family Practice*, *31*(3), 357–363. https://doi.org/10.1093/fampra/cmu016.

Blease, C., Colloca, L., & Kaptchuk, T. J. (2016). Are open-label placebos ethical? Informed consent and ethical equivocations. *Bioethics*, *30*(6), 407–414. https://doi.org/10.1111/bioe.12245.

Carvalho, C., Caetano, J. M., Cunha, L., Rebouta, P., Kaptchuk, T. J., & Kirsch, I. (2016). Open-label placebo treatment in chronic low back pain: A randomized controlled trial. *Pain*, *157*(12), 2766–2772. https://doi.org/10.1097/j.pain.0000000000000700.

Caspi, O., & Bootzin, R. R. (2002). Evaluating how placebos produce change. Logical and causal traps and understanding cognitive explanatory mechanisms. *Evaluation & the Health Professions*, *25*(4), 436–464.

Charlesworth, J. E. G., Petkovic, G., Kelley, J. M., Hunter, M., Onakpoya, I., Roberts, N., et al. (2017). Effects of placebos without deception compared with no treatment: A systematic review and meta-analysis. *Journal of Evidence-Based Medicine*, *10*(2), 97–107. https://doi.org/10.1111/jebm.12251.

Cohen, S., & Shapiro, H. (2013). "Comparable placebo treatment" and the ethics of deception. *The Journal of Medicine and Philosophy*, *38*(6), 696–709. https://doi.org/10.1093/jmp/jht052.

Colloca, L. (2014). *Ethics of placebos encyclopaedia of human biology* (3rd ed.). . Oxford, UK: Elsevier.

Colloca, L. (2017). Hypoalgesic placebo effects can occur with transparent disclosures. *Pain*, *158*(12), 2279–2280. https://doi.org/10.1097/j.pain.0000000000001041.

Colloca, L., Enck, P., & DeGrazia, D. (2016). Relieving pain using dose-extending placebos: A scoping review. *Pain*, *157*(8), 1590–1598. https://doi.org/10.1097/j.pain.0000000000000566.

Colloca, L., & Miller, F. G. (2011a). How placebo responses are formed: A learning perspective. *Philosophical Transactions of the Royal Society of London. Series B, Biological Sciences*, *366*(1572), 1859–1869. https://doi.org/10.1098/rstb.2010.0398.

Colloca, L., & Miller, F. G. (2011b). The nocebo effect and its relevance for clinical practice. *Psychosomatic Medicine*, *73*(7), 598–603. https://doi.org/10.1097/PSY.0b013e3182294a50.

Colloca, L., & Miller, F. G. (2011c). Role of expectations in health. *Current Opinion in Psychiatry*, *24*(2), 149–155. https://doi.org/10.1097/YCO.0b013e328343803b.

Colloca, L., Pine, D. S., Ernst, M., Miller, F. G., & Grillon, C. (2016). Vasopressin boosts placebo analgesic effects in women: A randomized trial. *Biological Psychiatry*, *79*(10), 794–802. https://doi.org/10.1016/j.biopsych.2015.07.019.

Comaroff, J. (1976). A bitter pill to swallow: Placebo therapy in general practice. *The Sociological Review*, *24*(1), 79–96.

Corsi, N., & Colloca, L. (2017). Placebo and nocebo effects: The advantage of measuring expectations and psychological factors. *Frontiers in Psychology*, *8*, 308. https://doi.org/10.3389/fpsyg.2017.00308.

Darragh, M., Yow, B., Kieser, A., Booth, R. J., Kydd, R. R., & Consedine, N. S. (2016). A take-home placebo treatment can reduce stress, anxiety and symptoms of depression in a non-patient population. *The Australian and New Zealand Journal of Psychiatry*, *50*(9), 858–865. https://doi.org/10.1177/0004867415621390.

Enck, P., Bingel, U., Schedlowski, M., & Rief, W. (2013). The placebo response in medicine: Minimize, maximize or personalize? *Nature Reviews. Drug Discovery*, *12*(3), 191–204. https://doi.org/10.1038/nrd3923.

Fassler, M., Gnadinger, M., Rosemann, T., & Biller-Andorno, N. (2009). Use of placebo interventions among Swiss primary care providers. *BMC Health Services Research*, *9*, 144. https://doi.org/10.1186/1472-6963-9-144.

Fassler, M., Meissner, K., Schneider, A., & Linde, K. (2010). Frequency and circumstances of placebo use in clinical practice—A systematic review of empirical studies. *BMC Medicine*, *8*, 15. https://doi.org/10.1186/1741-7015-8-15.

Friedman, R., Sedler, M., Myers, P., & Benson, H. (1997). Behavioral medicine, complementary medicine and integrated care. Economic implications. *Primary Care*, *24*(4), 949–962.

Fuchs, T., & Schlimme, J. E. (2009). Embodiment and psychopathology: A phenomenological perspective. *Current Opinion in Psychiatry*, *22*(6), 570–575. https://doi.org/10.1097/YCO.0b013e3283318e5c.

Goebel, M. U., Trebst, A. E., Steiner, J., Xie, Y. F., Exton, M. S., Frede, S., et al. (2002). Behavioral conditioning of immunosuppression is possible in humans. *The FASEB Journal*, *16*(14), 1869–1873. https://doi.org/10.1096/fj.02-0389com

Gold, A., & Lichtenberg, P. (2014). The moral case for the clinical placebo. *Journal of Medical Ethics*, *40*(4), 219–224. https://doi.org/10.1136/medethics-2012-101314.

Guo, J. Y., Wang, J. Y., & Luo, F. (2010). Dissection of placebo analgesia in mice: The conditions for activation of opioid and non-opioid systems. *Journal of Psychopharmacology*, *24*(10), 1561–1567. https://doi.org/10.1177/0269881109104848.

Harris, C. S., & Raz, A. (2012). Deliberate use of placebos in clinical practice: What we really know. *Journal of Medical Ethics*, *38*(7), 406–407. https://doi.org/10.1136/medethics-2012-100695.

Hassan, T., Fauzi, M., & Hasan, D. (2011). Perceptions of placebo therapy among health practitioners in jazan region of saudi arabia. *International Journal of Health Sciences*, *5*(1), 45–49.

Henriksen, K., & Hansen, E. H. (2004). The threatened self: General practitioners' self-perception in relation to prescribing medicine. *Social Science & Medicine*, *59*(1), 47–55. https://doi.org/10.1016/j.socscimed.2003.10.004.

Hoenemeyer, T. W., Kaptchuk, T. J., Mehta, T. S., & Fontaine, K. R. (2018). Open-label placebo treatment for cancer-related fatigue: A randomized-controlled clinical trial. *Scientific Reports*, *8*(1), 2784. https://doi.org/10.1038/s41598-018-20993-y. PubMed PMID, 29426869.

Holt, S., & Gilbey, A. (2009). Use of placebos by New Zealand doctors. *The New Zealand Medical Journal*, *122*(1298), 84–86.

Holt-Lunstad, J., Smith, T. B., & Layton, J. B. (2010). Social relationships and mortality risk: A meta-analytic review. *PLoS Medicine*, *7*(7), e1000316. https://doi.org/10.1371/journal.pmed.1000316.

Howick, J. (2017). The relativity of placebos: Defending a modified version of Grünbaum's scheme. *Synthese, 194*(4), 1363–1396.

Howick, J., Bishop, F. L., Heneghan, C., Wolstenholme, J., Stevens, S., Hobbs, F. D., et al. (2013). Placebo use in the United Kingdom: Results from a national survey of primary care practitioners. *PLoS One, 8*(3), e58247. https://doi.org/10.1371/journal.pone.0058247.

Howick, J., Friedemann, C., Tsakok, M., Watson, R., Tsakok, T., Thomas, J., et al. (2013). Are treatments more effective than placebos? A systematic review and meta-analysis. *PLoS One, 8*(5), e62599. https://doi.org/10.1371/journal.pone.0062599.

Hrobjartsson, A., & Norup, M. (2003). The use of placebo interventions in medical practice—A national questionnaire survey of Danish clinicians. *Evaluation & the Health Professions, 26*(2), 153–165.

Hull, S. C., Colloca, L., Avins, A., Gordon, N. P., Somkin, C. P., Kaptchuk, T. J., et al. (2013). Patients' attitudes about the use of placebo treatments: Telephone survey. *BMJ 347*, f3757. https://doi.org/10.1136/bmj.f3757.

Justman, S. (2013). Placebo: The lie that comes true? *Journal of Medical Ethics, 39*(4), 243–248. https://doi.org/10.1136/medethics-2012-101057.

Kaptchuk, T. J., Friedlander, E., Kelley, J. M., Sanchez, M. N., Kokkotou, E., Singer, J. P., et al. (2010). Placebos without deception: A randomized controlled trial in irritable bowel syndrome. *PLoS One, 5*(12), e15591. https://doi.org/10.1371/journal.pone.0015591.

Kelley, J. M., Kaptchuk, T., Cusin, C., Lipkin, S., & Fava, M. (2012). Open-label placebo for major depressive disorder: A pilot randomized controlled trial. *Psychotherapy and Psychosomatics, 81*, 312–314.

Kemmerer, D., Miller, L., Macpherson, M. K., Huber, J., & Tranel, D. (2013). An investigation of semantic similarity judgments about action and non-action verbs in Parkinson's disease: Implications for the embodied cognition framework. *Frontiers in Human Neuroscience,* 7, 146. https://doi.org/10.3389/fnhum.2013.00146.

Kermen, R., Hickner, J., Brody, H., & Hasham, I. (2010). Family physicians believe the placebo effect is therapeutic but often use real drugs as placebos. *Family Medicine, 42*(9), 636–642.

Linde, K., Friedrichs, C., Alscher, A., Blank, W. A., Schneider, A., Fassler, M., et al. (2013). Use of placebos and nonspecific and complementary treatments by German physicians—rationale and development of a questionnaire for a nationwide survey. *Forschende Komplementärmedizin, 20*(5), 361–367. https://doi.org/10.1159/000356230.

Locher, C., Nascimento, A. F., Kirsch, I., Kossowsky, J., Meyer, A., & Gaab, J. (2017). Is the rationale more important than deception? A randomized controlled trial of open-label placebo analgesia. *Pain, 158*, 2320–2328.

Louhiala, P. (2012). What do we really know about the deliberate use of placebos in clinical practice? *Journal of Medical Ethics, 38*(7), 403–405. https://doi.org/10.1136/medethics-2011-100420.

Martin, A. L., & Katz, J. (2010). Inclusion of authorized deception in the informed consent process does not affect the magnitude of the placebo effect for experimentally induced pain. *Pain, 149*(2), 208–215. https://doi.org/10.1016/j.pain.2009.12.004.

Meissner, K. (2005). Preferring patient-reported to observer-reported outcomes substantially influences the results of the updated systematic review on placebos by Hrobjartsson and Gotzsche. *Journal of Internal Medicine, 257*(4), 394. author reply 395–396. https://doi.org/10.1111/j.1365-2796.2005.01470.x.

Meissner, K., Hofner, L., Fassler, M., & Linde, K. (2012). Widespread use of pure and impure placebo interventions by GPs in Germany. *Family Practice, 29*(1), 79–85. https://doi.org/10.1093/fampra/cmr045.

Miller, F. G., & Kaptchuk, T. J. (2008). Deception of subjects in neuroscience: An ethical analysis. *The Journal of Neuroscience, 28*(19), 4841–4843. https://doi.org/10.1523/JNEUROSCI.1493-08.2008.

Miller, F. G., Wendler, D., & Swartzman, L. C. (2005). Deception in research on the placebo effect. *PLoS Medicine, 2*(9), e262. https://doi.org/10.1371/journal.pmed.0020262.

Nitzan, U., & Lichtenberg, P. (2004). Questionnaire survey on use of placebo. *BMJ, 329*(7472), 944–946. https://doi.org/10.1136/bmj.38236.646678.55.

Nizan, R., Barash, I., Valinsky, L., Lichter, A., & Manulis, S. (1997). The presence of hrp genes on the pathogenicity-associated plasmid of the tumorigenic bacterium Erwinia herbicola pv. Gypsophilae. *Molecular Plant-Microbe Interactions, 10*(5), 677–682. https://doi.org/10.1094/MPMI.1997.10.5.677.

O'Neil, C. C., & Miller, F. G. (2009). When scientists deceive: Applying the federal regulations. *The Journal of Law, Medicine & Ethics, 37*(2), 344–350. https://doi.org/10.1111/j.1748-720X.2009.00377.x.

Park, L. C., & Covi, L. (1965). Nonblind placebo trial: An exploration of neurotic patients' responses to placebo when its inert content is disclosed. *Archives of General Psychiatry, 12*, 36–45.

Pavlov, I. P. (1927). *Conditioned reflexes: An investigation of the physiological activity of the cerebral cortex*. London: Oxford University Press.

Perlis, M., Grandner, M., Zee, J., Bremer, E., Whinnery, J., Barilla, H., et al. (2015). Durability of treatment response to zolpidem with three different maintenance regimens: A preliminary study. *Sleep Medicine, 16*(9), 1160–1168. https://doi.org/10.1016/j.sleep.2015.06.015. PubMed PMID: 26298795. PubMed Central PMCID: PMC4709332.

Pollo, A., Amanzio, M., Arslanian, A., Casadio, C., Maggi, G., & Benedetti, F. (2001). Response expectancies in placebo analgesia and their clinical relevance. *Pain, 93*(1), 77–84.

Price, D. D., Finniss, D. G., & Benedetti, F. (2008). A comprehensive review of the placebo effect: Recent advances and current thought. *Annual Review of Psychology, 59*, 565–590. https://doi.org/10.1146/annurev.psych.59.113006.095941.

Raz, A., Campbell, N., Guindi, D., Holcroft, C., Dery, C., & Cukier, O. (2011). Placebos in clinical practice: Comparing attitudes, beliefs, and patterns of use between academic psychiatrists and nonpsychiatrists. *Canadian Journal of Psychiatry, 56*(4), 198–208.

Sandler, A. D., & Bodfish, J. W. (2008). Open-label use of placebos in the treatment of ADHD: A pilot study. *Child: Care, Health and Development, 34*(1), 104–110. https://doi.org/10.1111/j.1365-2214.2007.00797.x.

Sandler, A., Glesne, C., & Geller, G. (2008). Children's and parents' perspectives on open-label use of placebos in the treatment of ADHD. *Child: Care, Health and Development, 34*(1), 111–120. https://doi.org/10.1111/j.1365-2214.2007.00743.x.

Schaefer, M., Harke, R., & Denke, C. (2016). Open-label placebos improve symptoms in allergic rhinitis: A randomized controlled trial. *Psychotherapy and Psychosomatics, 85*(6), 373–374. https://doi.org/10.1159/000447242.

Schafer, S. M., Colloca, L., & Wager, T. D. (2015). Conditioned placebo analgesia persists when subjects know they are receiving a placebo. *The Journal of Pain, 16*(5), 412–420. https://doi.org/10.1016/j.jpain.2014.12.008.

Shah, K. N., Panchal, D. J., Vyas, B. M., & Patel, V. J. (2009). Use of placebo: Knowledge, attitude and practice among medical practitioners. *Indian Journal of Medical Sciences, 63*(10), 472–473. https://doi.org/10.4103/0019-5359.57640.

Shapiro, L. A. (2014). *The Routledge handbook of embodied cognition*. Routledge.

Sherman, R., & Hickner, J. (2008). Academic physicians use placebos in clinical practice and believe in the mind-body connection. *Journal of General Internal Medicine, 23*(1), 7–10. https://doi.org/10.1007/s11606-007-0332-z.

Siegel, S., Baptista, M. A., Kim, J. A., McDonald, R. V., & Weise-Kelly, L. (2000). Pavlovian psychopharmacology: The associative basis of tolerance. *Experimental and Clinical Psychopharmacology, 8*(3), 276–293.

Subkov, A. A., & Zilov, G. N. (1937). The role of conditioned reflex adaption in the origin of hyperergic reactions. *Bulletin de Biologie et Medecine Experimentale, 4*, 294–296.

Tam, N. T., Huy, N. T., Thoa le, T. B., Long, N. P., Trang, N. T., Hirayama, K., et al. (2015). Participants' understanding of informed consent in clinical trials over three decades: Systematic review and meta-analysis. *Bulletin of the World Health Organization, 93*(3), 186–198H. https://doi.org/10.2471/BLT.14.141390.

Thompson, J. J., Ritenbaugh, C., & Nichter, M. (2009). Reconsidering the placebo response from a broad anthropological perspective. *Culture, Medicine and Psychiatry, 33*(1), 112–152. https://doi.org/10.1007/s11013-008-9122-2.

Tilburt, J. C., Emanuel, E. J., Kaptchuk, T. J., Curlin, F. A., & Miller, F. G. (2008). Prescribing "placebo treatments": Results of national survey of US internists and rheumatologists. *BMJ, 337*, a1938. https://doi.org/10.1136/bmj.a1938.

CHAPTER THIRTEEN

Placebo and Psychotherapy: Differences, Similarities, and Implications

Jens Gaab*[,1], Cosima Locher*, Charlotte Blease[†,‡]

*Clinical Psychology and Psychotherapy, University of Basel, Basel, Switzerland
[†]School of Psychology, University College Dublin, Dublin, Ireland
[‡]Program in Placebo Studies and the Therapeutic Encounter, Harvard Medical School, Boston, MA, United States
[1]Corresponding author: e-mail address: jens.gaab@unibas.ch

Contents

Abstract

The placebo and psychotherapy are both effective psychological interventions. Next to being characterized by their own and specific controversies and debates, there is a persistent—and least for psychotherapy—looming notion that these two interventions share more than just the first letter. Based on Grünbaum's influential conceptualization of placebo, this chapter critically reviews both the time-honored claim that psychotherapy is a placebo as well as the argument that the placebo concept does not translate to psychotherapy. We conclude that there is an unwanted proximity between these two interventions and that empirical attempts to separate the "wheat from the chaff" in psychotherapy research face several distinctive challenges and thus are often methodologically comprised by the integrity of the placebo. However, drawing on recent, innovative research, we conclude that psychotherapy can be saved, i.e., shown to be distinct from the placebo, by employing study designs derived from the placebo research. We conclude that the placebo concept has profound implications for psychotherapy, psychotherapy research, and last but not least its ethical practice.

International Review of Neurobiology, Volume 138
ISSN 0074-7742
https://doi.org/10.1016/bs.irn.2018.01.013

1. PSYCHOTHERAPY AND PLACEBO: DEFINITIONS AND DISTINCTIONS

Psychotherapy is a generic label for a large and growing number of interventions, which share certain and defining characteristics, such as being intended to be therapeutic, being based on psychological principles and their derivative treatment methods, and being delivered by trained professionals. As such, psychotherapy—or more precisely psychotherapies—are well established, efficacious, and accepted treatments for an array of mental disorders and psychological problems (Goldfried, 2013). While the same can be said about the placebo or more precisely placebos and maybe with the exception that its use still *not* widely legitimated but nevertheless common in clinical practice (Fässler, Meissner, Schneider, & Linde, 2010, *BMC Medicine*), equating these interventions is seen unwarranted or as Irving Kirsch eloquently put it:

> *There is a problem with identifying psychotherapy with the placebo effect. A placebo is something that is sham, fake, false, inert, and empty. Psychotherapy is none of these. In this sense, it is different from medical placebos, and it does not deserve the pejorative connotations associated with the term.*
>
> **Cited from Kirsch (2005, p. 797)**

While there is little reason to doubt that relevant psychotherapy bodies as well as psychotherapists do their best to ensure the legitimacy of their professional actions, the aforementioned distinction between psychotherapy and placebo rests on the assumption that an intervention is only a placebo if given intentionally, since giving something that is inert and empty openly would not qualify as being sham or fake (Kaptchuk et al., 2010).

Defining the placebo can be seen as either a simple or a difficult matter. With regard to the former, *placebos are drugs, devices, or other treatments that are physically and pharmacologically inert* (and that) *do not, by definition, have any direct therapeutic effects on the body* (cited from Wager & Atlas, 2015). In this understanding, *placebo effects are thus brain–body responses to context information that promote health and well-being* (cited from Wager & Atlas, 2015). While this perspective focuses on the properties of placebos and elegantly avoids the conundrum of defining the placebo by its effects, it poses a problem as it by definition would also include psychotherapy as a placebo, as it is *pharmacologically inert* and it has no *direct therapeutic effects on the body*. And with regard to the latter, Adolf Grünbaum conceptualization (1981, 1986) *has been proposed by some authors as the best attempt thus far* (cited from Howick, 2016) and is *arguably the most sophisticated model in the literature* (cited from Alfano, 2015). Interestingly and relevant to the question at hand, Grünbaum's definition of what constitutes a placebo is based on the respective treatment theory for a given disorder as well as on the communication of the treatment provider and not on its properties or effects. Thus, on Grünbaum's model, a given treatment consists of a treatment theory-driven distinction between characteristic and incidental constituents, and *the generic distinction between placebos and non-placebos (depends on) whether the characteristic treatment factors do play a therapeutic role for (a given disorder) ... or not* (cited from Grünbaum, 1986, p. 33). A generic placebo is understood as an intervention containing no characteristic constituent for the disorder at hand on the basis of the respective treatment theory and which becomes an intended placebo when knowingly administered by the treatment provider to a patient, who remains otherwise ignorant of the true nature of the intervention. This theory-driven definition has the advantage to avoid the aforementioned trap to equating placebo and psychotherapy on the basis of their non-pharmacological properties.

With regard to the distinction between placebos and psychotherapy and according to Grünbaum's distinction between intended and inadvertent placebos, one might be tempted to easily rule out the possibility that psychotherapy is based on the intentional use of incidental treatment constituents, i.e., which do not play a therapeutic role for a given disorder. But this proposition rests on the assumption that the assumed characteristic constituents are truly responsible for an observed treatment effect, which implies that the underlying treatment theory correctly identifies and differentiates between incidental and characteristic treatment constituents. If this were not the case, psychotherapy would run the risk of qualifying as an inadvertent placebo.

The claim that psychotherapy is—or more precisely: psychotherap*ies* are—effective as a result of its incidental treatment constituents is far from new and dates back to the 1930s:

> *It is justifiable to wonder whether the factors alleged to be operating in a given therapy are identical with the factors that actually are operating, and whether the factors that actually are operating in several different therapies may not have much more in common than have the factors alleged to be operating.*
>
> **Cited from Rosenzweig (1936, p. 412)**

Interestingly, this assumption marked the beginning of the more recent, so-called *great psychotherapy debate* (Wampold & Imel, 2015) which has influenced psychotherapy research and practice today (Hofmann & Barlow, 2014; Laska & Wampold, 2014) and has recently been addressed prominently:

> *The old debate about whether or not psychotherapy and placebos have similar mechanisms consists of ascertaining whether psychotherapy is nothing but a placebo effect, and thus whether a placebo procedure is a very simple form of psychotherapy.*
>
> **Cited from Benedetti (2009, p. 141)**

Furthermore, different psychotherapy approaches have been related to placebo, such as Eysenck's infamous claim that *psychotherapy* (i.e., any psychotherapy excluding behavioral therapy) *works, as far as it does, by means of nonspecific or placebo effects* (cited from Eysenck, 1994, p. 490), the puzzling argument that due to the *significant epistemic pressures in the therapeutic encounter, the insight-oriented psychotherapies are highly susceptible to generating placebo insights, that is, illusions, deceptions, and adaptive self-misunderstandings that convincingly mimic veridical insight but have no genuine explanatory power* (cited from Jopling, 2001, p. 19), the proposition that *because psychotherapy is less burdened by doubts about the placebo effect that it was able to come to its aid when it was orphaned by medicine (and that) it is vain to expect something with so long a history as the placebo effect to disappear from the practices of healing* (cited from Justman, 2011, p. 95), and McNally's (1999) equation of modern-day eye movement desensitization and reprocessing with mid-18th century animal magnetism (Franklin et al., 1784/2002)—the former is an accredited psychological treatment by the Society of Clinical Psychology of the American Psychological Association.

Thus, there are strong reasons to address the theoretical as well as empirical similarities and differences between psychotherapy and placebo in order to elucidate their relationship and the possible practice implications for using psychological treatments.

2. EMPIRICAL APPROACHES TO CONTROL FOR INCIDENTAL TREATMENT CONSTITUENTS IN PSYCHOTHERAPY

In order to establish the effectiveness of specific factors of a given intervention in clinical trials, its incidental treatment constituents, such as expectancy and a caring patient–practitioner relationship, usually are controlled for with the administration of a placebo treatment; ideally placebos should be indistinguishable from the verum counterpart among patients as well as the treating clinicians and/or researchers. In order to control for incidental treatment constituents, the placebo should therefore mimic the verum in every respect except the characteristic factor hypothesized to be remedial. In psychotherapy outcome research, ideally, these control conditions would contain all and only treatment constituents which are considered incidental by the respective treatment theory. Due to this dependency on each respective treatment theory, control conditions in psychotherapy clearly differ in content and focus between trials. But while this does not automatically impose a psychotherapy-specific methodological problem, as medicinal placebos also differ in content and focus between trials, the problem in psychotherapy arises from the fact that neither single nor double blinding is possible in psychotherapy. Given that psychotherapy fully relies on the verbal interaction between psychotherapists and patients, it is hard to imagine how we might implement control conditions in psychotherapy trials without psychotherapists and patients being fully aware of the true nature of their treatment. Of course, it should be emphasized that breaking blind also regularly occurs in pharmacological placebo-controlled trials (Margraf, Ehlers, Roth, et al., 1991) and the knowledge among trial participants and clinicians that they are receiving or administering an active medication can have clear effects on trial outcomes (Rutherford et al., 2017) as can the use of active placebos, which mimic the side effects of active verum (Moncrieff, Wessely, & Hardy, 2004). In summary, the de facto transparency of the control condition can *play havoc with the logic of the design* (cited from Baskin, Tierney, Minami, & Wampold, 2003, p. 974). Where this is evident in biomedical clinical trials, the situation is accentuated in psychotherapy research. In principle, we argue that while the use of controls is conceivable in psychotherapy research, designing adequate placebos is fraught with many more challenges than are (typically) present in biomedical contexts.

We identify four interrelated challenges when it comes to the problem of designing adequate placebos in clinical trials in psychotherapy.

1. *Patients' expectancies*: Patients' perceptions about the credibility of therapists as well as of version of therapy may mediate response expectancies. These factors are not consistently empirically assessed nor adequately controlled for. Recognition of this problem can be traced back to Kazdin and Wilcoxon (1976) who stated with regard to claims of specificity of systematic desensitization—notably a founding tenet of behavior therapy—that *(the) vast majority of studies have not determined empirically whether desensitization and nonspecific treatment control conditions are equal in credibility and expectancy for improvement generated in the clients*. In a follow-up analysis of

Kazdin and Wilcoxon's (1976) seminal paper—which coincided with a substantial decline in scientific interest in systematic desensitization (McGlynn, Smitherman, & Gothard, 2004)—Locher, Hasler, and Gaab (2016) also concluded that the restrictions in control conditions are a prerequisite to find specific effects in this line of outcome research.

2. *Mimicking the structure of psychotherapy*: The failure to operationalize the structure of verum psychotherapy in control conditions poses a serious threat to the validity of outcome research. The differences may be subtle, but they can lead to clinically relevant effects. For example, Baskin et al. (2003) meta-analyzed 21 studies comparing a specific psychological treatment with placebo controls, i.e., that lacked the active ingredients of the specific treatment. Based on a careful rating of the structural qualities (such as number, length, format of sessions, and thematic restrictions), aggregated specific effects i.e., differences between psychotherapy and placebo conditions in studies with dissimilarly structured placebo control groups were $d = 0.47$, while those in studies with equivalently structured placebo control groups were negligible, i.e., $d = 0.15$.
3. *Researcher allegiance*: One possible cause of the subtle changes in the operationalization of placebo control conditions in psychotherapy research is researcher allegiance (Gerger & Gaab, 2016). Research has demonstrated that controlling for researcher allegiance reduced otherwise observed differences between active psychotherapy treatments and placebo control conditions to virtual zero (Cuijpers et al., 2012; Miller, Wampold, & Varhely, 2008).
4. *Differentiating between characteristic and incidental factors*: Finally, the problems with placebo control groups in psychotherapy outcome research might be considered a consequence of the challenges involved in differentiating between incidental and characteristic factors in psychotherapy. For example, Serfaty et al. (2009) fashioned a talking control condition to establish specific effect of cognitive-behavioral therapy for depressed older people in primary care. Although patients rated the talking control condition as acceptable (Serfaty, Csipke, Haworth, Murad, & King, 2011), it included the exclusion of dialogue about emotional issues. For example, *if the patient said, "My daughter does not like me as she never comes to visit me," the therapist would ask, "How many children do you have?"* (cited from Serfaty et al., 2009, p. 1334). Does this imply that talking about emotional issues is a characteristic constituent of psychotherapy?

The omission of reference to any dialogue referencing emotional subjects, or indeed, of emotional affect, would render many professional as well as normal social encounters highly unusual. Consider, for example, the range of emotional issues that are referenced outside of psychotherapy, including in close relation and friendships, with clerics, hairdressers, and barbers or even vendors in the local grocery store. Theoretically, placebo control groups should ideally only be restricted to characteristic treatment constituents—those considered specific according to the underlying treatment theory. Clearly, disclosing, addressing, and discussing personal and emotional relevant topics are not a specific psychotherapeutic feat but rather a genuine aspect of human interactions (Wampold, 2012). Based on these considerations the differentiation of placebo and psychotherapy from an empirical point of view is linked to the operationalization of the employed placebo and is dependent on how clinical researchers theoretically delineate between treatment incidental and characteristic constituents in therapy. The aforementioned problems have led prominent placebo and psychotherapy researchers to claim that *in evaluating the efficacy of psychotherapy, the placebo effect cannot and should not be controlled* (cited from Kirsch, Wampold, & Kelley, 2016, p. 121).

3. PLACEBO AND PSYCHOTHERAPY: UNWANTED PROXIMITY

But while the argument *not* to control for the placebo effect in psychotherapy research appears justified from a methodological perspective, it is deeply problematic insofar as it rules out the possibility of establishing the effectiveness of specific factors, i.e., on Grünbaum's account: the characteristic factors, in psychotherapy. In this regard, Cuijpers and Cristea (2015) pointed out that using waiting-list control groups results in larger effect sizes and listed this approach as a method that can help to *prove that your intervention is effective (even when it is not)* (cited from Cuijpers & Cristea, 2015, p. 3).

Things are further complicated by the fact that "common factors"—which include response expectancies; the therapeutic alliance; therapist empathy; therapists' positive regard for patients which are considered to be the major determinants of psychotherapy outcome across different psychotherapy modalities by some (see Cuijpers, van Straten, Andersson, & van Oppen, 2008; Wampold & Imel, 2015)—are often classified as "placebo" outside psychotherapy research. While there is still some controversy among researchers about the role of these so-called common factors, there is widespread agreement among psychotherapy researchers and psychotherapists that—whatever the role of specific treatment factors—common factors are significant mediators of outcome (Cuijpers, 2016; Huppert, Fabbro, & Barlow, 2006; Lambert & Barley, 2002; Marcus, O'Connell, Norris, & Sawaqdeh, 2014). However, it is important to emphasize that typically these common factors are considered incidental constituents of treatments as they are actively controlled for by the majority of researchers who undertake clinical research into psychological treatments.

Interestingly and relevant to the matter at hand, placebos and psychotherapy have independently been classified as meaning-transforming interventions (Frank, 1986; Moerman & Jonas, 2002). This indicates a rather disagreeable proximity between these treatments, which otherwise are associated with rather different connotations. In this perspective, both interventions are thought to provide an alternative meaning to the experience of symptoms, which would be delivered in the context of an empathetic and trustful clinical encounter. Interestingly, as it is part of the trick for placebos that the explanation given does not need to be true or verum, but only plausible, the same holds true for psychotherapy (Wampold, 2007).

However, these considerations are only consequential for the clinical application of psychotherapy if indeed psychotherapy is correctly conceived as a placebo—or put otherwise, if a psychotherapy placebo—i.e., an otherwise *sham, fake, false, inert, and empty (intervention)* (cited from Kirsch, 2005, p. 797), which would be deceptively administered as an active psychotherapy treatment—is possible and has clinically significant effects. In this regard, it is interesting to note that although placebos can and do take many forms, including acupuncture needles (Kaptchuk et al., 2008), surgical procedures (Jonas, Crawford, Colloca, et al., 2015), and even cleaning hotel rooms (Crum & Langer, 2007), the plethora of placebo research employs placebos with a more or less explicit medicinal meaning, such as pills, and that placebos with a psychotherapeutical meaning are comparably scarce. In this line of reasoning and according to the understanding of placebo effects as meaning responses (Moerman & Jonas, 2002) a psychotherapy placebo would not need to fully resemble "real" psychotherapy, but just do so sufficiently well so that patients would accept the administered placebo to have a psychotherapeutical meaning.

Given that the gold standard of clinical trials aims to establish the effectiveness of specific treatment components, how can one fashion a placebo that would lack only its specific components and yet would also convincingly mimic a verum treatment? One intriguing way to handle this problem has been proposed by Kim, Wollburg, and Roth (2012) and directly compares two opposing psychotherapeutical treatments, i.e., breathing trainings to either increase or decrease end-tidal partial pressure of carbon dioxide in patients with panic disorder. Notice that such an approach prevents patients, therapists, and researchers from identifying which of the administrated treatments is the placebo; yet it clearly rules out the possibility that both can be "verum," since both treatments have opposing characteristic constituents, i.e., hyper- or hypnocapic breathing. Thus, from the perspective of the opposing treatment (and its underlying treatment theory), either treatment could be considered a placebo. Interestingly, both treatments showed the same, impressively stable effects, both at postintervention and at follow-up, while a wait-list control group did not show a significant change. Furthermore, the quality of the patient-rated therapeutic alliance as well as patient-rated confidence that the therapy would produce improvement was the main and only predictors of observed remedial effects (Kim, Roth, & Wollburg, 2015). Finally, and relevant to this discussion, it is interesting to note that both interpersonal psychotherapy, which has been found to be effective for an array of mental disorders (Cuijpers, Donker, Weissman, Ravitz, & Cristea, 2016), and present-centered therapy, which is now a now empirically supported treatment (American Psychological Association, Society of Clinical Psychology, 2018), were originally devised as control conditions in a pharmacological and psychotherapy trials, respectively (Frost, Laska, & Wampold, 2014; Weissman, 2006).

4. THE IMPLICATIONS OF THE RELATIONSHIP BETWEEN PLACEBO AND PSYCHOTHERAPY

The study of Kim et al. (2015, 2012) clearly demonstrates that placebos with a psychotherapeutical meaning are both possible and clinically effective. Although we cannot infer that psychotherapy is placebo, neither can we preclude the possibility. Thus, and reframing Kirsch's aforementioned statement, psychotherapy is not *sham, fake, false, inert, and empty* – but it can be. This possibility has important implications, which in our view could be seen as both a challenge and an opportunity for psychotherapy research.

First, in order to both better understand the mechanisms driving the impressive effects of psychotherapy and to be able to comply with the moral principle of respect for autonomy and the patient's right to self-determination (Trachsel & Gaab, 2016), it is justified to on the one hand to investigate the role of those factors, which are considered incidental to most treatment theories, but which have been found to be underlying psychotherapy's impressive effects, and on the other hand—if necessary—to revise conventionally held theories about how these treatments work. If therapeutic alliance and treatment expectancies are highly relevant not only for placebo but also for psychotherapy effects, then how do they work and how does this translate to the different etiologies for different disorders? Clearly, psychotherapy could benefit from placebo research as this allows the controlled examination of treatment components (e.g., Kaptchuk et al., 2008) and could help to introduce new designs to test psychotherapy components for specificity. With regard to the latter, the open and hidden administration of a given treatment is one promising avenue. In open-hidden experiments an otherwise identical intervention is given to research participants who either know or do not know when (or if) the treatment will be administered or withdrawn. These experiments allow to examine the size of placebo effects within assumed verum treatments, without using placebos: any differences between open and hidden conditions are then understood to be placebo-like effects, since the intervention itself is not a placebo (e.g., Lund, Vase, Petersen, Jensen, & Finnerup, 2014). Using this approach, we recently observed differences in the effects of the psychotherapeutical method of expressive writing when administered open or hidden, which were of the same magnitude as the effects of this intervention itself (Tondorf et al., 2017).

Also, and importantly, these existing as well as future findings should be incorporated in existing psychotherapy treatment theories. For example, while psychotherapy treatment manuals and textbook in general contain some information on the importance of the therapeutic alliance and the plausibility and acceptance of the treatment rational to patients, this is seldom adequately incorporated in treatment theory or widely within psychotherapy education, nor is it typically communicated to patients (Locher, Meier, & Gaab, in press). Second and most important for the clinical practice of psychotherapy, patients have the right to be furnished with truthful information about how treatments work. Information about the nature of therapy—such as the importance of the therapeutic alliance, and patients' expectancies about treatment—is not morally trivial. Ethicists and clinical psychologists have argued that disclosure of so-called common factors (or in Grünbaum's terminology, the incidental features of treatment) is a moral obligation, one that may even positively influence treatment effects (Blease, 2015; Blease, Lilienfeld, & Kelley, 2016; Gaab, Blease, Locher, & Gerger, 2016; Trachsel & Gaab, 2016). It might be countered that informing patients about the role of the therapeutic alliance and of treatment expectancies as well as *not* informing patients about the role of specific methods and techniques—as they are *not* responsible for the majority of possible effects—would both diminish the effects of just those very factors, i.e., impair the therapeutic alliance as well as decrease the expectancy of a treatment effect. First, we argue that this is an illogical argument—one cannot deny the importance of expectancy and therapeutic alliance on the one hand and on the other hand plead not to disclose the lack of importance of so-called specific as this would have a negative effect on these very factors, i.e., expectancy and alliance. Also, it is inconsistent to defend the importance of one set of (so-called) characteristic treatment constituents yet deny the disclosure of other treatment constitutes—those which the evidence-based indicates may be better labeled as characteristic constituents. Second, this line of reasoning may also be empirically doubted. It is possible that disclosure of common factors (especially those relating to the importance of the therapeutic alliance, and empathy) may positively influence patients' decisions to stay in therapy, or to find another therapist or version of treatment, rather than simply to drop out and leave therapy for good (Blease, Kelley, & Trachsel, 2018; Gaab et al., 2016). In addition, disclosure of (so-called) incidental treatment constituents may not diminish treatment outcomes. The administration of open-label placebos shows that full disclosure of placebo factors may still lead to clinically relevant effects (Charlesworth et al., 2017)—indeed, research shows that these effects are similar to deceptively administrated placebos (Locher et al., 2017).

In conclusion, the relationship, similarities, and differences between psychotherapy and placebo are interesting and highly relevant for the understanding of these two psychological interventions and thus should not be further neglected by research, psychotherapy education, and clinical practice. And of course, psychotherapy needs to be anything but *sham, fake, false, inert, and empty* (cited from Kirsch, 2005, p. 797), but it also needs a dose of deeper self-insight. It needs to understand why something that is *sham, fake, false, inert, and empty* has such sizeable effects.

CONFLICT OF INTEREST

The authors declare that they have no financial conflict of interest concerning this chapter.

REFERENCES

Alfano, M. (2015). Placebo effects and informed consent. *The American Journal of Bioethics, 15*(10), 3–12.

American Psychological Association, Society of Clinical Psychology. (2018). *Research-supported psychological treatments.* Retrieved September 11, 2017, from http://www.div12.org/PsychologicalTreatments/disorders/ptsd_main.php.

Baskin, T. W., Tierney, S. C., Minami, T., & Wampold, B. E. (2003). Establishing specificity in psychotherapy: A meta-analysis of structural equivalence of placebo controls. *Journal of Consulting and Clinical Psychology, 71*, 973–979. https://doi.org/10.1037/0022-006X.71.6.973.

Benedetti, F. (2009). Placebo effects. *Understanding the mechanisms in health and disease.* New York: Oxford University Press.

Blease, C. R. (2015). Talking more about talking cures: Cognitive behavioural therapy and informed consent. *Journal of Medical Ethics, 41*, 750–755.

Blease, C. R., Kelley, J. M., & Trachsel, M. (2018). Informed consent in psychotherapy: Implications of evidence-based practice. *Journal of Contemporary Psychotherapy.* https://doi.org/10.1007/s10879-017-9372-9.

Blease, C. R., Lilienfeld, S. O., & Kelley, J. M. (2016). Evidence-based practice and psychological treatments: The imperatives of informed consent. *Frontiers in Psychology, 7*, Art. No.: 1170. https://doi.org/10.3389/fpsyg.2016.01170.

Charlesworth, J. E. G., Petkovic, G., Kelley, J. M., Hunter, M., Onakpoya, I., Roberts, N., et al. (2017). Effects of placebos without deception compared with no treatment: A systematic review and meta-analysis. *Journal of Evidence-Based Medicine, 10*(2), 97–107.

Crum, A. J., & Langer, E. J. (2007). Mind-set matters: Exercise and the placebo effect. *Psychological Science, 18*(2), 165–171.

Cuijpers, P. (2016). Are all psychotherapies equally effective in the treatment of adult depression? The lack of statistical power of comparative outcome studies. *Evidence-Based Mental Health, 19*, 39–42. https://doi.org/10.1136/eb-2016-102341.

Cuijpers, P., & Cristea, I. A. (2015). How to prove that your therapy is effective, even when it is not: A guideline. *Epidemiology and Psychiatric Sciences, 25.* https://doi.org/10.1017/S2045796015000864.

Cuijpers, P., Donker, T., Weissman, M. M., Ravitz, P., & Cristea, I. A. (2016). Interpersonal psychotherapy for mental health problems: A comprehensive meta-analysis. *American Journal of Psychiatry, 173*, 680–687.

Cuijpers, P., Driessen, E., Hollon, S. D., van Oppen, P., Barth, J., & Andersson, G. (2012). The efficacy of non-directive supportive therapy for adult depression: A meta-analysis. *Clinical Psychology Review, 32*, 280–291. https://doi.org/10.1016/j.cpr.2012.01.003.

Cuijpers, P., van Straten, A., Andersson, G., & van Oppen, P. (2008). Psychotherapy for depression in adults: A meta-analysis of comparative outcome studies. *Journal of Consulting and Clinical Psychology, 76*, 909. https://doi.org/10.1037/a0013075.

Eysenck, H. J. (1994). The outcome problem in psychotherapy: What have we learned? *Behaviour Research and Therapy, 32*, 477–495. https://doi.org/10.1016/0005-7967(94)90135-X.

Fässler, M., Meissner, K., Schneider, A., & Linde, K. (2010). Frequency and circumstances of placebo use in clinical practice—A systematic review of empirical studies. *BMC Medicine, 8*, 15. https://doi.org/10.1186/1741-7015-8-15.

Frank, J. D. (1986). Psychotherapy—The transformation of meanings: Discussion paper. *Journal of the Royal Society of Medicine, 79*, 341–346.

Franklin, B., Majault, Le Roy, Sallin, Bailly, J. S., D'Arcet, et al. (1784/2002). Report of the commissioners charged by the King with the examination of animal magnetism. 1784. *International Journal of Clinical and Experimental Hypnosis, 50*(4), 332–363. https://doi.org/10.1080/00207140208410109.

Frost, N. D., Laska, K. M., & Wampold, B. E. (2014). The evidence for present-centered therapy as a treatment for posttraumatic stress disorder. *Journal of Traumatic Stress, 27*, 1–8.

Gaab, J., Blease, C., Locher, C., & Gerger, H. (2016). Go open: A plea for transparency in psychotherapy. *Psychology of Consciousness: Theory, Research, and Practice, 3*(2), 175–198. https://doi.org/10.1037/cns0000063.

Gerger, H., & Gaab, J. (2016). Researcher allegiance as hidden moderator in psychotherapy outcome research. *Verhaltenstherapie, 26*, 41–45. https://doi.org/10.1159/000443543.

Goldfried, M. R. (2013). What should we expect from psychotherapy? *Clinical Psychology Review, 33*, 862–869. https://doi.org/10.1016/j.cpr.2013.05.003.

Grünbaum, A. (1981). The placebo concept. *Behaviour Research and Therapy, 19*, 157–167. https://doi.org/10.1016/0005-7967(81)90040-1.

Grünbaum, A. (1986). The placebo concept in medicine and psychiatry. *Psychological Medicine, 16*, 19–38. https://doi.org/10.1017/S0033291700002506.

Hofmann, S. G., & Barlow, D. H. (2014). Evidence-based psychological interventions and the common factors approach: The beginnings of a rapprochement? *Psychotherapy, 51*, 510–513. https://doi.org/10.1037/a0037045.

Howick, J. (2016). The relativity of 'placebos': Defending a modified version of Grünbaum's definition. *Synthese, 194*(4), 1363–1396.

Huppert, J. D., Fabbro, A., & Barlow, D. H. (2006). Evidence-based practice and psychological treatments in goodheart, CD. In A. E. Kazdin & R. J. Sternberg (Eds.), *Evidence-based psychotherapy: Where practice and research meet* (pp. 131–152). Washington, DC: American Psychological Association.

Jonas, W. B., Crawford, C., Colloca, L., et al. (2015). To what extent are surgery and invasive procedures effective beyond a placebo response? A systematic review with meta-analysis of randomized, sham controlled trials. *BMJ Open, 5*, e009655.

Jopling, D. A. (2001). Placebo insight: The rational- ity of insight-oriented psychotherapy. *Journal of Clinical Psychology, 57*, 19–36. 1 19::AID-JCLP4 3.0.CO;2-Z. https://doi.org/10.1002/1097-4679(200101)57.

Justman, S. (2011). From medicine to psychotherapy: The placebo effect. *History of the Human Sciences, 24*, 95–107. https://doi.org/10.1177/0952695110386655.

Kaptchuk, T. J., Friedlander, E., Kelley, J. M., Sanchez, M. N., Kokkotou, E., Singer, J. P., et al. (2010). Placebos without deception: A randomized controlled trial in irritable bowel syndrome. *PLoS One, 5*(12), e15591.

Kaptchuk, T. J., Kelley, J. M., Conboy, L. A., Davis, R. B., Kerr, C. E., Jacobson, E. E., et al. (2008). Components of placebo effect: Randomised controlled trial in patients with irritable bowel syndrome. *BMJ: British Medical Journal, 336*, 999–1003.

Kazdin, A. E., & Wilcoxon, L. A. (1976). Systematic desensitization and nonspecific treatment effects: A methodological evaluation. *Psychological Bulletin, 83*, 729–758. https://doi.org/10.1037/0033-2909.83.5.729.

Kim, S., Roth, W. T., & Wollburg, E. (2015). Effects of the therapeutic relationship expectancy, and credibility in breathing therapies for anxiety. *Bulletin of the Menninger Clinic, 79*(2), 116–130.

Kim, S., Wollburg, E., & Roth, W. T. (2012). Opposing breathing therapies for panic disorder: A randomized controlled trial of lowering vs raising end tidal P(CO2). *Journal of Clinical Psychiatry, 73*, 931–939.

Kirsch, I. (2005). Placebo psychotherapy: Synonym or oxymoron? *Journal of Clinical Psychology, 61*, 791–803. https://doi.org/10.1002/jclp.20126.

Kirsch, I., Wampold, B. E., & Kelley, J. (2016). Controlling for the placebo effect in psychotherapy: Noble quest or tilting at windmills? *Psychology of Consciousness: Theory, Research, and Practice, 3*(2), 121–131.

Lambert, M. J., & Barley, D. E. (2002). Research summary on the therapeutic relationship and psychotherapy outcome. Expectations and preferences. In J. C. Norcross (Ed.), *Psychotherapy relationships that work: Therapist contributions and responsiveness to patients* (pp. 17–32). London: Oxford University Press.

Laska, K. M., & Wampold, B. E. (2014). Ten things to remember about common factor theory. *Psychotherapy*, *51*(4), 519–524. https://doi.org/10.1037/a0038245.

Locher, C., Frey Nascimento, A., Kirsch, I., Kossowsky, J., Meyer, A., & Gaab, J. (2017). Is the rationale more important than deception? A randomized controlled trial of open-label placebo analgesia. *Pain*, *158*, 2320–2328. https://doi.org/10.1097/j.pain.0000000000001012.

Locher, C., Hasler, S., & Gaab, J. (2016). When do psychotherapeutic placebos work? A critical review on the example of systematic desensitization. *Verhaltenstherapie*, *26*, 9–20.

Locher, C., Meier, S. and Gaab, J. (in press). Psychotherapy: A world of meanings, *Perspectives in Biology & Medicine*.

Lund, K., Vase, L., Petersen, G. L., Jensen, T. S., & Finnerup, N. B. (2014). Randomised controlled trials may underestimate drug effects: Balanced placebo trial design. *PLoS One*, *9*(1), e84104.

Marcus, D. K., O'Connell, D., Norris, A. L., & Sawaqdeh, A. (2014). Is the Dodo bird endangered in the 21st century? A meta-analysis of treatment comparison studies. *Clinical Psychologist*, *34*, 519–530. https://doi.org/10.1016/j.cpr.2014.08.001.

Margraf, J., Ehlers, A., Roth, W. T., et al. (1991). How 'blind' are double-blind studies? *Journal of Consulting and Clinical Psychology*, *59*, 184–187.

McGlynn, F. D., Smitherman, T. A., & Gothard, K. D. (2004). Comment on the status of systematic desensitization. *Behavior Modification*, *28*, 194–205.

McNally, R. J. (1999). EMDR and mesmerism: A comparative historical analysis. *Journal of Anxiety Disorders*, *13*, 225–236. https://doi.org/10.1016/S0887-6185(98)00049-8.

Miller, S., Wampold, B., & Varhely, K. (2008). Direct comparisons of treatment modalities for youth disorders: A meta-analysis. *Psychotherapy Research*, *18*, 5–14. https://doi.org/10.1080/10503300701472131.

Moerman, D. E., & Jonas, W. B. (2002). Deconstructing the placebo effect and finding the meaning response. *Annals of Internal Medicine*, *136*, 471–476. https://doi.org/10.7326/0003-4819-136-6-200203190-00011.

Moncrieff, J., Wessely, S., & Hardy, R. (2004). Active placebos versus antidepressants for depression. *The Cochrane Database of Systematic Reviews*, (1). Art. No.: CD003012.pub2. https://doi.org/10.1002/14651858.CD003012.pub2.

Rosenzweig, S. (1936). Some implicit common factors in diverse methods of psychotherapy. *American Journal of Orthopsychiatry*, *6*, 412–415. https://doi.org/10.1111/j.1939-0025.1936.tb05248.x.

Rutherford, B. R., Wall, M. M., Brown, P. J., Choo, T. H., Wager, T. D., Peterson, B. S., et al. (2017). Patient expectancy as a mediator of placebo effects in antidepressant clinical trials. *The American Journal of Psychiatry*, *174*(2), 135–142.

Serfaty, M., Csipke, E., Haworth, D., Murad, S., & King, M. (2011). A talking control for use in evaluating the effectiveness of cognitive-behavioral therapy. *Behaviour Research and Therapy*, *49*, 433–440.

Serfaty, M. A., Haworth, D., Blanchard, M., Buszewicz, M., Murad, S., & King, M. (2009). Clinical effectiveness of individual cognitive behavioral therapy for depressed older people in primary care: A randomized controlled trial. *Archives of General Psychiatry*, *66*, 1332–1340.

Tondorf, T., Kaufmann, L.-K., Degel, A., Locher, C., Birkhäuer, J., Gerger, H., et al. (2017). Employing open/hidden administration in psychotherapy research: A randomized-controlled trial of expressive writing. *PLoS One*, *12*(11), e0187400. https://doi.org/10.1371/journal.pone.0187400.

Trachsel, M., & Gaab, J. (2016). Disclosure of incidental constituents of psychotherapy as a moral obligation for psychiatrists and psychotherapists. *Journal of Medical Ethics*, *42*(8), 493–495. Epub May 11, 2016. https://doi.org/10.1136/medethics-2015-102986.
Wager, T. D., & Atlas, L. Y. (2015). The neuroscience of placebo effects: Connecting context, learning and health. *Nature Reviews. Neuroscience*, *16*(7), 403–418. https://doi.org/10.1038/nrn3976 [Review].
Wampold, B. E. (2007). Psychotherapy: The humanistic (and effective) treatment. *American Psychologist*, *62*, 857–873. https://doi.org/10.1037/0003-066X.62.8.857.
Wampold, B. E. (2012). Humanism as a common factor in psychotherapy. *Psychotherapy (Chicago, IL)*, *49*(4), 445–449.
Wampold, B. E., & Imel, Z. E. (2015). *The great psychotherapy debate: The research evidence for what works in psychotherapy*. Routledge.
Weissman, M. M. (2006). A brief history of interpersonal psychotherapy. *Psychiatric Annals*, *36*(8), 553–557.

CHAPTER FOURTEEN

Expectation-Focused Psychotherapy to Improve Clinical Outcomes

Bettina K. Doering*, Julia A. Glombiewski†, Winfried Rief‡,1
*Catholic University Eichstätt-Ingolstadt, Eichstätt, Germany
†University of Koblenz–Landau, Landau, Germany
‡Philipps-University Marburg, Marburg, Germany
[1]Corresponding author: e-mail address: rief@staff.uni-marburg.de

Contents

Abstract

Placebo research has shown that patients' expectations are among the strongest predictors of treatment outcome regarding various medical conditions. Therefore, interventions focusing on the change of such expectations might be most effective. This could be of relevance not only for outcome expectations but also for disease-specific, problem-specific, and treatment-specific expectations.

A theoretical model is introduced to illustrate the conditions for expectation change and persistence. Strategies for patients regarding how to devalue learning effects in expectation violation situations are exemplified, and ways to optimize expectation-violating effects are highlighted.

Data from a large randomized controlled trial serve as a practical example and demonstrate the effect of a presurgery optimization of expectations in patients scheduled for heart surgery (coronary artery bypass grafting). Disability scores that were obtained 6 months postsurgery confirmed that patients who participated in preoperative psychological interventions, which aimed at optimizing expectations, demonstrate the best outcome after heart surgery. Practical aspects of the intervention are presented, and suggestions on implementing these procedures to optimize outcome in medical interventions are discussed.

Psychotherapy is conceptualized as an intervention that should target expectation violation of disorder- and treatment-specific expectations.

International Review of Neurobiology, Volume 138
ISSN 0074-7742
https://doi.org/10.1016/bs.irn.2018.02.004

1. INTRODUCTION

Patients' expectations in the context of medical and psychotherapeutic treatment are a many-facetted construct. They may refer to the outcome of a specific treatment or to the anticipated outcome of one's health behavior (Laferton, Kube, Salzmann, Auer, & Shedden-Mora, 2017) or they may even be an integral part of the medical condition, e.g., in psychiatric disorders (Rief et al., 2015).

During the last decades, placebo research has accumulated abounding evidence that patient expectations are a highly relevant determinant of treatment outcome. This evidence pertains to the influence of expectations on the courses of various medical conditions (Enck, Bingel, Schedlowski, & Rief, 2013), e.g., in cardiovascular diseases where expectations can positively influence long-term mortality (Barefoot et al., 2011). Patient expectations also influence the outcomes of highly specialized and standardized medical procedures, such as surgical treatments (Auer et al., 2015). Looking at specific surgical treatments, research in cardiovascular diseases demonstrates that expectations influence adjustment to illness and quality of life in patients undergoing a heart transplantation (Leedham, Meyerowitz, Muirhead, & Frist, 1995) and well-being after implantation of a cardioverter defibrillator (Habibović, Pedersen, van den Broek, & Denollet, 2014). In the domain of musculoskeletal disorders, preoperative expectations were a significant predictor of long-term postoperative walking speed after hip and knee arthroplasty (van den Akker-Scheek, Stevens, Groothoff, Bulstra, & Zijlstra, 2007) and patient expectations also proved an important independent predictor of improved functional outcomes and satisfaction following total joint arthroplasty (Mahomed et al., 2002). Placebo researchers have demonstrated the prominent role of expectations for outcomes in pharmacotherapy (Benedetti et al., 2003; Bingel et al., 2011). Thus, there is a critical need for the development of expectation-focused therapy in medical treatments.

Expectations are influenced by both associative and observational learning, and may also be responsive to the patient–provider interaction. Placebo research often focuses on patient expectations concerning treatment outcomes. However, other treatment-related expectations may also be relevant, such as personal control (i.e., expectations about the influence that the patients themselves have over the course of their disease) through, for example, modification of health behaviors. Expectations concerning outcomes and personal control pertain to both medical and psychotherapeutic contexts and should be targeted in order to optimize treatment success.

In this chapter, we will introduce a model that illustrates the formation and persistence of expectations. We will describe the results of a randomized controlled trial which focuses on a psychological intervention, including expectation manipulation in the medical context. This trial exemplifies how placebo mechanisms can be harnessed in everyday clinical practice in order to optimize treatment outcomes. We also discuss how expectation-focused therapy can potentially be used to minimize negative outcomes and the occurrence of nocebo effects. We will focus on expectations as core features of mental disorders and conceptualize psychotherapy as an interventional process that should concentrate on changing expectations in order to change the course of the disorder (Rief & Glombiewski, 2016). Our conclusion illustrates future directions of this approach.

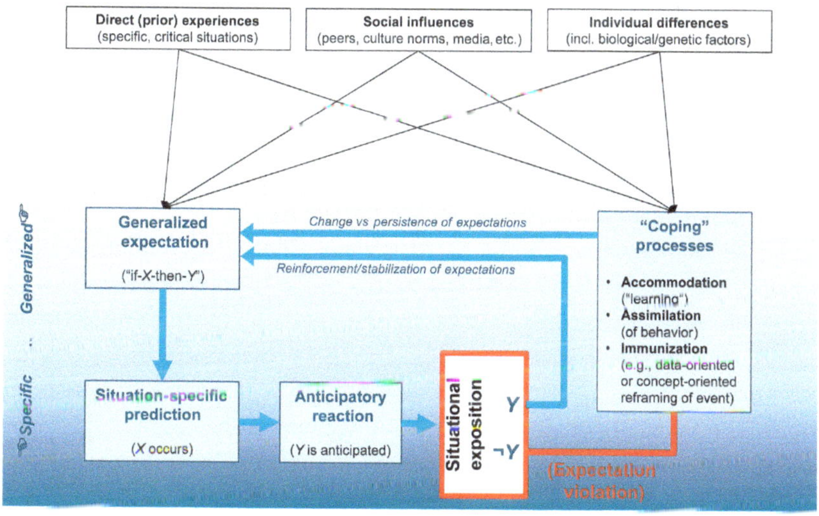

Fig. 1 The ViolEx model. This model illustrates the persistence and change of expectations in the face of confirming vs disconfirming situational evidence. *Adapted from Rief, W., Glombiewski, J. A., Gollwitzer, M., Schubö, A., Schwarting, R., & Thorwart, A. (2015). Expectancies as core features of mental disorders.* Current Opinion in Psychiatry, *28(5), 378–385.*

2. FORMATION, PERSISTENCE, AND CHANGE OF EXPECTATIONS

In 2015, several authors from different psychological subdisciplines developed the "ViolEx" model, which describes factors that might contribute to expectation change or expectation persistence (Rief et al., 2015; see Fig. 1). According to the ViolEx model, when thinking about optimizing patient expectations, it is important to consider that patients enter treatment with preexisting expectations. These may have been formed through past experiences (i.e., through learning/conditioning), social influences (e.g., media, peers), and individual differences, both biological and genetic.

In a recent study, we were able to demonstrate as a proof of principle that side effects can be acquired through classical conditioning (Rheker, Winkler, Doering, & Rief, 2017). In this study, healthy participants were randomized to an acquisition period taking either amitriptyline or placebo. Participants taking amitriptyline reported significantly more antidepressant-specific side effects after acquisition compared to the placebo group. After evocation, participants who underwent the conditioning procedure with amitriptyline reported significantly more antidepressant-specific side effects than those who never received amitriptyline, even though both groups received a placebo pill. Our results indicate that side effects can be learned using a conditioning paradigm. They may thus influence expectations toward and outcomes of subsequent treatment approaches.

Expectations are the basis for, and simultaneously the result of, situation-specific predictions, i.e., if *x* occurs, then *y* will follow (e.g., "If I undergo the surgery, I will be impaired and have to stay in the hospital for several weeks in order to recover"). Importantly, the presence of expectations alters how individuals deal with a given situation. In such a case, anticipatory reactions could occur (e.g., If a patient holds such a negative expectation toward the surgery, he may delay treatment, thereby worsening his/her condition and inadvertently contributing to a negative outcome). The anticipatory reaction could also be a physiological reaction such as general anxiety or, more specifically, something such as elevated blood pressure contributing to potential adverse events of treatment. In this example, the expectation could become a self-fulfilling prophecy. Once the patient in this example undergoes the surgery, the situational exposition occurs, i.e., the expectation is tested. If the consequences of the surgery concur with the previous expectation, the expectation will be reinforced and stabilized. It may also become more generalized and thus be applied to other medical treatments as well.

If, however, the predicted outcome does not occur (e.g., the patient is sent home after only a short stay at a hospital and is encouraged to be active), the expectation is violated. Expectation violation can lead to a change in the respective expectation ("the surgery did not lead to impairment"). The greater the discrepancy between the expected outcome and the reality, the more likely the expectation will be changed. However, in many instances expectations are maintained in the presence of obviously discouraging evidence (e.g., patient feels disabled although his doctors did not see any reason to hospitalize him). This fact points to the crucial role of information processing after the situational exposition. There are various facets of information processing that may work against expectation change. Contradictory evidence may be ignored, which is known as selective attention (e.g., regarding the postsurgery discomforts), thus maintaining the preexisting expectation. Evidence may be reappraised so that the preexisting expectation is confirmed by a post hoc evaluation (immunization), and the contradictory evidence can be discarded as irrelevant (e.g., "They sent me home because of insurance or other financial reasons, but I am still severely impaired"). Expectation violations can also be redefined as the "exception to the rule" (e.g., "The week after surgery I felt unusually well due to the weather, this does not mean that it will persist").

The ViolEx model is applicable not only to the medical context but also to the psychological interventions in mental illnesses such as depression, anxiety, obsessive–compulsive disorder, and chronic pain. Interventions such as exposure for anxiety explicitly address expectations (e.g., "When I experience elevated heart rate it is a sign of a heart attack") in order to change those expectations through expectation violation. In order to achieve this goal, specific measurement instruments are needed that assess these disease-specific situations. In depression, for example, questionnaires should address depression-specific dysfunctional expectations. Importantly, the questionnaire items should be situation-specific and falsifiable so that they can be tested by the individual using behavioral experiments and may therefore facilitate cognitive restructuring. Our own group has developed and validated an instrument to assess these expectations in depression (Kube, D'Astolfo, Glombiewski, Doering, & Rief, 2016). Since these expectations are disease-specific, future efforts should focus on providing similar assessment tools for other disorders.

However, a significant number of patients do not change their expectations despite successful exposures, probably due to "immunization" and other processes as described in the ViolEx model (Kube, Rief, & Glombiewski, 2017). Thus, to foster expectation change in medical contexts or psychological disorders, it is not only necessary to provide situations or information that put the respective expectation to the test, but it is equally important to generate a situational environment that is conducive to a realistic evaluation of the corrective experience, for example, an empathic patient–provider interaction. Therapists should try to counteract biased information processing that could maintain unrealistic expectations in spite of contradictory evidence.

3. THE MEDICAL CONTEXT: OPTIMIZATION OF PATIENT EXPECTATIONS

The above-mentioned model makes specific recommendations for expectation change in medical contexts. It stresses the importance of both providing information while monitoring patients' information processing and establishing an empathic patient–provider relationship. The respective relevance of these both components, however, needs to be tested. This was accomplished in a recent study in patients undergoing coronary artery bypass graft (CABG) surgery. The PSY-HEART trial (PSYchological preoperative interventions to improve outcome in HEART surgery patients) was based on observational evidence that preexisting patient expectations are associated with improved outcome of CABG surgery (Juergens, Seekatz, Moosdorf, Petrie, & Rief, 2010; Ronaldson et al., 2014; Scheier et al., 1999). PSY-HEART set out to investigate whether (1) presurgical patient expectations could be modified by a short psychological intervention (EXPECT) aimed at developing realistic expectations about beneficial effects of the surgery and the recovery process, and (2) whether this expectation manipulation intervention would lead to better postsurgery outcomes. Two control groups were established for the randomized controlled trial. The first control group received the standard medical care (SMC) presurgery. The second control group (SUPPORT) additionally received an unspecific supportive therapy in which expectations were not addressed. However, patients in this group received the same amount of empathic patient–provider interaction, thus controlling for another important mechanism of placebo responses. In total, 124 patients were randomized to the three conditions. The primary outcome was disability for 6 months after the surgery. Among secondary outcomes were quality of life (QoL) and mental health, but also biological markers of the recovery process (e.g., interleukin 6 and 8). Outcomes were assessed at baseline (presurgery) and after 6 months (postsurgery). Changes in expectation (i.e., the effect of the interventions on expectations) were assessed directly after the intervention. The trial protocol and the trial results have already been published (Auer et al., 2017; Laferton, Auer, Shedden Mora, Moosdorf, & Rief, 2013; Rief et al., 2017; Salzmann et al., 2017). The following section will therefore discuss PSY-HEART from the perspective of expectation-focused psychotherapy.

The expectation manipulation intervention encompassed two in-person sessions (50 min each) and three phone calls (20 min each). One of the phone calls was scheduled after the surgery ("booster call"); all other interventions took place before the surgery. The interventions targeted both outcome expectations and expectations about personal control. At the start of the intervention, patients' knowledge and expectations concerning their condition and its course were assessed. The therapist provided information about the surgery and its potential benefits. Importantly, this information was not overly optimistic; instead, it aimed to foster realistic treatment outcome expectations. In order to encourage patients to develop very individualized and specific outcome expectations, patients wrote a personalized recovery action plan (e.g., which positive activities they would be able to enjoy at which stage of the recovery process). They were encouraged to imagine this vividly in an imagination exercise. Thus, both psychoeducation as a more cognitive approach of expectation manipulation and imagination as a more emotional approach were implemented in EXPECT. In order to positively influence expectations about personal control, patients were informed about possible bodily symptoms that could occur after the surgery, for example, suffering from gag reflexes due to the respiratory tube. Therapists and patients jointly developed ideas and plans on how the patient could cope with these symptoms. Patients also discussed health behavior changes with their therapist that could help them to speed up their recovery, thus also fostering expectations of personal control. In the "booster call" after surgery, patients' expectations were reviewed and patients were reminded of specific intervention content, if necessary.

The intervention successfully impacted patient expectations. Compared to the two control groups, patients in the EXPECT intervention group demonstrated more positive expectations to personally control the disease after the intervention, while no significant change in expectations occurred in the SMC group or the SUPPORT group. This also indicates that providing only an empathic patient–provider interaction (SUPPORT group) was not sufficient to positively influence expectations. It seems that specifically tailored interventions are necessary to induce this expectation change. Concerning the primary outcome, patients in the EXPECT group demonstrated significant improvements in disability scores 6 months postsurgery compared to the SMC group. Thus, a very short psychological intervention improved postoperative outcome. Within the SUPPORT group patients benefitted from the intervention as well as demonstrated improved disability scores. The comparison between the SUPPORT and EXPECT conditions evidenced a trend for EXPECT to outperform SUPPORT; however, this trend was not statistically significant. Positive effects of EXPECT were also evident on secondary outcome measures, such as QoL, subjective ability to work, and immune responses. The SUPPORT group did also show improvements in some outcome measures compared to the SMC group; however, no consistent superiority over SMC could be ascertained for SUPPORT. While the results are very encouraging and speak for the importance of changing expectations, two aspects of the trial merit further discussion. First, the EXPECT intervention targeted multiple expectations simultaneously, such as outcome expectation and expectations of personal control with regard to the recovery process and with regard to coping with potential side effects. The results do not allow a dismantling of this effect, i.e., it remains unclear whether all these expectations need to be addressed to achieve a positive outcome. To investigate this issue further, it is also necessary to develop more sophisticated instruments to assess different expectation constructs (Laferton et al., 2017). Second, the results also demonstrate a positive effect for only providing an empathic patient–provider relationship, i.e., the SUPPORT group. This could in part be due to contamination effects, since the same therapists conducted all interventions. Alternatively, this could be the effect of the placebo mechanism of the empathic patient–provider relationship. If this is the case, it would be interesting to find out whether there is a differential indication for SUPPORT and EXPECT. As a whole, the trial demonstrates that focusing on patient expectations and changing them to a realistic, but positive view on the surgery and the recovery process can optimize treatment outcomes after CABG. This is especially notable taking the very limited duration of the intervention into account. Incorporating both psychoeducational elements and more emotion-focused techniques such as imagination exercises seems helpful. The intervention was conducted in an individual setting. Group formats would increase the cost-effectiveness of the intervention; however, they would need to be tested for their acceptance and feasibility. Implementing this short intervention within the practical confinements of a medical setting proved feasible and benefits patients as well.

While in PSY-HEART the optimization of expectations aimed at improving treatment outcome, similar interventions can also be designed to change negative treatment expectations in order to avoid nocebo effects. A recent clinical trial in patients with breast cancer who were about to receive adjuvant endocrine therapy (AET) illustrates the relevance of such interventions (Nestoriuc et al., 2016). This prospective cohort study used a minimal intervention that consisted of only one session of approximately 15 min. In this session, a trained professional provided standardized verbal and written information on mechanisms, benefits, and potential side effects of AET before treatment start and discussed this information with the patient. Side effect expectations and relevant outcomes were assessed at baseline, 3 months after beginning of treatment, and 2 years later. Pretreatment expectations were significantly associated with self-reported side effects and QoL. Additionally, patients with more negative expectations at baseline reported more side effects after 2 years of endocrine treatment. This trial demonstrates the need to optimize not only positive treatment expectations but also to focus on negative treatment expectations, such as side effect expectations, in order to optimize medical treatment. A more comprehensive expectation-focused intervention, i.e., a side effect prevention training, is currently being investigated (Shedden-Mora et al., 2017; von Blanckenburg, Schuricht, Albert, Rief, & Nestoriuc, 2013).

4. EXPECTATION-FOCUSED PSYCHOTHERAPY

Expectations concerning treatment outcome or personal control are relevant not only to medical treatments but also to psychotherapy and should therefore be optimized in both fields. However, psychotherapy needs to consider that, in addition to the treatment-related expectations, in the case of mental disorders, disease-specific expectations exist that are inherent to the disorder and contribute to its occurrence and maintenance. Thus, changing these disease-specific expectations should be the goal of psychotherapy. An example is exposure treatment for chronic pain. According to the fear avoidance model (Vlaeyen & Linton, 2000), chronic pain occurs when movements are avoided due to the expectation of harm. Exposure treatment for chronic back pain aims at correcting those expectations by exposing patients to feared movements which do not lead to feared consequences such as spinal cord damages or pain elevation. This experience helps individuals with chronic pain to correct their expectations concerning movement, which results in higher levels of activities and lower levels of disability (Woods & Asmundson, 2008).

Disease-specific expectations can also be illustrated using the example of depression: negative expectations and hopelessness concerning the future are typical of the disease (Korn, Sharot, Walter, Heekeren, & Dolan, 2014) and are also mirrored in the cognitive triad (Beck, 1987). Depressive core beliefs also translate into situation-specific dysfunctional expectations, e.g., "When I ask someone for help, I will be rejected" (Kube et al., 2016). Interestingly, such expectations often persist even if the patient encounters multiple situations in which he/she receives social support instead of experiencing social rejection. To give another example, disease-specific expectations in anxiety disorders are characterized by catastrophic thinking and negative expectations concerning the consequences of feared situations (Boehme et al., 2014; Sebastiani, D'Alessandro, & Gemignani, 2014). Motivated by their fear, many patients avoid the respective stimuli. However, even if they confront the phobic stimuli repeatedly, and even if this exposure is accompanied by a therapist who minimizes strategies of covert avoidance, disease-specific expectations may persist. Craske and colleagues have convincingly argued that expectation violation should be optimized in exposure therapy (Craske, Treanor, Conway, Zbozinek, & Vervliet, 2014) by considering the principles of inhibitory learning. Additionally, factors hindering expectation change after expectation violation (e.g., immunization), as described in the ViolEx model, should be actively addressed by psychotherapists. Since expectations are such an integral characteristic of many mental disorders and since they are an important predictor of treatment success in psychotherapy (Wampold, Minami, Tierney, Baskin, & Bhati, 2005), psychotherapeutic interventions should aim to assess patient expectations and optimize them. Expectation-focused psychological interventions should emphasize disorder-specific expectations, develop situational tests to check the credibility of these expectations, and evaluate expectations by comparing expectations with the experience during those tests.

Considering different forms of psychotherapy, it seems that they acknowledge and address patient expectations, however, to a different degree and with varying disease specificity. Some forms of psychotherapy use the therapeutic relationship to violate dysfunctional patient expectations regarding interpersonal contexts and to provide corrective emotional experiences (McCullough, 2003; Young, Klosko, & Weishaar, 2003). Modern cognitive behavioral therapy targets disease-specific expectations routinely, e.g., in anxiety disorders and obsessive–compulsive disorders. Exposure, behavioral experiments, or cognitive interventions aim at modifying such expectations. We would argue, however, that an expectation-focused approach is applicable to a broader range of conditions and would also benefit, e.g., the treatment of depression (Kube et al., 2017) or chronic pain. Limitations of this approach concern probably mental disorders which are predominantly linked to an ongoing evaluation of past events (e.g., very strong feelings of guilt). Although many of these evaluations also trigger expectations ("I cannot continue to live with these feelings of guilt"), a reevaluation of the experience may be reached more easily by discussing the past event. Another potential limitation concerns skills deficits (e.g., social competence deficits), which may also be more amenable to change via skills training instead of expectation focusing. Psychological interventions would be more effective and more cost efficient, if they focused on modifying future relevant cognitions. Additionally, establishing positive outcome expectations should be one of the first goals of every psychological treatment.

5. FUTURE DIRECTIONS

The significance of expectation-focused interventions thus pertains to both medical treatments and psychotherapeutic treatments. With regard to medical treatments, placebo and nocebo research has convincingly demonstrated the importance of expectations for both positive and negative treatment outcomes. Different forms of expectations represent potential targets for an intervention. PSY-HEART has demonstrated that a short expectation-focused intervention can change presurgery expectations in patients undergoing CABG (Rief et al., 2017). Importantly, the intervention addressed both global outcome expectations concerning the treatment and disease-specific expectations (i.e., personal control with regard to the recovery process and with regard to coping with potential side effects). Other expectation-focused interventions are currently being evaluated (von Blanckenburg et al., 2013). Before we can systematically implement such approaches in medicine, future research must establish the general framework of such interventions.

We need to identify the medical treatments that are most promising for an application of expectation-focused interventions. One relevant aspect concerns the side effects associated with a given treatment. Treatment invasiveness or perceived treatment risk could be other relevant aspects. Next, future trials need to corroborate the existing evidence, but also to expand our knowledge concerning further questions regarding expectation-focused interventions. Do all patients benefit equally from expectation-focused interventions or are there some for whom sufficient time to express and discuss emotions with an empathetic therapist (such as the SUPPORT condition in PSY-HEART) would be more helpful? Is there a subgroup of patients for whom side effect prevention training could backfire because they would prefer not to know too much about potential side effects (Barsky, Saintfort, Rogers, & Borus, 2002; Nestoriuc et al., 2016)? More information about such limitations is clearly necessary. Additionally, future studies could evaluate more generic treatment modules. Such modules could focus on frequent themes of expectations that apply to various medical treatments, e.g., how to address expectation of side effects or how to optimize expectation of treatment benefit. Standardizing the intervention would also facilitate dissemination. While trained psychologists have delivered the expectation-focused intervention in our trials, training other health care professionals in this approach could increase its availability tremendously.

6. CONCLUSIONS

Mechanisms of expectation change concerning treatment outcome or personal control are relevant to both medical treatments (e.g., heart surgery) and psychotherapy (e.g., depression and chronic pain) and should be optimized in both fields. The ViolEx model offers a theoretical framework for understanding processes that hinder successful expectation change despite expectation violation.

REFERENCES

Auer, C. J., Glombiewski, J. A., Doering, B. K., Winkler, A., Laferton, J. A., Broadbent, E., et al. (2015). Patients' expectations predict surgery outcomes: A meta-analysis. *International Journal of Behavioral Medicine*, *23*(1), 49–62.

Auer, C. J., Laferton, J. A. C., Shedden-Mora, M. C., Salzmann, S., Moosdorf, R., & Rief, W. (2017). Optimizing preoperative expectations leads to a shorter length of hospital stay in CABG patients: Further results of the randomized controlled PSY-HEART trial. *Journal of Psychosomatic Research*, *97*, 82–89.

Barefoot, J. C., Brummett, B. H., Williams, R. B., Siegler, I. C., Helms, M. J., Boyle, S. H., et al. (2011). Recovery expectations and long-term prognosis of patients with coronary heart disease. *Archives of Internal Medicine*, *171*(10), 929–935.

Barsky, A. J., Saintfort, R., Rogers, M. P., & Borus, J. F. (2002). Nonspecific medication side effects and the nocebo phenomenon. *JAMA*, *287*(5), 622–627.

Beck, A. T. (1987). Cognitive models of depression. *Journal of Cognitive Psychotherapy: An International Quarterly*, *1*, 5–37.

Benedetti, F., Maggi, G., Lopiano, L., Lanotte, M., Rainero, I., Vighetti, S., et al. (2003). Open versus hidden medical treatments: The patient's knowledge about a therapy affects the therapy outcome. *Prevention & Treatment*, *6*(1), 1a.

Bingel, U., Wanigasekera, V., Wiech, K., Ni Mhuircheartaigh, R., Lee, M. C., Ploner, M., et al. (2011). The effect of treatment expectation on drug efficacy: Imaging the analgesic benefit of the opioid remifentanil. *Science Translational Medicine*, *3*(70), 70ra14.

Boehme, S., Ritter, V., Tefikow, S., Stangier, U., Strauss, B., Miltner, W. H., et al. (2014). Brain activation during anticipatory anxiety in social anxiety disorder. *Social Cognitive and Affective Neuroscience*, *9*(9), 1413–1418.

Craske, M. G., Treanor, M., Conway, C. C., Zbozinek, T., & Vervliet, B. (2014). Maximizing exposure therapy: An inhibitory learning approach. *Behaviour Research and Therapy*, *58*, 10–23.

Enck, P., Bingel, U., Schedlowski, M., & Rief, W. (2013). The placebo response in medicine: Minimize, maximize or personalize? *Nature Reviews Drug Discovery*, *12*, 191–204.

Habibović, M., Pedersen, S. S., van den Broek, K. C., & Denollet, J. (2014). Monitoring treatment expectations in patients with an implantable cardioverter-defibrillator using the EXPECT-ICD scale. *Europace*, *16*(7), 1022–1027.

Juergens, M. C., Seekatz, B., Moosdorf, R. G., Petrie, K. J., & Rief, W. (2010). Illness beliefs before cardiac surgery predict disability, quality of life, and depression 3 months later. *Journal of Psychosomatic Research*, *68*(6), 553–560.

Korn, C. W., Sharot, T., Walter, H., Heekeren, H. R., & Dolan, R. J. (2014). Depression is related to an absence of optimistically biased belief updating about future life events. *Psychological Medicine*, *44*(3), 579–592.

Kube, T., D'Astolfo, L., Glombiewski, J. A., Doering, B. K., & Rief, W. (2016). Focusing on situation-specific expectations in major depression as basis for behavioural experiments—Development of the Depressive Expectations Scale. *Psychology and Psychotherapy: Theory, Research and Practice*, *90*(3), 336–352.

Kube, T., Rief, W., & Glombiewski, J. A. (2017). On the maintenance of expectations in major depression—Investigating a neglected phenomenon. *Frontiers in Psychology*, *8*, 9.

Laferton, J. A. C., Auer, C. J., Shedden Mora, M., Moosdorf, R., & Rief, W. (2013). Enhancing the efficacy of heart surgery by optimizing patients' preoperative expectations: Study protocol of a randomized controlled trial. *American Heart Journal*, *165*, 1–7.

Laferton, J. A., Kube, T., Salzmann, S., Auer, C. J., & Shedden-Mora, M. C. (2017). Patients' expectations regarding medical treatment: A critical review of concepts and their assessment. *Frontiers in Psychology*, *8*, 233.

Leedham, B., Meyerowitz, B. E., Muirhead, J., & Frist, W. H. (1995). Positive expectations predict health after heart transplantation. *Health Psychology*, *14*(1), 74.

Mahomed, N. N., Liang, M. H., Cook, E. F., Daltroy, L. H., Fortin, P. R., Fossel, A. H., et al. (2002). The importance of patient expectations in predicting functional outcomes after total joint arthroplasty. *Journal of Rheumatology*, *29*(6), 1273–1279.

McCullough, J. P. (2003). Treatment for chronic depression using cognitive behavioral analysis system of psychotherapy (CBASP). *Journal of Clinical Psychology*, *59*(8), 833–846.

Nestoriuc, Y., von Blanckenburg, P., Schuricht, F., Barsky, A. J., Hadji, P., Albert, U. S., et al. (2016). Is it best to expect the worst? Influence of patients' side-effect expectations on endocrine treatment outcome in a 2-year prospective clinical cohort study. *Annals of Oncology*, *27*(10), 1909–1915.

Rheker, J., Winkler, A., Doering, B. K., & Rief, W. (2017). Learning to experience side effects after antidepressant intake—Results from a randomized, controlled, double-blind study. *Psychopharmacology, 234*(3), 329–338.

Rief, W., & Glombiewski, J. A. (2016). Erwartungsfokussierte Psychotherapeutische Interventionen (EFPI). *Verhaltenstherapie, 26*(1), 47–54.

Rief, W., Glombiewski, J. A., Gollwitzer, M., Schubö, A., Schwarting, R., & Thorwart, A. (2015). Expectancies as core features of mental disorders. *Current Opinion in Psychiatry, 28*(5), 378–385.

Rief, W., Shedden-Mora, M. C., Laferton, J. A., Auer, C., Petrie, K. J., Salzmann, S., et al. (2017). Preoperative optimization of patient expectations improves long-term outcome in heart surgery patients: Results of the randomized controlled PSY-HEART trial. *BMC Medicine, 15*(1), 4.

Ronaldson, A., Poole, L., Kidd, T., Leigh, E., Jahangiri, M., & Steptoe, A. (2014). Optimism measured pre-operatively is associated with reduced pain intensity and physical symptom reporting after coronary artery bypass graft surgery. *Journal of Psychosomatic Research, 77*(4), 278–282.

Salzmann, S., Euteneuer, F., Laferton, J. A. C., Auer, C. J., Shedden-Mora, M. C., Schedlowski, M., et al. (2017). Effects of preoperative psychological interventions on catecholamine and cortisol levels after surgery in coronary artery bypass graft patients: The randomized controlled PSY-HEART trial. *Psychosomatic Medicine, 79*(7), 806–814.

Scheier, M. F., Matthews, K. A., Owens, J. F., Schulz, R., Bridges, M. W., Magovern, G. J., et al. (1999). Optimism and rehospitalization after coronary artery bypass graft surgery. *Archives of Internal Medicine, 159*(8), 829–835.

Sebastiani, L., D'Alessandro, L., & Gemignani, A. (2014). Does fear expectancy prime fear? An autonomic study in spider phobics. *International Journal of Psychophysiology, 91*(3), 178–185.

Shedden-Mora, M., Heisig, S., Von Blanckenburg, P., Witzel, I., Rief, W., Albert, U., et al. (2017). Optimizing expectations to prevent side effects of adjuvant endocrine treatment in breast cancer: A randomized controlled trial. *Journal of Psychosomatic Research, 97*, 167–168.

van den Akker-Scheek, I., Stevens, M., Groothoff, J. W., Bulstra, S. K., & Zijlstra, W. (2007). Preoperative or postoperative self-efficacy: Which is a better predictor of outcome after total hip or knee arthroplasty? *Patient Education & Counseling, 66*(1), 92–99.

Vlaeyen, J. W. S., & Linton, S. J. (2000). Fear-avoidance and its consequences in chronic musculoskeletal pain: A state of the art review. *Pain, 85*(3), 317–332.

von Blanckenburg, P., Schuricht, F., Albert, U.-S., Rief, W., & Nestoriuc, Y. (2013). Optimizing expectations to prevent side effects and enhance quality of life in breast cancer patients undergoing endocrine therapy: Study protocol of a randomized controlled trial. *BMC Cancer, 13*(1), 426.

Wampold, B. E., Minami, T., Tierney, S. C., Baskin, T. W., & Bhati, K. S. (2005). The placebo is powerful: Estimating placebo effects in medicine and psychotherapy from randomized clinical trials. *Journal of Clinical Psychology, 61*(7), 835–854.

Woods, M. P., & Asmundson, G. J. (2008). Evaluating the efficacy of graded in vivo exposure for the treatment of fear in patients with chronic back pain: A randomized controlled clinical trial. *Pain, 136*(3), 271–280.

Young, J. E., Klosko, J. S., & Weishaar, M. E. (2003). *Schema therapy: A practitioner's guide.* Guilford Press.

CHAPTER FIFTEEN

Nocebo Effects: Neurobiological Mechanisms and Strategies for Prevention and Optimizing Treatment

Julian Kleine-Borgmann[1], Ulrike Bingel
University Hospital Essen, University of Duisburg–Essen, Essen, Germany
[1]Corresponding author: e-mail address: julian.kleine-borgmann@uk-essen.de

Contents

Abstract

Psychosocial and contextual factors, such as patient–physician relationship, prior treatment experience, and treatment expectation, can either improve or compromise treatment efficacy. These phenomena are commonly specified as placebo and nocebo effects. As placebo and nocebo effects can influence symptom development, adverse event rate, and treatment efficacy, it is crucial to be aware of these effects and to develop strategies for prevention to optimize treatment outcomes. While experimental studies have made substantial progress in elucidating the psychosocial and neurobiological mechanisms underlying placebo effects, the detailed mechanisms of nocebo effects remain largely unexplored. A better understanding of these mechanisms promises to facilitate the development of easy-to-use strategies in clinical care to improve overall treatment outcomes and patient satisfaction.

1. INTRODUCTION

The psychosocial context is an essential and inevitable part of most medical treatments. Evidence gained in the recent past highlights the

International Review of Neurobiology, Volume 138
ISSN 0074-7742
https://doi.org/10.1016/bs.irn.2018.02.005

importance of the various mechanisms involved in shaping this therapeutic context, such as patients' expectations, learning mechanisms, and the patient–physician relationship. Positive physiological or psychological effects induced by the treatment context are referred to as placebo effects or placebo responses and are known to represent complex psychoneurobiological phenomena (Benedetti et al., 2005; Enck et al., 2013). Importantly, placebo effects are not restricted to placebo treatments; rather, we know that they can substantially modulate the efficacy and tolerability of active pharmacological therapies, for example.

On the other hand, the negative effects the treatment context induces—defined as nocebo effects (nocere, lat. = to hurt)—can impair medical (treatment) outcomes on various levels including the development and maintenance of new symptoms (Crichton & Petrie, 2015), as well as the efficacy and tolerability of medical treatments including pharmacotherapy. However, despite their potentially detrimental effects on treatment outcomes, the neurobiological and neurochemical mechanisms involved in nocebo effects are largely unknown and much less studied than placebo effects. In this chapter, we highlight the levels at which different nocebo effects interfere negatively with medical and treatment outcomes, and propose strategies for preventing or even overcoming their influence so as to optimize treatment outcomes.

2. PSYCHOSOCIAL MECHANISMS

From a psychological point of view, nocebo effects are induced and/or moderated by several factors that ultimately converge to negative treatment outcome expectations. These include, among others, negative prior information regarding a treatment or health outcome provided by healthcare professionals, the media or a social peer, or negative personal experiences as evidence following treatment failure. These latter learning mechanisms are not restricted to first-hand experiences. Negative information can also be gathered from social learning, like watching other patients develop or acquire worsening symptoms, or the knowledge of a higher exposure dose (Blasini, Corsi, Klinger, & Colloca, 2017). One of the best known influences on nocebo effects are verbally induced expectations, which even can compromise pharmacological treatments. Dworkin et al. (1983) reported on one of the first investigations on the impact of verbal suggestions on pharmacological treatments such as the analgesic drug nitrous oxide (N_2O) that was often used in settings of dentistry, childbirth, and outpatient surgery. In their studies, Dworkin et al. combined an administration of 33% N_2O mixture with various verbal explanations. While the control group received objective information about the mechanisms, common use, and application of N_2O, the experimental study group heard alternative information about N_2O as a consciousness-modifying drug, which proved able to increase subjective sensitivity rather than have an analgesic effect. Experimental pain was induced electrically in a tooth pulp test at baseline (while breathing ambient air) and while breathing 33% N_2O mixture. Surprisingly, their group with experimentally induced altered expectations exhibited the opposite effect of N_2O, namely, lower pain thresholds and pain tolerance as hyperalgesia. Although that study has some limitations (such as the lack of a placebo-controlled condition), it provides evidence for the importance of verbal information in the development of nocebo effects and following investigations have confirmed this mechanism (Amanzio & Benedetti, 1999; Flaten, Simonsen, & Olsen, 1999). In addition to expectations induced by verbal information, associative learning of one's own and vicarious therapy experiences represents another key mechanism in the development of nocebo effects. Interestingly, while learning induced by a conditioning procedure can enhance placebo effects induced by verbal instructions, no such effect has been observed in conjunction with nocebo hyperalgesia, where verbal instructions and/or conditioning lead to similar magnitudes of nocebo hyperalgesia (Colloca, Sigaudo, & Benedetti, 2008). Whether this observation reflects a true difference regarding the mechanisms underlying placebo and nocebo effects and is applicable beyond the field of pain remains unknown. Other factors (i.e., personality factors) and biomarkers (e.g., genetic polymorphisms) that may also predict placebo or nocebo effects are important subjects of current research (also reviewed in Blasini et al., 2017).

3. NEUROBIOLOGICAL CORRELATES

Despite their potentially detrimental clinical consequences, the mechanisms underlying nocebo effects including nocebo hyperalgesia have received very little attention so far. Initial fMRI studies have shown that nocebo hyperalgesia is associated with increased activity in pain-processing areas (Keltner et al., 2006; Sawamoto et al., 2000), but the specific mechanisms underlying this amplification of nociceptive information have not been adequately explored. Placebo analgesia is known to be modulated, at least in part, by descending inhibitory pathways involving the endogenous opioidergic and dopaminergic system (Benedetti et al., 2005). While these also play a role in nocebo hyperalgesia (Geuter & Buchel, 2013; Scott et al., 2006), as shown by positron emission tomography studies providing evidence of reduced dopaminergic and opioidergic neurotransmission in pain-processing and pain-regulatory networks during nocebo interventions (Scott et al., 2008), additional mechanisms may play a role. Benedetti et al. (1997) emphasized the importance of anxiety and stress as basic components of nocebo effects, and provided evidence for the involvement of cholecystokinin (CCK), an anxiogenic and panicogenic transmitter, in nocebo hyperalgesia. In their randomized, controlled trial, patients were told they would receive a pain-increasing infusion as a nocebo intervention after thoracoscopy while in reality, saline was infused. Afterward, proglumide, a CCK-A/B antagonist, or the opioid antagonist naloxone was infused in randomized order. The CCK-A/B antagonist significantly reduced the induced nocebo hyperalgesic effect, highlighting the role of anxiety-related mechanisms in nocebo effects. In contrast, the opioid antagonist naloxone exhibited no impact on the nocebo effect. Further investigations revealed activation of the hypothalamic–pituitary–adrenal axis in nocebo hyperalgesia and strengthened the psychoneurochemical relationship between anxiety and nocebo hyperalgesia. The anxiolytic benzodiazepine diazepam and the previously mentioned CCK were both able to reduce nocebo hyperalgesia-induced pain intensity (but not basal pain intensities) (Benedetti et al., 2006). Furthermore, neuroimaging studies have linked nocebo hyperalgesia with enhanced activity and functional connectivity in the hippocampus and amygdala (Bingel et al., 2011; Kong et al., 2008), lending further support for anxiety-related circuitry's role in nocebo hyperalgesia. Initial fMRI studies investigated placebo analgesia and nocebo hyperalgesia within the same experimental setting and revealed partially distinct and nonoverlapping brain areas involved in the neural processing of placebo analgesia and nocebo hyperalgesia (Freeman et al., 2015). In short, although placebo and nocebo effects are commonly understood as a continuum on the same scale, recent research points toward both shared and potentially distinct mechanisms of the two. Indeed, verbal suggestions, conditioning, and social learning seem to be common concepts involved in the development of placebo and nocebo effects. However, the distinct psychoneurobiological mechanisms such as the involvement of neurotransmitters and anxiety-related networks may differ between the two. Future studies will have to elucidate the distinct psychoneurobiological mechanisms underlying placebo and nocebo effects and equally importantly, individual psychological and biological predictors of an individual's susceptibility/vulnerability to developing such responses. This is not only true for the field of pain; it is of even greater relevance to nocebo effects in other physiological systems.

4. SYMPTOM DEVELOPMENT

A well-known nocebo effect is symptom augmentation, as Benedetti, Durando, and Vighetti (2014) showed recently in a study investigating nocebo and placebo effects in hypobaric hypoxia headache. As a symptom of acute mountain sickness, hypobaric hypoxia headache results from a rise in the concentration of prostaglandins due to hypoxia at high altitudes (Benedetti et al., 2014). In this trial, negative information (i.e., the induction of a negative expectation) from other study volunteers (social peers) about a potentially emerging hypobaric hypoxia headache caused a significant increase in the prevalence and magnitude of reported headache at 3500 m altitude. Salivary cyclooxygenase activity, a core factor in prostaglandin synthesis, was enhanced at the same time. Interestingly, in the absence of hypobaric conditions (i.e., at 1500 m altitude) the negative expectation revealed no influence on the prevalence and pain intensity of headache. One could conclude that inducing a negative expectation led to stimulation of prostaglandin synthesis in hypobaric and hypoxic conditions and could therefore augment or even cause headache symptoms.

Crichton and Petrie (2015) investigated the effects of patient education regarding nocebo effects on symptoms supposedly originating from infrasound produced by wind mill farms in New Zealand. Subjects were randomly assigned to two groups receiving different information regarding the potential mechanisms underlying symptom development due to wind mill farm exposure. One group was informed about the nature of nocebo effects and their potential contribution to symptom development, while a second group received a biological explanation of how infrasound causes adverse effects. Subjects were then experimentally exposed to infrasound and audible wind mill sound during two sessions. Between the sessions, they received audiovisual instruction about nocebo effects or biological mechanisms. The participants' symptoms and mood were recorded during each session. The results show that the knowledge of nocebo effects potentially contributing to symptom occurrence lowered symptom scores and enhanced mood back to baseline levels during the second session, while information on putative biological mechanisms raised symptom scores and lowered mood. This study corroborates the importance of nocebo effects in shaping symptom perception and of patient education as fundamental strategy to prevent the same.

Although the impact of nocebo effects on symptom development can vary among different physiological systems and medical conditions, these two examples highlight the crucial contribution of nocebo effects to symptom development (and persistence) and to their perception.

5. TREATMENT EFFICACY AND COURSE OF DISEASE

A recent meta-analysis involving studies with experimental, acute (e.g., procedural or postoperative), and chronic pain (e.g., chronic back pain) settings reveals that different positive (placebo-)expectations—either verbally induced, by conditioning or by imagery—can lead to pain relief, especially in case of acute pain, therefore improving pain therapy and its outcome (Peerdeman et al., 2016). In contrast, negative expectations are not just able to augment symptoms. It has long been suggested that they also influence the efficacy of drug treatment (Lyerly et al., 1964). Several experimental studies have strengthened the hypothesis of nocebo effects modulating pharmacological treatments. In a randomized, controlled fMRI trial, Bingel et al. (2011) investigated how expectations can influence the analgesic efficacy of the potent μ-opioid remifentanil in an experimental heat pain paradigm. As expected, the positive expectation enhanced remifentanil's analgesic efficacy (placebo effect). More interesting was the observation that the negative treatment expectation neutralized its analgesic effect, which was associated with blood-oxygen-level-dependent activity in the hippocampus. For a schematic overview of involved brain regions, see Fig. 1. Prior treatment experiences, particularly those that are negative, can also impact treatment responses, which can be transferred to another treatment as Kessner et al. (2013) demonstrate. Their randomized, controlled fMRI trial employed two subsequent placebo treatment conditions during a heat pain paradigm. The subjects were randomized to a positive and negative treatment history group, where either a positive or negative treatment experience with an analgesic patch was induced experimentally via a conditioning procedure. All participants were then exposed to

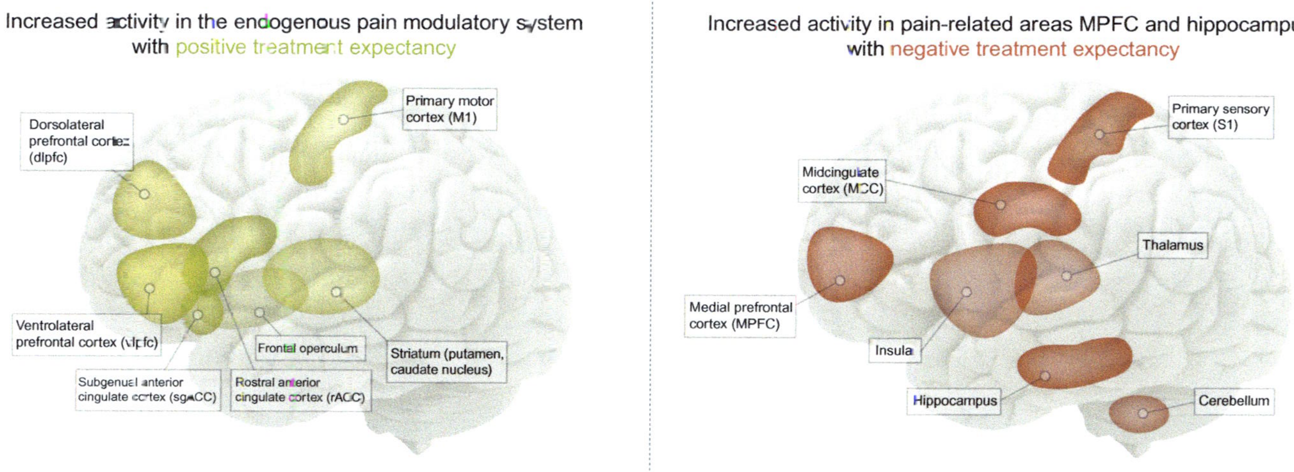

Fig. 1 Brain activity related to positive and negative treatment expectancy. This illustrates changes in pain-related brain activity during experimentally induced positive and negative treatment expectation, in a simplified scheme. *Modified in accordance with Benedetti, F., et al. (2005). Neurobiological mechanisms of the placebo effect.* The Journal of Neuroscience, 25*(45), 10390–10402.*

the second treatment condition during which an ointment instead of a patch was used. Interestingly, the prior experience with the first treatment substantially modulated the response to the second treatment, indicating a nocebo effect induced by associative learning that travels over time and over condition. As recently shown by Zunhammer et al. (2017), these negative carryover effects seem to generalize across treatments and cannot be countered by changing the drug administration route, e.g., from topical to oral administration. These new findings underline the maleficence of nocebo and carryover effects in treatment failure.

6. UNWANTED/ADVERSE EVENTS

Aside from the harm nocebo effects have on treatment efficacy, negative expectations or prior experiences may also facilitate the occurrence and development of side effects during drug and other therapies. This is well demonstrated by Rief et al. (2009), who studied the reported adverse effects in 12,742 patients randomized to the placebo arm of RCTs investigating tricyclic antidepressants (TCA) and selective serotonin reuptake inhibitors. Although all patients analyzed in this meta-analysis received placebo, the others presented very high rates of up to 19% of unwanted adverse events (Rief et al., 2009), particularly in the placebo arm of trials investigating TCA. This difference in unwanted adverse events is probably due to the negative expectation the informed consent procedure induced. Furthermore, the authors conclude that the very assessment of side effects itself may contribute to the rate of reported adverse events. Another example is the recently published results of the Lipid-Lowering Arm of the Anglo-Scandinavian Cardiac Outcomes Trial (ASCOT-LLA), which illustrates that the informed consent procedure after the act of unblinding both the physician and the patient may lead to higher rates of adverse events (AEs, e.g., musculoskeletal-related AEs) (Gupta et al., 2017). Basically, side effects can influence patient compliance and thus treatment efficacy.

Heisig et al. (2016) investigated the association between treatment expectation and side effects under endocrine therapy in breast cancer patients. Via standardized questionnaires, the authors assessed side effect expectations, treatment efficacy expectations, and the treatment necessity-concern balance. As expected, patients with high expectations of side effects revealed at the same time lower expectations of treatment efficacy. Moreover, negative necessity beliefs were associated with both lower treatment efficacy expectations and less intention to adhere. In sum, prior side effect experience and expectations influence negatively the expectation of treatment efficacy, and may also impair adherence and treatment outcome. In such a situation, proper patient information and education is crucial to shape treatment expectations, which ultimately promise to heighten treatment efficacy while avoiding nocebo effects in the form of nonpharmacologically induced side effects or nonadherence.

In sum, there is accumulating experimental and clinical evidence that nocebo effects can induce, aggravate, or prolong symptoms, shape the course of disease, and modulate the efficacy and tolerability of treatments, thereby impairing overall treatment outcomes. These insights have profound clinical implications and call for a systematic approach to prevent or overcome the detrimental consequences of nocebo effects on health outcomes (Bingel & Placebo Competence Team, 2014; Enck et al., 2013).

Knowledge of the psychobiological mechanisms underlying these effects offers the potential to systematically modulate and thereby prevent or reduce the negative effects caused by nocebo effects. Although nocebo effects have been much less investigated than placebo effects, negative treatment expectations induced by verbal information, and one's own or observed prior treatment experiences appear to play a major role. And importantly, both of these can be modulated to benefit the patient, as elegantly shown in a recent prospective three-arm randomized clinical trial investigating the effect of an experimentally induced positive expectation in patients scheduled for coronary artery bypass graft surgery (Rief et al., 2017). In that study, 124 patients were randomized to three groups: the first contained a psychological presurgery intervention to optimize outcome expectations by evaluating possible aims in the future after surgery and informing about realistic postsurgery symptoms differentiated from complications. The second group underwent a psychological control intervention focusing on emotional support and general advice, and the third was randomized to standard medical care. The group's results show that the psychological intervention significantly improved outcome expectation and importantly, that this expectation modulation leads to significantly larger improvements in disability and quality of life after 6 month postsurgery. Notably, these findings were corroborated on the physiological level by lower levels of proinflammatory cytokine concentrations compared to the standard medical care group at the same time. In conclusion, the systematic modulation of mechanisms underlying placebo and nocebo effects—here the improvement in outcome expectations—and patient education aiming to enable the patient to differentiate between "normal" symptoms and severe complications led to a significant and clinically relevant improvement in postsurgery outcome. Although such approaches need to be validated and replicated in other patient groups, this study highlights the potential of systematically modulating patients' expectations and learning experiences to enhance the placebo component and minimize the nocebo component of medical treatments, thereby improving the overall treatment outcome.

In clinical settings, there are various easy hands-on strategies to prevent nocebo effects. First, the essential basis is the awareness of all healthcare professionals of the existence of placebo and nocebo effects, and that both influence treatment efficacy and outcome (Fig. 2). All the other strategies for preventing nocebo effects address the mechanisms of placebo and nocebo effects known to date. For example, by paying attention to the patient–physician relationship and helping patients understand the therapeutic strategy, the patient's expectations can be modulated, thus leading to more efficient therapy. Specifically, there is no need to hide any information, but it may help to first focus on a given treatment's desired effects (e.g., pain relief, or preventing pain chronification) before discussing potential side effects. For the latter, communication with the physician should contribute to a realistic evaluation of side effects' frequency by providing specific additional information such as percentages that could be graphically displayed to enhance clarity. In addition to informing about side effects, potential coping strategies should be delineated regarding how to deal with certain side effects in advance of adverse events to strengthen perceived self-efficacy and treatment compliance. More approaches are listed in Table 1.

Taken together, nocebo effects impair treatment and health outcomes on several levels. Negative expectations and learned behavior are both modulated by the therapeutic context and patient–physician relationship. Therapy failure and side effects can significantly detract from patients' compliance and moreover harm subsequent therapeutic approaches. Deeper insights into the psychoneurobiological mechanisms underlying nocebo responses will lay the foundation upon which systematic strategies can be developed to prevent or reduce nocebo effects and their detrimental effect on health and treatment outcomes. Such approaches will require concerted efforts on the part of healthcare professionals at all levels including medical education, and promise to improve treatment quality and health care.

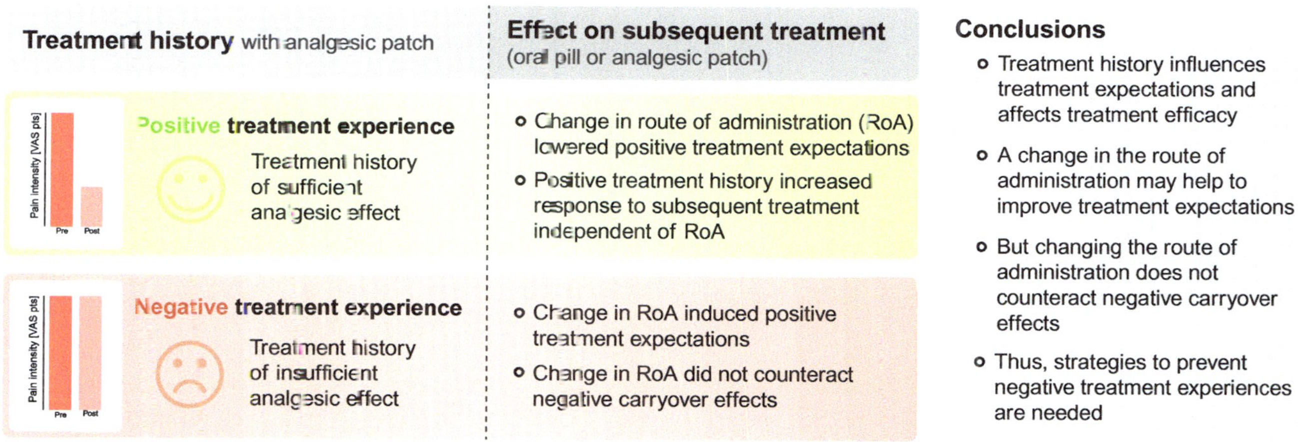

Fig. 2 The effect of prior treatment experience on treatment outcome. Treatment history influences the subsequent treatment efficacy even when changing the route of administration (e.g., from analgesic patch to oral pill). Positive prior treatment experiences lead to higher subsequent treatment responses, while negative treatment experiences lead to reduced subsequent treatment responses. Changing the route of administration between treatment approaches can help to restore a positive treatment expectation, but does not counter the detrimental effects of prior treatment failure. *Modified in accordance with Crichton, F., & Petrie, K. J. (2015). Health complaints and wind turbines: The efficacy of explaining the nocebo response to reduce symptom reporting.* Environmental Research, 140, *449–455; Enck, P., et al. (2013). The placebo response in medicine: Minimize, maximize or personalize?* Nature Reviews. Drug Discovery, 12*(3), 191–204.*

Table 1 Strategies for Avoiding Nocebo Effects and Optimizing Treatment

Expectation and Learning-Based Mechanisms	Patient–Physician-Relation
• Determine patients' treatment and side effects expectations • Assess prior treatment (failure) experiences • Explain the desired treatment effects before addressing potential risks • Graphically illustrate adverse event rate • Help differentiate between pharmacologically induced adverse events and nocebo effects • Use illustrative methods like videos or figures and everyday language • Provide evidence-based information via web based or others rather than unproven anxiety-increasing sources	• Determine the patient's level of knowledge • Provide adequate information on diagnosis, disease, treatment, treatment goals, and side effects • Adopt an empathic and authentic communication style • Double-check patient's understanding of treatment goals (e.g., let the patient summarize) to prevent misunderstandings • Acknowledge and answer patients' concerns to strengthen compliance

REFERENCES

Amanzio, M., & Benedetti, F. (1999). Neuropharmacological dissection of placebo analgesia: Expectation-activated opioid systems versus conditioning-activated specific subsystems. *The Journal of Neuroscience*, *19*(1), 484–494.

Benedetti, F., Durando, J., & Vighetti, S. (2014). Nocebo and placebo modulation of hypobaric hypoxia headache involves the cyclooxygenase-prostaglandins pathway. *Pain*, *155*(5), 921–928.

Benedetti, F., et al. (1997). Blockade of nocebo hyperalgesia by the cholecystokinin antagonist proglumide. *Pain*, *71*(2), 135–140.

Benedetti, F., et al. (2005). Neurobiological mechanisms of the placebo effect. *The Journal of Neuroscience*, *25*(45), 10390–10402.

Benedetti, F., et al. (2006). The biochemical and neuroendocrine bases of the hyperalgesic nocebo effect. *The Journal of Neuroscience*, *26*(46), 12014–12022.

Bingel, U., & Placebo Competence Team (2014). Avoiding nocebo effects to optimize treatment outcome. *JAMA*, *312*(7), 693–694.

Bingel, U., et al. (2011). The effect of treatment expectation on drug efficacy: Imaging the analgesic benefit of the opioid remifentanil. *Science Translational Medicine*, *3*(70) . 70ra14.

Blasini, M., Corsi, N., Klinger, R., & Colloca, L. (2017). Nocebo and pain: An overview of the psychoneurobiological mechanisms. *Pain Reports*, *2*(2), e585. http://doi.org/10.1097/PR9.0000000000000585.

Colloca, L., Sigaudo, M., & Benedetti, F. (2008). The role of learning in nocebo and placebo effects. *Pain*, *136*(1–2), 211–218.

Crichton, F., & Petrie, K. J. (2015). Health complaints and wind turbines: The efficacy of explaining the nocebo response to reduce symptom reporting. *Environmental Research*, *140*, 449–455.

Dworkin, S. F., et al. (1983). Cognitive reversal of expected nitrous oxide analgesia for acute pain. *Anesthesia and Analgesia, 62*(12), 1073–1077.

Enck, P., et al. (2013). The placebo response in medicine: Minimize, maximize or personalize? *Nature Reviews. Drug Discovery, 12*(3), 191–204.

Flaten, M. A., Simonsen, T., & Olsen, H. (1999). Drug-related information generates placebo and nocebo responses that modify the drug response. *Psychosomatic Medicine, 61*(2), 250–255.

Freeman, S., et al. (2015). Distinct neural representations of placebo and nocebo effects. *Neuroimage, 112*, 197–207.

Geuter, S., & Buchel, C. (2013). Facilitation of pain in the human spinal cord by nocebo treatment. *The Journal of Neuroscience, 33*(34), 13784–13790.

Gupta, A., et al. (2017). Adverse events associated with unblinded, but not with blinded, statin therapy in the Anglo-Scandinavian Cardiac Outcomes Trial-Lipid-Lowering Arm (ASCOT-LLA): A randomised double-blind placebo-controlled trial and its non-randomised non-blind extension phase. *Lancet, 389*(10088), 2473–2481.

Heisig, S. R., et al. (2016). What do women with breast cancer expect from their treatment? Correlates of negative treatment expectations about endocrine therapy. *Psychooncology, 25*(12), 1485–1492.

Keltner, J. R., et al. (2006). Isolating the modulatory effect of expectation on pain transmission: A functional magnetic resonance imaging study. *The Journal of Neuroscience, 26*(16), 4437–4443.

Kessner, S., et al. (2013). The effect of treatment history on therapeutic outcome: An experimental approach. *JAMA Internal Medicine, 173*(15), 1468–1469.

Kong, J., et al. (2008). A functional magnetic resonance imaging study on the neural mechanisms of hyperalgesic nocebo effect. *The Journal of Neuroscience, 28*(49), 13354–13362.

Lyerly, S. B., et al. (1964). Drugs and placebos: The effects of instructions upon performance and mood under amphetamine sulphate and chloral hydrate. *Journal of Abnormal Psychology, 68*, 321–327.

Peerdeman, K. J., et al. (2016). Relieving patients' pain with expectation interventions: A meta-analysis. *Pain, 157*(6), 1179–1191.

Rief, W., et al. (2009). Differences in adverse effect reporting in placebo groups in SSRI and tricyclic antidepressant trials: A systematic review and meta-analysis. *Drug Safety, 32*(11), 1041–1056.

Rief, W., et al. (2017). Preoperative optimization of patient expectations improves long-term outcome in heart surgery patients: Results of the randomized controlled PSY-HEART trial. *BMC Medicine, 15*(1), 4.

Sawamoto, N., et al. (2000). Expectation of pain enhances responses to nonpainful somatosensory stimulation in the anterior cingulate cortex and parietal operculum/posterior insula: An event-related functional magnetic resonance imaging study. *The Journal of Neuroscience, 20*(19), 7438–7445.

Scott, D. J., et al. (2006). Variations in the human pain stress experience mediated by ventral and dorsal basal ganglia dopamine activity. *The Journal of Neuroscience, 26*(42), 10789–10795.

Scott, D. J., et al. (2008). Placebo and nocebo effects are defined by opposite opioid and dopaminergic responses. *Archives of General Psychiatry, 65*(2), 220–231.

Zunhammer, M., et al. (2017). The effects of treatment failure generalize across different routes of drug administration. *Science Translational Medicine, 9*(393).

CHAPTER SIXTEEN

Nocebo Effects and Experimental Models in Visceral Pain

Sigrid Elsenbruch[1], Franziska Labrenz
Institute of Medical Psychology and Behavioral Immunobiology, University Hospital Essen, University of Duisburg-Essen, Essen, Germany
[1]Corresponding author: e-mail address: sigrid.elsenbruch@uk-essen.de

Contents

Abstract

Despite its clinical relevance and the potential to extend insights into the processing and modulation of pain derived from investigations of placebo phenomena, the nocebo effect has received comparably little attention over the past decades. Research from experimental and clinical studies is only beginning to unravel the behavioral, functional, and psychoneurobiological mechanisms underlying the nocebo effect. Herein, we summarize current evidence regarding nocebo effects in the field of pain, with a particular emphasis on visceral pain. We provide an overview over behavioral and neuroimaging findings on the impact of expectations and learning and propose promising future directions to help fostering the transition of experimental research from bench to bedside.

International Review of Neurobiology, Volume 138
ISSN 0074-7742
https://doi.org/10.1016/bs.irn.2018.01.010

1. INTRODUCTION

Even though we are only beginning to unravel the nocebo phenomenon, its scientific and clinical relevance undoubtedly rivals that of the placebo effect. Over the past decades, the field of pain has driven interdisciplinary research efforts to elucidate psychological and neurobiological mechanisms as well as clinical implications of placebo effects. To advance knowledge about nocebo effects, experimental and clinical pain research is proving to be similarly fruitful, complementing knowledge emerging from other research areas. Within the broad field of pain with its many facets and subspecialties, visceral pain arguably plays an important yet often underestimated role. From a clinical perspective, both acute and chronic visceral pain are highly prevalent and afflict patients in diverse clinical subspecialties ranging from abdominal surgery over gynecology, general medicine to gastroenterology, psychosomatic medicine, and psychiatry. Beyond the indisputable clinical relevance of visceral pain, scientific efforts to elucidate visceral perception and interoception address fascinating research questions about perceptions, emotions, and cognitions at the interface of the neurosciences, neurogastroenterology, and psychology.

Together, studies conducted in the context of both visceral and somatic pain research have fundamentally shaped current knowledge that expectations and learning—the basic mechanisms underlying both placebo and nocebo effects—are relevant for clinical practice and treatment trials as they impact patient-reported outcomes, including the symptom of pain. This rapidly evolving field further inspires new interdisciplinary experimental approaches and transdisciplinary knowledge contributing to a greater appreciation regarding the crucial role of psychological factors in the pathophysiology of many clinical conditions, including those characterized by chronic visceral pain. In this chapter, we review existing knowledge about psychological and neurobiological mechanisms underlying nocebo effects in the context of visceral pain. Of note, besides visceral pain, a different yet highly connected and clinically relevant line of research has addressed nocebo effects for nausea (Elsenbruch & Enck, 2015), a topic which we only briefly touch upon as it is beyond the scope of this review. In its first part, we give a brief background into visceral pain and placebo knowledge as it has emerged over the past decade driven primarily by early patient-oriented studies. We then summarize existing human research studies that have implemented designs built on negative suggestions or learning/conditioning procedures to induce nocebo effects in the context of visceral pain. In an extended outlook, which draws on knowledge not only from within the visceral but also the somatic pain field, we highlight connected areas of research which have interesting conceptual and mechanistic overlap to nocebo effects. We thereby hope to encourage novel translational research strategies extending existing paradigms and concepts to incorporate and address the putative roles of predictability, context, and learning phenomena like overshadowing in nocebo research. Whenever possible, we attempt to integrate evidence from patients with chronic visceral pain. We acknowledge gaps in mechanistic knowledge which continue to exist even though this rapidly evolving transdisciplinary research field strives to quickly move from bench to bedside.

2. VISCERAL PAIN AND PLACEBO RESEARCH

Existing knowledge about placebo and more recently nocebo effects in visceral pain comes primarily from studies conducted in the context of functional gastrointestinal disorders such as the irritable bowel syndrome (IBS). With a pooled global prevalence of 11.2% in the general population (Enck et al., 2016), IBS is the most common functional gastrointestinal disorder (Canavan, West, & Card, 2014) with primary symptoms of recurrent abdominal pain or discomfort and altered bowel habits in the absence of identifiable organic pathology. Although the pathophysiology remains incompletely understood, it is assumed to rely on the interaction between biological and psychological factors along the brain–gut axis, involving hyperalgesia and hypervigilance for interoceptive, visceral signals as key centrally mediated mechanisms. Psychological alterations comprise affective comorbidity, abnormal illness behavior, heightened stress, and maladaptive cognitions, all of which are subsumed in a biopsychosocial model (Elsenbruch, 2011). In IBS and other gastrointestinal conditions, high placebo response rates have initially drawn rather negative attention as a "nuisance" hampering clinical trials (Elsenbruch & Enck, 2015). Yet, it was also in patients with IBS that the first brain imaging studies on placebo analgesia have been accomplished (Lieberman et al., 2004; Price, Craggs, Nicholas Verne, Perlstein, & Robinson, 2007; Price et al., 2009), paving the way for many brain imaging studies in both healthy individuals and patients with different chronic pain conditions. Furthermore, treatment approaches that incorporate placebo elements have successfully been employed in IBS (Ballou & Keefer, 2017; Dorn et al., 2007; Kaptchuk et al., 2010), inspiring placebo-based treatment studies in patients with other chronic pain conditions (Benedetti, Arduino, et al., 2006; Klinger, Soost, Flor, & Worm, 2007; Peerdeman et al., 2016).

Compared to placebo, nocebo effects remain poorly studied not only in the context of visceral pain or disturbed visceroception (Elsenbruch & Enck, 2015) but also in the broader somatic pain field (Schedlowski, Enck, Rief, & Bingel, 2015; Vase, Skyt, & Hall, 2016), despite their likely contribution to the reporting of adverse events and side effects as well as to the generation or worsening of symptoms (Elsenbruch & Enck, 2015). As such, nocebo effects are likely relevant for the pathophysiology of chronic pain irrespective of pain modality or affected bodily region, as they conceivably contribute to hyperalgesia and hypervigilance. Given accumulating evidence supporting pain modality-specific behavioral and neural processing (Aziz et al., 2000; Cervero, 2010; Koenen et al., 2017), dedicated visceral pain studies in clinically relevant pain models are needed in order to complement and extend knowledge derived from somatic pain models. In the comparably small field of placebo/nocebo research on visceral pain, first experimental research on treatment expectations has been conducted using a clinically relevant visceral pain model, i.e., pressure-controlled rectal or esophageal distensions. Graded distensions of the rectum, esophagus, or even the stomach with pressure-controlled inflatable balloon systems constitute a well-validated and widely used experimental model to assess visceroception and visceral pain in healthy individuals and patients with chronic visceral pain like IBS or functional dyspepsia (Mayer et al., 2008). This method allows the determination of sensory and discomfort/pain thresholds and in evoked pain studies the controlled and finely tuned application of distensions of varying intensity and duration that closely resemble aversive visceral sensations experienced by patients as well as sensations commonly (but less frequently and/or less intensely) experienced by healthy individuals.

3. NEGATIVE SUGGESTIONS

Verbal suggestions and learning paradigms constitute the two paradigms which have been used to generate and explore both placebo and nocebo effects in laboratory settings (Chavarria et al., 2017). This distinction has been the basis for different experimental study designs on placebo and nocebo effects in humans, and herein we adhere to this distinction in order to structure our review. However, it is becoming increasingly clear that expectations induced by suggestions and those resulting from learning processes are conceptually intertwined and difficult to disentangle experimentally, at least in humans (Enck, Chae, & Elsenbruch, 2017). Clearly, prior experience and learning processes shape and interact with treatment suggestions within any given clinical or experimental context, giving rise to expectancies (Rief & Petrie, 2016).

It is well established that positive treatment expectations arising from the psychosocial treatment context, especially written information or verbal suggestions delivered by healthcare professionals, the media, or peers, demonstrably shape the experience and neural processing of pain, irrespective of pain modality. This holds true not only for placebo analgesia but also for nocebo hyperalgesia mediated by negative expectations induced by suggestions and/or learning processes. Several studies on visceral placebo analgesia have implemented positive treatment suggestions in healthy humans and in patients with IBS (reviewed in Elsenbruch & Enck, 2015). With respect to nocebo effects induced by negative suggestions, studies implementing visceral pain models are scarce and have exclusively been conducted in healthy humans. Our group has studied nocebo effects using two different types of negative suggestions. On the one hand, we have implemented suggestions focused on negative *treatment-related* effects, and on the other hand, we have also employed negative *pain-related* suggestions. For example, in two studies (Roderigo et al., 2017; Schmid et al., 2013), we induced negative treatment expectations using deceptive written and verbal information regarding an intravenous (i.v.) administration of a supposed opioid antagonist. As a control, we provided participants with truthful information that an inert substance (saline) was administered. Both studies revealed evidence of nocebo hyperalgesia for visceral pain (Fig. 1), i.e., greater perceived pain induced by rectal distensions when participants expected a treatment that supposedly enhances pain when compared to when they expected an ineffective drug, which was modulated by acute stress in one study (Roderigo et al., 2017).

To complement and extend these findings, we employed negative suggestions that were focused directly on pain rather than treatment. In a first behavioral study (Elsenbruch, Schmid, et al., 2012), we randomized healthy women to negative, neutral, or positive information in a 2 day paradigm involving three series of rectal distensions (Fig. 2A). Herein, after a baseline series of distensions, all groups received i.v. administration of saline, and the nocebo group was additionally informed about sensitization processes (supposedly) occurring over the course of repeated distensions. In addition, on day 1 of the study, distension pressures were surreptitiously increased slightly to evoke a negative pain experience supporting our information regarding

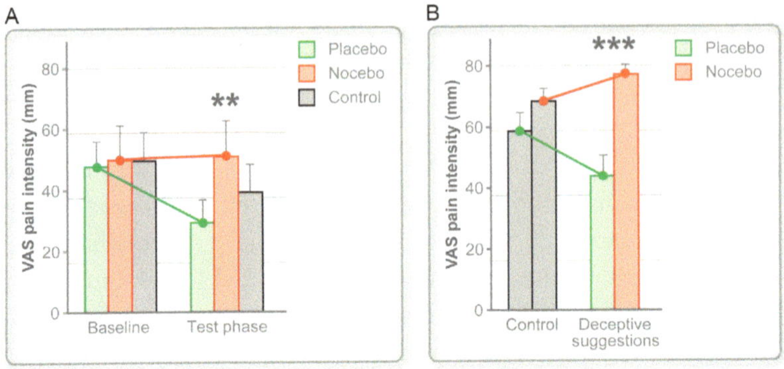

Fig. 1 Placebo analgesia and nocebo hyperalgesia in visceral pain studies implementing deceptive treatment suggestions. Changes in perceived pain intensity in response to painful rectal distensions, assessed with visual analog scales (VAS; 0–100 mm with end points labeled 0 = not painful and 100 = very painful), from two exemplary studies on visceral placebo and nocebo effects in healthy volunteers. (A) The study by Roderigo et al. (2017) implemented a between-group study design with a baseline and a test. Participants were randomly assigned to groups receiving deceptive positive ("spasmolytic drug" = placebo group), deceptive negative ("opioid antagonist" = nocebo group), or truthful neutral ("saline" = control group) treatment information about an intravenous administration which in reality always contained saline. Compared to the control group, pain ratings increased in the nocebo group indicating hyperalgesia and decreased in the placebo group indicating analgesic effects. (B) The brain imaging study by Schmid et al. (2013) implemented the same treatment suggestions in a combined within- and between-group study design. Volunteers received neutral suggestions (control) and either deceptive positive (placebo group) or deceptive negative treatment suggestions (nocebo group). The order of neutral and deceptive suggestions was randomized. Compared to neutral suggestions (control), negative suggestions led to nocebo hyperalgesia, while positive suggestions induced placebo analgesia, along with changes in neural activation (not shown here). Data are shown as mean ± standard error of the mean (SEM). $^{***}P < .001$; $^{**}P < .01$.

pain sensitization. On day 2, i.e., the test day, distension pressures were identical to baseline pressures. In the nocebo group, this procedure successfully induced enhanced pain ratings on the test day (Fig. 2B). Of note, given that this was an approach which combined suggestions with a manipulated experience of greater pain intensity, it is not possible to disentangle effects of suggestions and learning in this first proof-of-concept study. To clarify this and to address neural mechanisms, in a subsequent functional brain imaging study, we solely delivered negative suggestions regarding pain sensitization vs no such suggestions and without any i.v. catheter (Schmid, Bingel, et al., 2015). Herein, we also observed significant nocebo effects, i.e., greater pain ratings in the group with instructions of sensitization compared to a group

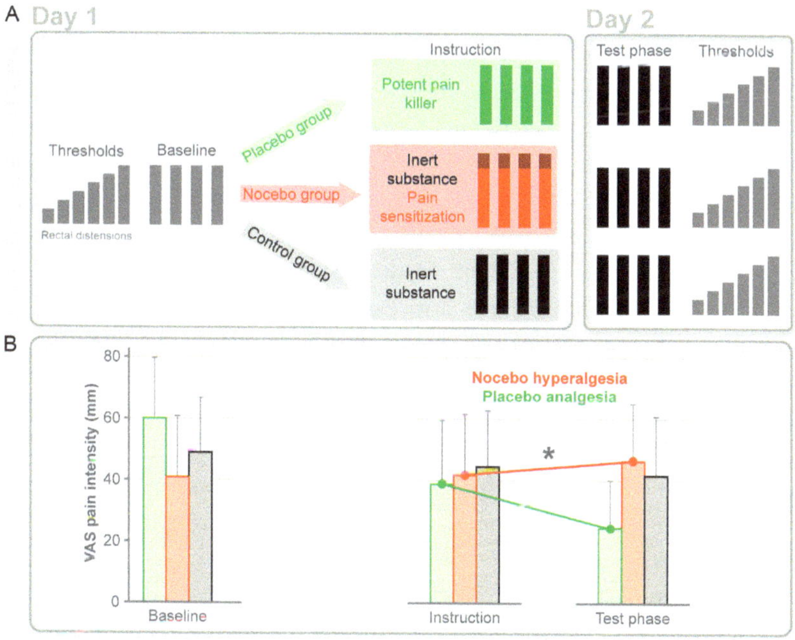

Fig. 2 Study design and pain ratings in a 2-day placebo/nocebo visceral pain study. In the behavioral study by Elsenbruch, Schmid, et al. (2012), healthy women were randomly assigned to a placebo, nocebo, or control group. On the first day, rectal sensory and pain thresholds were determined, and individually calibrated painful distensions were delivered (baseline). Different treatment suggestions (placebo group: positive; nocebo group: negative; control group: neutral) were delivered and another series of rectal distensions was accomplished. In the nocebo group, distension pressures were slightly increased (+2 mmHg). During the test on the second day, all groups received distensions at identical pressures as during baseline. Participants rated each distension on a visual analog scale (VAS: 0–100 mm). The nocebo group revealed a significant increase in perceived pain intensity from instruction to test phase indicating nocebo hyperalgesia, while the placebo group demonstrated placebo analgesia. Data are shown as mean ± standard error of the mean (SEM). $^*P < .05$.

without negative pain-related suggestions, but effects were markedly smaller than in the previous study. Since the studies differed in many other methodological aspects, it is not possible however to deduce that combining negative pain-related suggestions with a painful experience induces larger nocebo responses and suggestions or conditioning alone. Of note, studies employing negative pain-related suggestions and/or implementing negative pain-related experiences do not implement a "placebo pill" or other sham intervention. Hence, one could argue that the observed "nocebo effects" are mislabeled as such. While a discussion of definition of nocebo effects is beyond the scope herein, we argue that mechanistically such approaches are valuable to understand the mechanisms of nocebo effects that undoubtedly (also) occur outside of a direct treatment context, and are clinically highly relevant especially in patients with waxing and waning symptoms of pain, such as in IBS. Together, these data support that negative suggestions are capable of producing nocebo hyperalgesia in response to experimental visceral pain stimuli in healthy individuals, calling for more research to clarify if these effects also exist in patients and to discern both neurobiological mechanisms and clinical implications.

From a methodological perspective, we emphasize that it is crucial to implement control group(s) in order to properly quantify nocebo effects and distinguish them from other effects like habituation, sensitization, and order effects. Therefore, our approach to utilize truthful information regarding i.v. administration of saline as an inert substance has many advantages and arguably lacks better alternatives. At the same time, however, it may not be unequivocal and clearly should not be regarded as a "natural history" control group, i.e., a completely untreated group that receives pain stimuli without any type of information relating to treatment. First, receiving an i.v. infusion with the information that it contains saline may in itself induce explicit or implicit treatment expectations or conditioned responses, depending on prior knowledge and experience. Hence, strictly speaking, this group (or condition in within-group study designs) is a control group but not necessarily one completely free of treatment expectations. Participants could, despite the information that the substance is inert, have positive expectations—in the sense of an active placebo—which if true would effectively increase the difference between the nocebo and control group. On the other hand, it is also conceivable that the information that the substance is "inert" results in negative expectancies. Participants may perceive the treatment as "inefficacious"—knowing that other participants receive an active treatment (in between-group designs where participants are randomized) or knowing that they themselves will or have received pain stimuli with (supposed) drugs. Hence, suggestions about saline could also induce negative expectations, which would reduce the nocebo effect and lead to an underestimation of its effect size. In studies implementing deceptive treatment information in combination with i.v. administration (as in our studies), the only way to clarify this would be to include an additional control group who receives no i.v. catheter, which we did not do primarily given that we aimed to control for possible stress or discomfort that placement of an i.v.

catheter reportedly can induce itself and feasibility considerations—after all, these first studies in visceral pain were proof-of-concept studies designed at a time when experimental nocebo studies in the visceral pain field were lacking and primarily focused on elucidating brain mechanisms (Schmid et al., 2013) or interactions with acute stress (Roderigo et al., 2017), as explained in detail later.

4. CONDITIONING: FROM PAIN-RELATED FEAR TO NOCEBO HYPERALGESIA?

The role of pain-related fear is increasingly recognized in the pathophysiology and treatment of chronic pain (Vlaeyen, 2015). Pain-related fear in response to pain-predictive cues is established through associative learning and can hence be readily investigated using classical conditioning procedures, which are essentially based on fear conditioning (Elsenbruch & Wolf, 2015). Whether conditioned pain-related fear elicits (or contributes to) nocebo effects in visceral pain remains unclear given the small number of studies available, but conceptual overlap is intriguing. Further, studies emerging from the somatic pain field support nocebo effects induced by conditioning (Bąbel et al., 2017; Colagiuri, Quinn, & Colloca, 2015; Jensen et al., 2012; Jensen, Kirsch, Odmalm, Kaptchuk, & Ingvar, 2015; Reicherts et al., 2017) involving conscious as well as unconscious mechanisms. In the context of visceral pain, different types of innovative conditioning protocols using esophageal or rectal distensions as conditioned stimuli (CS) or unconditioned stimuli (US) have emerged in recent years (Ceunen et al., 2016; Gramsch et al., 2014; Icenhour et al., 2015; Labrenz et al., 2016; Yágüez et al., 2005), but very few of those have thus far directly addressed nocebo effects induced by conditioning.

Using the rectal distension model in healthy volunteers, our group tested effects of two types of conditioning protocols on nocebo effects at both the behavioral and neural levels. In one study (Labrenz et al., 2016), a conditioned group underwent differential fear conditioning using one visual symbol (CS+) that was consistently paired with painful rectal distensions as US, while another visual symbol remained unpaired (CS−). In a subsequent test phase, painful US were presented that were signaled by either CS+ or CS−. The other study (Icenhour et al., 2017) implemented conditioned cues signaling either distensions of high intensity (CS high) or low intensity (CS low), followed by a test phase with low-intensity distensions signaled by either CS high or CS low. Both studies showed no effects at the behavioral level, i.e., no evidence of increased perceived pain intensity in response to distensions, arguing against nocebo hyperalgesia induced by conditioned pain-related fear. However, neuroimaging data provided evidence that central changes occurred as a result of conditioning, suggesting hypervigilance rather than hyperalgesia, as explained later (see Section 5).

Likewise, it was demonstrated that fear toward nonpainful gastrointestinal sensations can be effectively induced through associative learning by implementing interoceptive conditioning procedures with esophageal balloon distensions (Ceunen et al., 2016). Moreover, classically conditioned fear to nonpainful sensations was shown to generalize to approximate distension pressures (Zaman, Weltens, et al., 2016) and to induce a perception bias to experience visceral sensations as more intense, aversive, or painful (Zaman, Madden, et al., 2016). As such, interoceptive fear learning may promote hypervigilance and visceral hypersensitivity and is hence highly relevant for nocebo effects. Since patients with functional gastrointestinal disorders are characterized by hypervigilance, excessive fear, and distress (Labus et al., 2004; Spiegel et al., 2011), the strong associability of unpleasant and benign visceral sensations might be an ecologically valid target for future research and an important factor contributing to nocebo-induced hyperalgesia.

5. BRAIN MECHANISMS

Neuroimaging studies have started to elucidate the neural mechanisms underlying visceral nocebo effects along the brain–gut axis. Based on knowledge from placebo studies, the general hypothesis was that negative expectations amplify pain processing in neural networks mediating sensory-discriminatory aspects of visceral pain, such as the thalamus, insula, and somatosensory cortex, as well as modify activity in regions involved in pain modulation by emotional or cognitive factors, such as the cingulate cortex, prefrontal cortices, and the amygdala. This was essentially confirmed in functional magnetic resonance imaging studies in healthy volunteers who received either positive or negative treatment suggestions (Schmid et al., 2013) or negative pain-related suggestions (Schmid, Bingel, et al., 2015). Negative treatment expectations demonstrably increased pain-induced neural activation in the insula when compared to neutral expectations and led to enhanced activity in the somatosensory cortex when compared directly to positive expectations (Schmid et al., 2013) (Fig. 3). In nocebo responders who showed hyperalgesia in response to negative

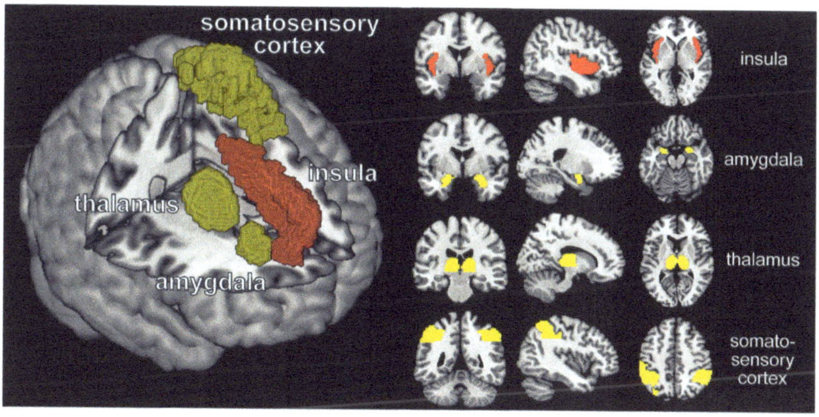

Fig. 3 Schematic depiction of neuroimaging findings on visceral nocebo responses induced by negative treatment or pain-related suggestions. Overview of brain regions involved in sensory-discriminative aspects of pain processing and cognitive-emotional pain modulation underlying nocebo effects in two of our visceral pain studies (Schmid, Bingel, et al., 2015; Schmid et al., 2013). Upregulation of pain-induced neural activation in the insula was found in both studies (Schmid, Bingel, et al., 2015; Schmid et al., 2013). Additionally, significant effects in the somatosensory cortex were found when placebo and nocebo groups were directly contrasted (Schmid et al., 2013), and nocebo responders compared to controls demonstrated higher pain-induced neural activity in the amygdala and thalamus (Schmid, Bingel, et al., 2015).

pain-related suggestions, greater pain-induced upregulation of the thalamus, insula, and amygdala was observed, which correlated significantly with increased perceived pain intensity (Schmid, Bingel, et al., 2015) (Fig. 3). The thalamus finding is in line with its established role in the processing of nociceptive information and coding of pain intensity (Bingel & Tracey, 2008) as well as its involvement in attentional processes as an aspect of cognitive pain regulation (Peyron, García-Larrea, Grégoire, & Costes, 1999). The upregulation of the insula in both our studies is consistent with three previous reports from other groups utilizing somatic pain models (Bingel et al., 2011; Freeman et al., 2015; Kong et al., 2008). Intriguingly, as a function of negative expectations, the insular cortex revealed increased functional connectivity with the midcingulate cortex extending to the posterior cingulate cortex during painful stimulation. The midcingulate cortex is involved in a number of pain related processes (Misra & Coombes, 2015; Pereira et al., 2010), including attention to behavioral stimuli and cognitive pain control (Shackman et al., 2011), and has recently been linked to central sensitization in visceral pain in healthy volunteers (Hamaguchi et al., 2013) and disturbed pain modulation in patients with chronic abdominal pain (Elsenbruch et al., 2010; Schmid, Langhorst, et al., 2015; Tillisch, Mayer, & Labus, 2011; Vogt, 2013). Together, these findings support the role of the insula across pain modalities and nocebo paradigms, making this one of the most robust findings in the otherwise heterogeneous brain imaging literature on nocebo hyperalgesia. Since the insula is also found highly active during placebo intervention in IBS patients (Hubbard, Labus, et al., 2011; Lee et al., 2012; Lieberman et al., 2004), it is tempting to speculate that the insula is a central hub for cognitive and emotional aspects of pain modulation in both acute and chronic visceral pain.

With respect to conditioning, the only two existing brain imaging studies employing visceral conditioning protocols to test for nocebo effects (Icenhour et al., 2017; Labrenz et al., 2016) support that conditioned pain-predictive cues do alter the neural processing of visceral stimuli, albeit in the absence of behavioral effects, as mentioned earlier. For example, neural activation in response to nonpainful rectal distensions evoked greater activation in dorsolateral prefrontal cortex, caudate nucleus, pregenual anterior, and ventral posterior cingulate cortex when preceded by visual cues that were previously conditioned to signal high pain compared to cues conditioned to signal low pain intensity (Icenhour et al., 2017). These findings may reflect hypervigilance and alterations in attention and are hence principally relevant to altered pain processing, but their role specifically in nocebo effects remains to be determined. Clearly, these studies only provide initial insight into the putative role of classical conditioning in nocebo responses for visceral pain.

In general, inferences in conditioning studies are limited by confounding effects of prior conditioning and habituation. Indeed, the majority of classical conditioning procedures are vulnerable to criticism of missing sham or yoked procedures that are urgently needed to disentangle the mutual interference between pain processing and associative learning and to elucidate differences in threat predictability, as explained in detail in the following section.

6. FUTURE DIRECTIONS

6.1 Anxiety and Stress

A role of trait or state negative emotions as mediators and/or moderators of placebo and especially nocebo effects has been proposed (Flaten, Aslaksen, Lyby, & Bjorkedal, 2011) based on experimental evidence that however almost exclusively comes from research in somatic pain models. Briefly, studies have suggested a role of dispositional fear (Aslaksen & Lyby, 2015), dispositional anxiety (Colloca, Pine, Ernst, Miller, & Grillon, 2016) as well as state anxiety (Elsenbruch, Schmid, et al., 2012; Vase, Robinson, Verne, & Price, 2005; Woo, 2015), and subjective stress levels (Aslaksen & Lyby, 2015; Aslaksen, Zwarg, Eilertsen, Gorecka, & Bjørkedal, 2015; Bjørkedal & Flaten, 2012) as mediators of expectancy effects. There also exists first evidence that neuroendocrine mediators of the hypothalamic–pituitary–adrenal (HPA) axis and sympathetic nervous systems may play a role in nocebo responses (Aslaksen et al., 2015; Benedetti, Amanzio, et al., 2006; Colloca et al., 2016). In contrast to this seemingly converging body of evidence emerging from the somatic pain field, existing placebo and nocebo studies in visceral pain have provided little or no support for a role of trait or state negative emotions as mediators of placebo or nocebo effects. In our line of studies, only one study (Elsenbruch, Schmid, et al., 2012) showed increased state anxiety in the nocebo group, while all other studies (Elsenbruch, Kotsis, et al., 2012; Icenhour et al., 2017; Labrenz et al., 2016; Roderigo et al., 2017; Schmid, Bingel, et al., 2015; Schmid, Langhorst, et al., 2015; Schmid et al., 2013) found no effects of suggestions on state anxiety or current tension in groups that received positive (placebo) or negative (nocebo) suggestions, and traits reflecting greater anxiety or stress symptoms did not correlate with the magnitude of placebo or nocebo responses.

Given this background, in a proof-of concept study we recently tested the impact of *acute* psychosocial stress on the magnitude of placebo or nocebo effects in a large sample of healthy volunteers (Roderigo et al., 2017). By utilizing a combined protocol of acute psychosocial stress followed by positive (placebo), negative (nocebo), or neutral (control) treatment suggestions, we could demonstrate that stress amplified the placebo and nocebo effects for rectal pain intensity and urge to defecate. Hence, acute stress may constitute a moderator of expectancy effects in the visceral system, with interesting clinical implications. After all, patients with chronic visceral pain and other burdening gastrointestinal symptoms such as rectal urgency demonstrably show enhanced emotional and HPA axis responses to psychological stress (Elsenbruch, 2011; Grinsvall, Törnblom, Tack, Van Oudenhove, & Simren, 2015), large pain reductions during placebo analgesia (Vase et al., 2005; Vase, Robinson, Verne, & Price, 2003), and altered neural modulation during placebo analgesia (Lee et al., 2012; Schmid, Langhorst, et al., 2015). It is also conceivable that stress impacts associative learning and memory processes, including conditioned pain-related learning and extinction which are demonstrably altered in patients with IBS (Icenhour et al., 2015; Labus et al., 2013). Applying existing findings regarding effects of stress or HPA axis mediators such as cortisol on memory consolidation and reconsolidation to the field of gastrointestinal disorders, one could postulate that stress results in a reactivation of the pain-related memory trace and/or facilitates its reconsolidation, ultimately making the pain-related fear memory more permanent. This process may contribute to the maintenance of pain-related fear and hypervigilance and thereby to maladaptive avoidance behavior as part of a vicious circle maintained by stress and fear (Elsenbruch & Wolf, 2015). While it is tempting to speculate that interactions between stress and negative expectancy effects—whether induced by suggestions or conditioning—could be increased in patients, the translation from healthy individuals tested in experimental laboratory settings to clinical populations is not unequivocal. Clearly, it is a challenging yet indispensable future research task to elucidate mediators and moderators of placebo and nocebo responses in patients experiencing chronic visceral pain in order to delineate the putative role of stress in nocebo effects in laboratory research, everyday clinical practice as well as randomized-controlled clinical trials (Elsenbruch & Enck, 2015; Enck, Bingel, Schedlowski, & Rief, 2013; Schedlowski et al., 2015).

6.2 Uncertainty and Predictability

Evidence from the somatic pain field has established that uncertainty (or unpredictability) is disruptive and may promote hyperalgesia (Hubbard, Ornitz, et al., 2011; Vlaeyen, 2015; Yoshida, Seymour, Koltzenburg, & Dolan, 2013). Indeed, unpredictable pain is generally assumed to incapacitate regulatory pain-inhibitory central mechanisms and through persistent prediction error signaling to result in elevated anxiety and pain spreading (Grupe & Nitschke, 2013). As such, it is highly relevant to control for and clarify putative effects of predictability in pain-related learning and nocebo effects in the context of pain. To do so, in one of our fear-conditioning studies (Labrenz et al., 2016), we included a control group in addition to a conditioned group. Specifically, participants were randomly assigned to a classically conditioned or sham-conditioned group, both operationalized through different CS–US contingencies but with identical number of stimuli. While the conditioned group received contingent CS–US pairings in a 75% reinforcement schedule, the sham-conditioned control group received noncontingently presented CS and US to prevent differential fear conditioning. As intended, the sham group did not reveal differential changes in CS valence but rather perceived both cues as equally aversive and pain predictive. Interestingly, during acquisition the sham-conditioned group demonstrated higher engagement of the somatosensory cortex, prefrontal, orbitofrontal, and cingulate cortices, parahippocampus, thalamus, insula, and periaqueductal gray when compared to the conditioned group, strongly supporting the role of pain predictability in shaping processing of the US. Based on these initial results and the fact that patients with chronic pain typically experience episodes of unpredictable pain, it appears fruitful to clarify the putative role of predictably in nocebo effects.

6.3 Context

Specifically in light of fear generalization and processing biases, the question of learning and retrieval context arises (Vervliet, Baeyens, Van den Bergh, & Hermans, 2013). Since contextual information can be dissociated from discrete stimuli representing CS and US and takes on an important role in information retrieval and resolving of ambiguous situations (Maren, Phan, & Liberzon, 2013), effects of contextual learning or context-dependent extinction arise for nocebo effects. In the field of visceral pain, the only existing brain imaging study addressing effects of a context change on the extinction and retrieval of pain-related memories demonstrated contextual factors to profoundly affect neural mechanisms underlying extinction learning (Icenhour et al., 2015). Specifically, a context change reduced differentiation of new pain-related safety and danger signal properties during extinction and may therefore substantially contribute to generalization processes impacting nocebo effects in pain.

An interesting new approach to investigate placebo- and nocebo-induced nausea has recently been introduced using galvanic vestibular stimulation (GVS) (Park et al., 1999; Quinn, MacDougall, & Colagiuri, 2015). By delivering electrical currents to the vestibular end organs, individuals experience feelings of motion sickness or undergo vehicle settings which provide an advantageous extension to prior conditioning studies in placebo and nocebo research. By combining instructions and conditioning using GVS, Quinn et al. (2015) could demonstrate higher nausea symptoms during test in the experimental groups. Specifically, in comparison with the group that was conditioned and tested within one context, the effect of increased nausea in the context-change group indicates generalization across contexts. In this line of evidence, the experimental implementation of suggestion and conditioning was further addressed, e.g., by Quinn, Livesey, and Colagiuri (2017), showing a comparable effect in reducing nausea via placebo responses. However, there is yet no empirical evidence on the effects of a combined approach and no evidence for similar effects and mechanisms of instruction and conditioning in nocebo responses.

6.4 Overshadowing, Latent Inhibition, and Blocking

Based on long-standing knowledge regarding the principles of classical conditioning, several promising targets for future conditioning studies on nocebo-related mechanisms are conceivable—many of them with interesting and obvious clinical perspectives. One goal with implications for the therapy of chronic visceral pain and other visceral symptoms would be the reduction of conditioned fear responses. This could be achieved by preexposure to the CS (latent inhibition), simultaneous presentation of a more salient CS (overshadowing), or pairing with a CS previously established to signal the US (blocking). With respect to latent inhibition, the initial, unreinforced presentation of a cue attenuates distinct features of novelty and salience, presumably resulting in less attention devoted to the cue upon later exposure, hence attenuating the strength of the conditioned response during reinforcement. This has already been demonstrated in nocebo nausea and was found effective even when participants were fully informed about the preexposure to treatment (Quinn et al., 2017). Likewise, in an approach to prevent anticipatory and to reduce posttreatment nausea, cancer patients received a saliently tasting beverage before chemotherapy that reduced overall nausea symptoms (Stockhorst et al., 1998). Overshadowing is a complex phenomenon of cue competition ranking among a variety of examples in which behavioral responses toward one stimulus are diminished through the concurrent presentation of a second stimulus. Another prominent example is the blocking effect, which has been shown to diminish conditioned analgesia (Dafters, Hetherington, & McCartney, 1983; Ross, 1985) and could be used in refined and extended experimental designs in visceral pain and nausea. For example, the specificity of either latent inhibition and overshadowing or a combined approach was tested in nausea-based context conditioning in a healthy population (Hall, Stockhorst, Enck, & Klosterhalfen, 2015). While the latent inhibition group showed the highest and the overshadowing group the lowest response in nausea symptoms, ratings of the combined and control group were in between these extremes. However, taking into account different baseline measures further analyses revealed that all three treatment groups were effective in alleviating nausea when compared to the control group. Consequently, latent inhibition and overshadowing are promising candidates for future development of conditioning techniques that could prevent the development of anticipatory nausea in patients.

7. CONCLUSIONS

In conclusion, research on the nocebo effect has begun to unravel the functional and neurobiological mechanisms underlying visceral hyperalgesia and hypervigilance. Emerging research from the pain field and beyond supports the crucial role of negative expectations and learning processes in shaping clinical outcomes. Further, nocebo effects—or rather nocebo-relevant mechanisms—are of pivotal significance beyond their treatment implications. They may indeed also play a role in the transition from acute to chronic pain as well as in the maintenance and perpetuation of symptom chronicity—consistent with a biopsychosocial disease model of chronic pain. At the same time, remaining gaps in knowledge call for more mechanistic research in order to integrate and clarify the contribution of different mediators and moderators, offering fascinating future directions in this rapidly evolving field. Only a more refined understanding of nocebo mechanisms will ultimately allow the development and testing of effective ways to reduce these effects in clinical settings.

ACKNOWLEDGMENTS

Funding for the authors' visceral pain studies was provided by the German Research Foundation (Deutsche Forschungsgemeinschaft, DFG) within the research unit FOR 1328 "Expectation and conditioning as basic processes of the placebo and nocebo response" (EL 236/8-1; 236/8-2), the research unit FOR 1581 "Extinction Learning: Neural Mechanisms, Behavioral Manifestations, and Clinical Implications" (EL 236/9-1; 236/9-2), and the Collaborative Research Centre 1280 "Extinction Learning."

REFERENCES

Aslaksen, P. M., & Lyby, P. S. (2015). Fear of pain potentiates nocebo hyperalgesia. *Journal of Pain Research*, *8*, 703–710.

Aslaksen, P. M., Zwarg, M. L., Eilertsen, H.-I. H., Gorecka, M. M., & Bjørkedal, E. (2015). Opposite effects of the same drug. *Pain*, *156*, 39–46.

Aziz, Q., Thompson, D. G., Ng, V. W., Hamdy, S., Sarkar, S., Brammer, M. J., et al. (2000). Cortical processing of human somatic and visceral sensation. *The Journal of Neuroscience: The Official Journal of the Society for Neuroscience*, *20*, 2657–2663.

Bąbel, P., Bajcar, E. A., Adamczyk, W., Kicman, P., Lisińska, N., Świder, K., et al. (2017). Classical conditioning without verbal suggestions elicits placebo analgesia and nocebo hyperalgesia. *PLoS One*, *12*, 1–17.

Ballou, S., & Keefer, L. (2017). The impact of irritable bowel syndrome on daily functioning: Characterizing and understanding daily consequences of IBS. *Neurogastroenterology and Motility: The Official Journal of the European Gastrointestinal Motility Society, 29*, 1–7.

Benedetti, F., Amanzio, M., Vighetti, S., & Asteggiano, G. (2006). The biochemical and neuroendocrine bases of the hyperalgesic nocebo effect. *Journal of Neuroscience, 26*, 12014–12022.

Benedetti, F., Arduino, C., Costa, S., Vighetti, S., Tarenzi, L., Rainero, I., et al. (2006). Loss of expectation-related mechanisms in Alzheimer's disease makes analgesic therapies less effective. *Pain, 121*, 133–144.

Bingel, U., & Tracey, I. (2008). Imaging CNS modulation of pain in humans. *Physiology, 23*, 371–380.

Bingel, U., Wanigasekera, V., Wiech, K., Ni Mhuircheartaigh, R., Lee, M. C., Ploner, M., et al. (2011). The effect of treatment expectation on drug efficacy: Imaging the analgesic benefit of the opioid remifentanil. *Science Translational Medicine, 3*, 70ra14.

Bjørkedal, E., & Flaten, M. A. (2012). Expectations of increased and decreased pain explain the effect of conditioned pain modulation in females. *Journal of Pain Research, 5*, 289–300.

Canavan, C., West, J., & Card, T. (2014). The epidemiology of irritable bowel syndrome. *Clinical Epidemiology, 6*, 71–80.

Cervero, F. (2010). Visceral versus somatic pain: Similarities and differences. *Digestive Diseases, 27*, 3–10.

Ceunen, E., Zaman, J., Weltens, N., Sarafanova, E., Arijs, V., Vlaeyen, J. W. S., et al. (2016). Learned fear of gastrointestinal sensations in healthy adults. *Clinical Gastroenterology and Hepatology, 14*, 1552–1558.e2.

Chavarria, V., Vian, J., Pereira, C., Data-Franco, J., Fernandes, B. S., Berk, M., et al. (2017). The placebo and nocebo phenomena: Their clinical management and impact on treatment outcomes. *Clinical Therapeutics, 39*, 477–486.

Colagiuri, B., Quinn, V. F., & Colloca, L. (2015). Nocebo hyperalgesia, partial reinforcement, and extinction. *The Journal of Pain: Official Journal of the American Pain Society, 16*, 995–1004.

Colloca, L., Pine, D. S., Ernst, M., Miller, F. G., & Grillon, C. (2016). Vasopressin boosts placebo analgesic effects in women: A randomized trial. *Biological Psychiatry, 79*, 794–802.

Dafters, R., Hetherington, M., & McCartney, H. (1983). Blocking and sensory preconditioning effects in morphine analgesic tolerance: Support for a Pavlovian conditioning model of drug tolerance. *The Quarterly Journal of Experimental Psychology. B, Comparative and Physiological Psychology, 35*, 1–11.

Dorn, S. D., Kaptchuk, T. J., Park, J. B., Nguyen, L. T., Canenguez, K., Nam, B. H., et al. (2007). A meta-analysis of the placebo response in complementary and alternative medicine trials of irritable bowel syndrome. *Neurogastroenterology and Motility: The Official Journal of the European Gastrointestinal Motility Society, 19*, 630–637.

Elsenbruch, S. (2011). Abdominal pain in irritable bowel syndrome: A review of putative psychological, neural and neuro-immune mechanisms. *Brain, Behavior, and Immunity, 25*, 386–394.

Elsenbruch, S., & Enck, P. (2015). Placebo effects and their determinants in gastrointestinal disorders. *Nature Reviews. Gastroenterology & Hepatology, 12*, 472–485.

Elsenbruch, S., Kotsis, V., Benson, S., Rosenberger, C., Reidick, D., Schedlowski, M., et al. (2012). Neural mechanisms mediating the effects of expectation in visceral placebo analgesia: An fMRI study in healthy placebo responders and nonresponders. *Pain, 153*, 382–390.

Elsenbruch, S., Rosenberger, C., Bingel, U., Forsting, M., Schedlowski, M., & Gizewski, E. R. (2010). Patients with irritable bowel syndrome have altered emotional modulation of neural responses to visceral stimuli. *Gastroenterology, 139*, 1310–1319, e4.

Elsenbruch, S., Schmid, J., Bäsler, M., Cesko, E., Schedlowski, M., & Benson, S. (2012). How positive and negative expectations shape the experience of visceral pain: An experimental pilot study in healthy women. *Neurogastroenterology and Motility: The Official Journal of the European Gastrointestinal Motility Society*, *24*, 914-e460.

Elsenbruch, S., & Wolf, O. T. (2015). Could stress contribute to pain-related fear in chronic pain? *Frontiers in Behavioral Neuroscience*, *9*, 340.

Enck, P., Aziz, Q., Barbara, G., Farmer, A. D., Fukudo, S., Mayer, E. A., et al. (2016). Irritable bowel syndrome. *Nature Reviews Disease Primers*, *2*, 16014.

Enck, P., Bingel, U., Schedlowski, M., & Rief, W. (2013). The placebo response in medicine: Minimize, maximize or personalize? *Nature Reviews Drug Discovery*, *12*, 191–204.

Enck, P., Chae, Y., & Elsenbruch, S. (2017). Novel designs and paradigms to study the placebo response in gastroenterology. *Current Opinion in Pharmacology*, *37*, 72–79.

Flaten, M. A., Aslaksen, P. M., Lyby, P. S., & Bjorkedal, E. (2011). The relation of emotions to placebo responses. *Philosophical Transactions of the Royal Society B: Biological Sciences*, *366*, 1818–1827.

Freeman, S., Yu, R., Egorova, N., Chen, X., Kirsch, I., Claggett, B., et al. (2015). Distinct neural representations of placebo and nocebo effects. *NeuroImage*, *112*, 197–207.

Gramsch, C., Kattoor, J., Icenhour, A., Forsting, M., Schedlowski, M., Gizewski, E. R., et al. (2014). Learning pain-related fear: Neural mechanisms mediating rapid differential conditioning, extinction and reinstatement processes in human visceral pain. *Neurobiology of Learning and Memory*, *116*, 36–45.

Grinsvall, C., Törnblom, H., Tack, J., Van Oudenhove, L., & Simren, M. (2015). Psychological factors selectively upregulate rectal pain perception in hypersensitive patients with irritable bowel syndrome. *Neurogastroenterology and Motility*, *27*, 1772–1782.

Grupe, D. W., & Nitschke, J. B. (2013). Uncertainty and anticipation in anxiety: An integrated neurobiological and psychological perspective. *Nature Reviews. Neuroscience*, *14*, 488–501.

Hall, G., Stockhorst, U., Enck, P., & Klosterhalfen, S. (2015). Overshadowing and latent inhibition in nausea-based context conditioning in humans: Theoretical and practical implications. *The Quarterly Journal of Experimental Psychology*, *218*, 1–12.

Hamaguchi, T., Kano, M., Kanazawa, M., Itoh, M., Yanai, K., & Fukudo, S. (2013). Effects of preceding stimulation on brain activation in response to colonic distention in humans. *Psychosomatic Medicine*, *75*, 453–462.

Hubbard, C. S., Labus, J. S., Bueller, J., Stains, J., Suyenobu, B., Dukes, G. E., et al. (2011). Corticotropin-releasing factor receptor 1 antagonist alters regional activation and effective connectivity in an emotional-arousal circuit during expectation of abdominal pain. *Journal of Neuroscience*, *31*, 12491–12500.

Hubbard, C. S., Ornitz, E., Gaspar, J. X., Smith, S., Amin, J., Labus, J. S., et al. (2011). Modulation of nociceptive and acoustic startle responses to an unpredictable threat in men and women. *Pain*, *152*, 1632–1640.

Icenhour, A., Kattoor, J., Benson, S., Boekstegers, A., Schlamann, M., Merz, C. J., et al. (2015). Neural circuitry underlying effects of context on human pain-related fear extinction in a renewal paradigm. *Human Brain Mapping*, *36*, 3179–3193.

Icenhour, A., Labrenz, F., Ritter, C., Theysohn, N., Forsting, M., Bingel, U., et al. (2017). Learning by experience? Visceral pain-related neural and behavioral responses in a classical conditioning paradigm. *Neurogastroenterology and Motility: The Official Journal of the European Gastrointestinal Motility Society*, *29*, 1–11.

Jensen, K. B., Kaptchuk, T. J., Kirsch, I., Raicek, J., Lindstrom, K. M., Berna, C., et al. (2012). Nonconscious activation of placebo and nocebo pain responses. *Proceedings of the National Academy of Sciences of the United States of America*, *109*, 15959–15964.

Jensen, K., Kirsch, I., Odmalm, S., Kaptchuk, T., & Ingvar, M. (2015). Classical conditioning of analgesic and hyperalgesic pain responses without conscious awareness. *Proceedings of the National Academy of Sciences of the United States of America, 112*, 7863–7867.

Kaptchuk, T. J., Friedlander, E., Kelley, J. M., Sanchez, M. N., Kokkotou, E., Singer, J. P., et al. (2010). Placebos without deception: A randomized controlled trial in irritable bowel syndrome. *PLoS One, 5*, e15591.

Klinger, R., Soost, S., Flor, H., & Worm, M. (2007). Classical conditioning and expectancy in placebo hypoalgesia: A randomized controlled study in patients with atopic dermatitis and persons with healthy skin. *Pain, 128*, 31–39.

Koenen, L. R., Icenhour, A., Forkmann, K., Pasler, A., Theysohn, N., Forsting, M., et al. (2017). Greater fear of visceral pain contributes to differences between visceral and somatic pain in healthy women. *Pain, 158*, 1599–1608.

Kong, J., Gollub, R. L., Polich, G., Kirsch, I., Laviolette, P., Vangel, M., et al. (2008). A functional magnetic resonance imaging study on the neural mechanisms of hyperalgesic nocebo effect. *The Journal of Neuroscience: The Official Journal of the Society for Neuroscience, 28*, 13354–13362.

Labrenz, F., Icenhour, A., Schlamann, M., Forsting, M., Bingel, U., & Elsenbruch, S. (2016). From Pavlov to pain: How predictability affects the anticipation and processing of visceral pain in a fear conditioning paradigm. *NeuroImage, 130*, 104–114.

Labus, J. S., Bolus, R., Chang, L., Wiklund, I., Naesdal, J., Mayer, E. A., et al. (2004). The visceral sensitivity index: Development and validation of a gastrointestinal symptom-specific anxiety scale. *Alimentary Pharmacology & Therapeutics, 20*, 89–97.

Labus, J. S., Hubbard, C. S., Bueller, J., Ebrat, B., Tillisch, K., Chen, M., et al. (2013). Impaired emotional learning and involvement of the corticotropin-releasing factor signaling system in patients with irritable bowel syndrome. *Gastroenterology, 145*, 1253–1261.e3.

Lee, H. F., Hsieh, J. C., Lu, C. L., Yeh, T. C., Tu, C. H., Cheng, C. M., et al. (2012). Enhanced affect/cognition-related brain responses during visceral placebo analgesia in irritable bowel syndrome patients. *Pain, 153*, 1301–1310.

Lieberman, M. D., Jarcho, J. M., Berman, S., Naliboff, B. D., Suyenobu, B. Y., Mandelkern, M., et al. (2004). The neural correlates of placebo effects: A disruption account. *NeuroImage, 22*, 447–455.

Maren, S., Phan, K. L., & Liberzon, I. (2013). The contextual brain: Implications for fear conditioning, extinction and psychopathology. *Nature Reviews. Neuroscience, 14*, 417–428.

Mayer, E. A., Bradesi, S., Chang, L., Spiegel, B. M., Bueller, J. A., & Naliboff, B. D. (2008). Functional GI disorders: From animal models to drug development. *Gut, 57*(3), 384–404.

Misra, G., & Coombes, S. A. (2015). Neuroimaging evidence of motor control and pain processing in the human midcingulate cortex. *Cerebral Cortex, 25*, 1906–1919.

Park, B. R., Kim, M. S., Lee, M. Y., Kim, Y. K., Choi, S. C., & Nah, Y. H. (1999). Effects of galvanic stimulation of the mastoid process on the gastric motility induced by caloric stimulation. *Auris, Nasus, Larynx, 26*, 263–268.

Peerdeman, K. J., van Laarhoven, A. I. M., Keij, S. M., Vase, L., Rovers, M. M., Peters, M. L., et al. (2016). Relieving patients' pain with expectation interventions: A meta-analysis. *Pain, 157*, 1179–1191.

Pereira, M. G., de Oliveira, L., Erthal, F. S., Joffily, M., Mocaiber, I. F., Volchan, E., et al. (2010). Emotion affects action: Midcingulate cortex as a pivotal node of interaction between negative emotion and motor signals. *Cognitive, Affective, & Behavioral Neuroscience, 10*, 94–106.

Peyron, R., García-Larrea, L., Grégoire, M., & Costes, N. (1999). Haemodynamic brain responses to acute pain in humans. *Brain*, 1765–1779.

Price, D. D., Craggs, J., Nicholas Verne, G., Perlstein, W. M., & Robinson, M. E. (2007). Placebo analgesia is accompanied by large reductions in pain-related brain activity in irritable bowel syndrome patients. *Pain, 127*, 63–72.

Price, D. D., Craggs, J. G., Zhou, Q., Verne, G. N., Perlstein, W. M., & Robinson, M. E. (2009). Widespread hyperalgesia in irritable bowel syndrome is dynamically maintained by tonic visceral impulse input and placebo/nocebo factors: Evidence from human psychophysics, animal models, and neuroimaging. *NeuroImage, 47*, 995–1001.

Quinn, V. F., Livesey, E. J., & Colagiuri, B. (2017). Latent inhibition reduces nocebo nausea, even without deception. *Annals of Behavioral Medicine, 51*, 432–441.

Quinn, V. F., MacDougall, H. G., & Colagiuri, B. (2015). Galvanic vestibular stimulation: A new model of placebo-induced nausea. *Journal of Psychosomatic Research, 78*, 484–488.

Reicherts, P., Wiemer, J., Gerdes, A. B. M., Schulz, S. M., Pauli, P., & Wieser, M. J. (2017). Anxious anticipation and pain: The influence of instructed vs conditioned threat on pain. *Social Cognitive and Affective Neuroscience, 12*(4), 544–554.

Rief, W., & Petrie, K. J. (2016). Can psychological expectation models be adapted for placebo research? *Frontiers in Psychology, 7*, 1876.

Roderigo, T., Benson, S., Schöls, M., Hetkamp, M., Schedlowski, M., Enck, P., et al. (2017). Effects of acute psychological stress on placebo and nocebo responses in a clinically relevant model of visceroception. *Pain, 158*, 1489–1498.

Ross, R. T. (1985). Blocking and unblocking of conditioned analgesia. *Learning and Motivation, 16*, 173–189.

Schedlowski, M., Enck, P., Rief, W., & Bingel, U. (2015). Neuro-bio-behavioral mechanisms of placebo and nocebo responses: Implications for clinical trials and clinical practice. *Pharmacological Reviews, 67*, 697–730.

Schmid, J., Bingel, U., Ritter, C., Benson, S., Schedlowski, M., Gramsch, C., et al. (2015). Neural underpinnings of nocebo hyperalgesia in visceral pain: A fMRI study in healthy volunteers. *NeuroImage, 120*, 114–122.

Schmid, J., Langhorst, J., Gaß, F., Theysohn, N., Benson, S., Engler, H., et al. (2015). Placebo analgesia in patients with functional and organic abdominal pain: A fMRI study in IBS, UC and healthy volunteers. *Gut, 64*, 418–427.

Schmid, J., Theysohn, N., Gaß, F., Benson, S., Gramsch, C., Forsting, M., et al. (2013). Neural mechanisms mediating positive and negative treatment expectations in visceral pain: A functional magnetic resonance imaging study on placebo and nocebo effects in healthy volunteers. *Pain, 154*, 2372–2380.

Shackman, A. J., Salomons, T. V., Slagter, H. A., Fox, A. S., Winter, J. J., & Davidson, R. J. (2011). The integration of negative affect, pain and cognitive control in the cingulate cortex. *Nature Reviews. Neuroscience, 12*, 154–167.

Spiegel, B. M. R., Khanna, D., Bolus, R., Agarwal, N., Khanna, P., & Chang, L. (2011). Understanding gastrointestinal distress: A framework for clinical practice. *The American Journal of Gastroenterology, 106*, 380–385.

Stockhorst, U., Wiener, J. A., Klosterhalfen, S., Klosterhalfen, W., Aul, C., & Steingrüber, H. J. (1998). Effects of overshadowing on conditioned nausea in cancer patients: An experimental study. *Physiology and Behavior, 64*, 743–753.

Tillisch, K., Mayer, E. A., & Labus, J. S. (2011). Quantitative meta-analysis identifies brain regions activated during rectal distension in irritable bowel syndrome. *Gastroenterology, 140*, 91–100.

Vase, L., Robinson, M. E., Verne, G. N., & Price, D. D. (2003). The contributions of suggestion, desire, and expectation to placebo effects in irritable bowel syndrome patients. An empirical investigation. *Pain, 105*, 17–25.

Vase, L., Robinson, M. E., Verne, G. N., & Price, D. D. (2005). Increased placebo analgesia over time in irritable bowel syndrome (IBS) patients is associated with desire and expectation but not endogenous opioid mechanisms. *Pain, 115*, 338–347.

Vase, L., Skyt, I., & Hall, K. T. (2016). Placebo, nocebo, and neuropathic pain. *Pain, 157*(Suppl. 1), S98–105.

Vervliet, B., Baeyens, F., Van den Bergh, O., & Hermans, D. (2013). Extinction, generalization, and return of fear: A critical review of renewal research in humans. *Biological Psychology, 92*, 51–58.

Vlaeyen, J. W. S. (2015). Learning to predict and control harmful events: Chronic pain and conditioning. *Pain, 156*(Suppl. 1), S86–93.

Vogt, B. A. (2013). Inflammatory bowel disease: Perspectives from cingulate cortex in the first brain. *Neurogastroenterology and Motility, 25*, 93–98.

Woo, K. Y. (2015). Unravelling nocebo effect: The mediating effect of anxiety between anticipation and pain at wound dressing change. *Journal of Clinical Nursing, 24*, 1975–1984.

Yágüez, L., Coen, S., Gregory, L. J., Amaro, E., Altman, C., Brammer, M. J., et al. (2005). Brain response to visceral aversive conditioning: A functional magnetic resonance imaging study. *Gastroenterology, 128*, 1819–1829.

Yoshida, W., Seymour, B., Koltzenburg, M., & Dolan, R. J. (2013). Uncertainty increases pain: Evidence for a novel mechanism of pain modulation involving the periaqueductal gray. *The Journal of Neuroscience: The Official Journal of the Society for Neuroscience, 33*, 5638–5646.

Zaman, J., Madden, V. J., Iven, J., Wiech, K., Weltens, N., Ly, H. G., et al. (2016). Biased intensity judgements of visceral sensations after learning to fear visceral stimuli: A drift diffusion approach. *Journal of Pain*, 1–12.

Zaman, J., Weltens, N., Ly, H. G., Struyf, D., Vlaeyen, J. W. S., Van den Bergh, O., et al. (2016). Influence of interoceptive fear learning on visceral perception. *Psychosomatic Medicine, 78*, 248–258.

CHAPTER SEVENTEEN

Using Learning Strategies to Inhibit the Nocebo Effect

Veronica F. Quinn[1], Ben Colagiuri
School of Psychology, The University of Sydney, Sydney, NSW, Australia
[1]Corresponding author: e-mail address: vqui6056@uni.sydney.edu.au

Contents

Abstract

Learning is a key mechanism underpinning the development of the nocebo effect. The learning literature has cataloged and explored numerous ways in which the environment can be manipulated to prevent, reduce, or eradicate learning. Knowledge of these processes could be used to both inhibit the development of nocebo effects and reduce already established nocebo learning. This review describes the available evidence on how such learning strategies have, or could be, applied to reduce the nocebo effect in both healthy participants and patients to date. These learning strategies include overshadowing, latent inhibition, extinction, and contingency degradation. These strategies represent important new avenues for investigation and should be used by researchers to design and test interventions to reduce nocebo effects.

International Review of Neurobiology, Volume 138
ISSN 0074-7742
https://doi.org/10.1016/bs.irn.2018.01.011

1. INTRODUCTION

The nocebo effect is a pervasive problem that occurs when features of a treatment context elicit adverse health reactions. While less studied than its positive counterpart, the placebo effect, there is considerable evidence for the nocebo effect across a range of conditions, including pain and nausea (Barsky, Saintfort, Rogers, & Borus, 2002; Colagiuri, Schenk, Kessler, Dorsey, & Colloca, 2015; Colloca & Miller, 2011; Colloca, Siguado, & Benedetti, 2008). These nocebo effects are not only inherently unpleasant to the individuals who experience them but can also cause substantial social and economic burden. As such, it is essential to develop strategies to counteract the substantial harm caused by the nocebo effect. Given that learning is one of the mechanisms playing a critical role in establishing the nocebo effect, in this review, we explore learning strategies that could be used to inhibit the nocebo effect.

2. LEARNING AS A MECHANISM OF THE NOCEBO EFFECT

Learning is one of the key mechanisms that drives the development of the nocebo effect. Learning accounts of the nocebo effect are often couched in terms of classical conditioning (hereafter "conditioning"), the process by which organisms learn about relations among cues and events in their environment (Rescorla, 1988). According to conditioning accounts, if an individual is exposed to treatment cues and then experiences an adverse outcome, then the treatment cues can become associated with the adverse outcome, such that exposure to the treatment cues alone can cause the individual to experience the same adverse reaction. This conditioning process has been shown to lead to nocebo effects in many conditions, including nausea (Klosterhalfen, Kellermann, Braun, et al., 2009; Klosterhalfen, Ruttgers, Krumrey, et al., 2000; Quinn & Colagiuri, 2016; Quinn, Livesey, & Colagiuri, 2017; Quinn, MacDougall, & Colagiuri, 2015), pain (Colagiuri, Quinn, & Colloca, 2015; Colloca, Petrovic, Wager, Ingvar, & Benedetti, 2010; Jensen, Kirsch, Odmalm, Kaptchuk, & Ingvar, 2015), itch (Bartels, van Laarhoven, Haverkamp, et al., 2014), and hormonal and cortisol secretion (Benedetti, Pollo, Lopiano, et al., 2003).

We recently developed a new model of conditioned nausea to examine the contribution of learning to the nocebo effect (see Fig. 1). In this model,

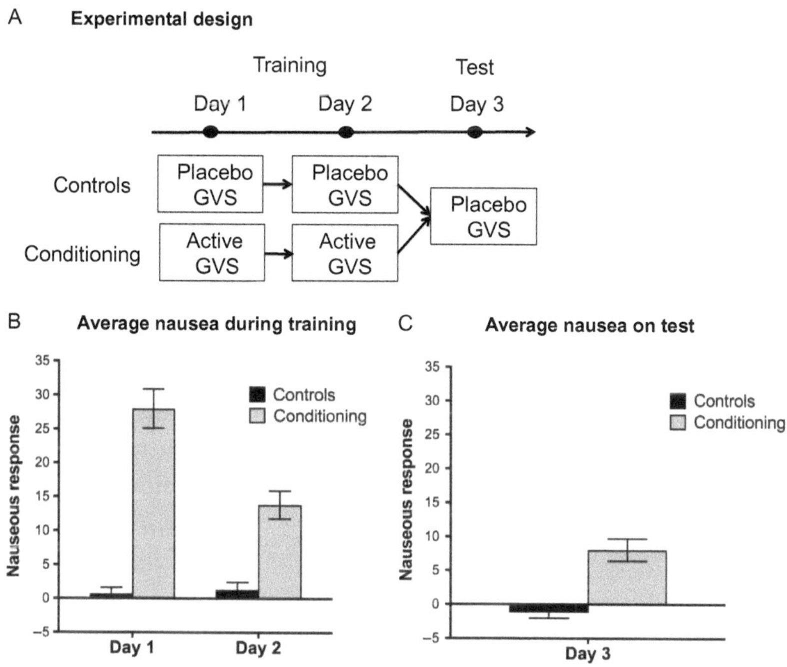

Fig. 1 Simplified version of the conditioning paradigm presented in Quinn et al. (2015). (A) The type of GVS received by the control and conditioning groups across the 3 days. (B) The conditioning group reporting significantly more nausea after receiving active GVS compared with controls that received placebo GVS. In (C), on test both groups receive placebo GVS, but the conditioning group reports higher ratings of nausea: evidence of conditioned nausea.

healthy participants were first exposed to 2 days of active galvanic vestibular stimulation (GVS), which causes them to experience nausea by disrupting their vestibular system, akin to motion sickness. On test on the third day, these same participants were exposed to placebo GVS which mimics the sensation of active GVS but does not typically induce nausea in and of itself. Critically, those participants who received prior conditioning with active GVS reported significant nocebo nausea when exposed to placebo GVS on the third day, whereas no such nausea occurred in control participants who only ever received placebo GVS. These types of findings highlight the importance of an individual's history with a given treatment. Where a treatment has previously led to adverse outcomes, the features of the treatment context may later come to evoke similar adverse outcomes simply as a result of conditioning between those features and the adverse outcome.

Having a reliable model in which nocebo effects can be induced via conditioning means that we can then use this same model to explore ways of inhibiting their development.

From the outset, however, it is important to emphasize that conditioning does not preclude conscious deliberation. That is, evidence of conditioned nocebo effects does not mean that these effects are not cognitively mediated. Early accounts described conditioning as a passive, low-level automatic process whereby the conditioned cue was essentially encoded as if it were the unconditioned stimulus, referred to as the "stimulus substitution" model of conditioning (Mackintosh, 1974). Modern accounts instead view conditioning as an active process whereby the individual learns to anticipate events in its environment (Rescorla, 1988), which can involve conscious deliberation. In fact, some modern learning accounts have gone as far as to propose that all human conditioning is mediated via conscious deliberation (e.g., "propositional account"; Mitchell, De Houwer, & Lovibond, 2009). While counterexamples to learning accounts that require conscious deliberation exist both in the field of placebo research (e.g., Benedetti et al., 2003) and in learning more generally (e.g., Colagiuri & Livesey, 2016; Perruchet, Cleeremans, & Destrebecqz, 2006), the very fact that such propositional accounts exist highlights the fact that conditioning and conscious deliberation are far from mutually exclusive. This review focuses on studies involving explicit conditioning manipulations, but there are a number of other studies in which nocebo effects are induced or inhibited via verbal suggestions instead (Benedetti et al., 2003; Voudouris, Peck, & Coleman, 1990). Their omission from this review is not to say that the mechanism through which verbal suggestions induce the nocebo effect and the mechanism underpinning conditioning manipulations necessarily differs. Indeed, a number of authors have suggested that they share the same mechanism, such as expectancy (Kirsch, 1997). However, untangling these differences is beyond the scope of the current manuscript and continues to be explored elsewhere (Kirsch, 2004; Stewart-Williams & Podd, 2004).

Importantly, by understanding the nocebo effect as a learning phenomenon in the modern sense, researchers and clinicians may be able to capitalize on the large body of learning literature in other areas to identify strategies to inhibit the nocebo effect. The purpose of this review is to describe such learning strategies, the available evidence for them, and how they might be applied as to inhibit the nocebo effect in clinical practice.

3. LEARNING STRATEGIES TO INHIBIT THE NOCEBO EFFECT

We will detail four of the most applicable learning strategies: extinction, overshadowing, latent inhibition, and contingency degradation. Extinction is a reactive strategy that attempts to reverse any existing learning leading to the nocebo effect. Overshadowing, latent inhibition, and contingency degradation reflect preventive strategies that are intended to impair the development of the learning that ultimately produces the nocebo effect. The evidence for their efficacy in reducing nocebo effects has been summarized in Table 1 and is explored in more detail in the following section.

3.1 Extinction

On face value, extinction is likely the most obvious way of attempting to reduce the nocebo effect. Extinction refers to the fact that conditioned responding typically diminishes when a cue is presented without its associated outcome. For example, if a rat is conditioned to expect an electric shock after a light is presented, then the rat will freeze when the light is presented again. However, after several displays of the light with no shock, eventually the rat will stop freezing to the light. That is, the fear to the light will have extinguished. As such, it should be possible to extinguish the nocebo effect simply by presenting the nocebo itself in the absence of any aversive outcome. In a clinical setting, this would involve presenting treatment cues without any associated adverse effect like pain or nausea so that conditioning, and the nocebo effect, diminishes.

However, so far, research from nocebo hyperalgesia suggests that the extinction of nocebo effects may be more difficult to obtain. Three studies of nocebo hyperalgesia conditioned in healthy participants have discovered that despite the withdrawal of the aversive outcome, the nocebo effect persisted (Colagiuri & Quinn, 2018; Colagiuri, Quinn, et al., 2015; Colloca et al., 2008). As shown in Fig. 2A, in one of our recent studies, we randomized healthy volunteers to receive nocebo conditioning or no conditioning (there was also a placebo conditioning that is omitted here) and then examined their extinction once the associated outcome was removed (Colagiuri & Quinn, 2018). In the nocebo conditioning, a sham electrode was paired with a surreptitious increase in electrical shocks compared with when no treatment was applied. In the control group, the sham electrode and no treatment were noncontingently paired with high and low shocks.

Table 1 Summary of the Current Evidence for Promising Learning Strategies in Reducing Nocebo Effects

Strategy	Description	Experimental Evidence	Clinical Evidence
Facilitating extinction	Extinction refers to the "unlearning" of treatment–outcome relationships after the treatment cues are presented without the accompanying aversive outcome	Three studies (Colagiuri, Quinn, et al., 2015; Colloca et al., 2008; Voudouris et al., 1990) have suggested that extinction of nocebo learning may not occur even when the aversive outcome is no longer present. One of these studies (Voudouris et al., 1990) also demonstrated that reducing anxiety may facilitate extinction processes	None
Overshadowing	More salient cues presented at the same time as conditioning between treatment and aversive outcomes can act as "scapegoats" that prevent treatment–outcome learning	Two studies (Pavlov, 1927; Symonds & Hall, 1999) found evidence that overshadowing may have reduced post rotation nausea, with a suggestion that anticipatory nausea could also be reduced using overshadowing strategies	A small pilot study (Dunsmoor, Niv, Daw, & Phelps, 2015) suggested that overshadowing could reduce nocebo nausea in chemotherapy
Latent inhibition	Pre-exposure to treatment-related cues prior to experiencing them with an aversive outcome can retard subsequent treatment–outcome conditioning	Two studies (Hall, Stockhorst, Enck, & Klosterhalfen, 2015; Symonds & Hall, 1999) found conflicting evidence as to the efficacy of pre-exposure reducing rotation-induced nocebo nausea. Two other experiments (Quinn et al., 2017) found that LI can reduce GVS-induced nocebo nausea if treatment-related cues are pre-exposed	None
Contingency degradation	Involves breaking down the contingency between treatment-relevant cues and aversive outcomes to prevent or reduce nocebo learning	Evidence in reducing nocebo hyperalgesia (Colagiuri, Quinn, et al., 2015)	None

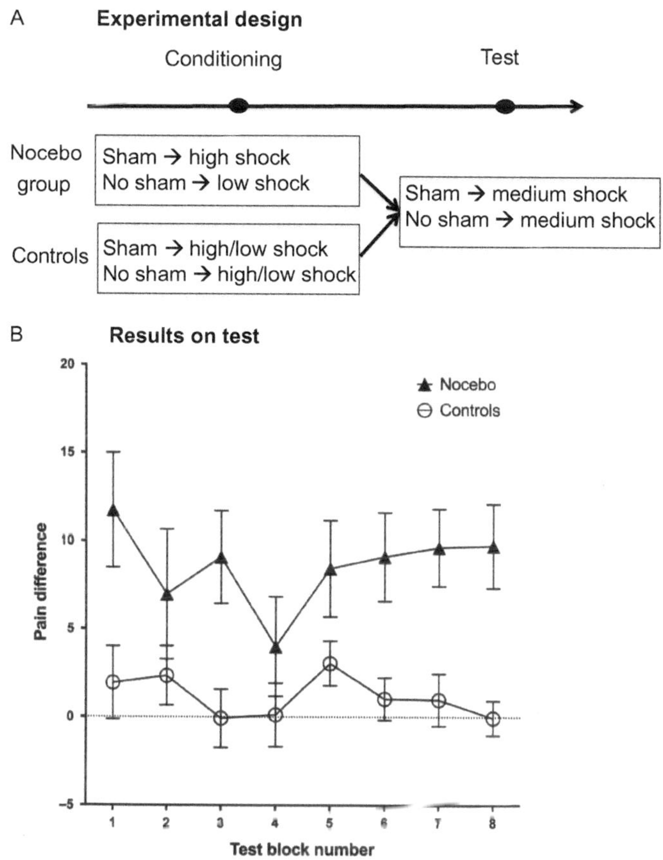

Fig. 2 Experimental design (A) and results (B) of a study exploring the rate of extinction of nocebo hyperalgesia in healthy participants.

In a later test phase, all participants received identical electrical shocks, with and without the sham electrode applied. As shown in Fig. 2B, there was clear evidence of a nocebo effect with the nocebo group demonstrating significant hyperalgesia when the sham was applied during the test. However, it was also clear that this nocebo hyperalgesia failed to extinguish over the entire 16 trials of the test session (8 blocks). Furthermore, the results revealed an interesting asymmetry between nocebo and placebo effects (not shown here) because despite the placebo conditioning leading to placebo analgesia, this placebo effect extinguished over the course of the test phase. As such, it appears that nocebo hyperalgesia is much more resistant to extinction than its positive counterpart and may persist indefinitely.

If nocebo effects do fail to extinguish for pain and other conditions, then, this means that nocebo effects could be very difficult to disrupt, once established. As such, it is important for researchers in the field more broadly to explore the extinction of nocebo effects across a range of symptom modalities and determine whether the failure to extinguish is specific to learning about pain or more general. If not, then simply presenting conditioned treatment cues without their respective outcomes is unlikely to be enough to reverse nocebo effects.

3.1.1 Anxiety and Extinction

While nocebo effects may not readily extinguish using standard extinction procedures, it may be possible to use knowledge from the animal and human learning literatures to facilitate extinction and hence reduce nocebo learning. In particular, anxiety and autonomic arousal have been proposed as potential mediators of failure to extinguish. We have reviewed this relationship elsewhere (Colagiuri & Quinn, 2018), but in brief, the increased autonomic arousal experienced by individuals when anticipating hyperalgesia may impair their ability to learn that a cue is no longer leading to this outcome. In the experiment depicted in Fig. 2, we discovered evidence that autonomic arousal was elevated throughout the extinction phase in the nocebo but not placebo group, and that autonomic arousal completely mediated the effect of nocebo vs placebo training on extinction (Colagiuri & Quinn, 2018). This suggests that interventions designed to reduce anticipatory anxiety may be effective at reducing the persistence of nocebo hyperalgesia and any other nocebo effects that are found not to extinguish. Trialing the use of relaxation procedures, distraction techniques, or pharmacological agents such as beta blockers could shed light on the causality of this effect and provide a promising method of reducing the burden of pain in those with persistent nocebo hyperalgesia. Interestingly, the animal literature suggests that benzodiazepines should not be used in this context as they may actually prevent extinction learning by disrupting the representations of these cues (Hart, Harris, & Westbrook, 2009), but that the beta blocker propranolol could reduce the physiological symptoms of anxiety without impairing this extinction learning (Rodriguez-Romaguera, Sotres-Bayon, Mueller, & Quirk, 2009).

3.1.2 Reemergence of Learning Following Extinction

A further problem in terms of extinction is that even when initially successful in reducing conditioned responding, these responses can later reemerge. Rather than simply being the undoing of old cue–outcome learning, it is most commonly thought that extinction represents new "cue–no outcome" learning (Bouton, 2002; Bouton & Bolles, 1979). Unfortunately, this means that even after extinction, this old learning can reemerge. This can happen when some time has passed after the extinction learning (spontaneous recovery), when extinguished cues are presented outside of the context in which they are extinguished (contextual renewal), or when the treatment is presented with the side effect again (rapid reacquisition). Particularly in the context of aversive stimuli such as fear, nausea, and anxiety, retaining the initial acquisition learning is likely to have been adaptive, as the survival costs of not anticipating aversive stimuli are high (Dunsmoor et al., 2015). But, this obviously presents a problem in the context of the nocebo effect reemerging even if it can first be extinguished.

Taken together, the current evidence suggests that despite being readily observed for many other learning phenomena, extinction of the nocebo effect is difficult to achieve. In the studies having assessed extinction of nocebo effects to date, there has been very little evidence that the nocebo effect diminishes over time. While it may be possible to induce extinction of nocebo effects with additional interventions, such as those targeting anticipatory anxiety and arousal, we propose that a better approach to combatting the nocebo effect is to prevent the learning leading to its establishment before it occurs. Attention now turns to such preventive strategies.

3.2 Overshadowing

Overshadowing is a learning strategy based on the principles of cue competition that could prevent or diminish nocebo learning from occurring. Cue competition refers to the fact that when multiple cues are present, learning about each of the cues develops preferentially based on the salience of each cue and any prior learning about them. In overshadowing, this type of cue competition occurs when two or more cues are presented simultaneously with a given outcome. Despite the fact that all of the cues are paired equally frequently with the given outcome, more learning about the outcome occurs to the more salient cue. Overshadowing was described as early as Pavlov (1927), who observed that when a weaker and more intense tone were presented simultaneously as conditioned stimuli, the conditioned response to the more intense tone was much greater when each was presented individually later (Pavlov, 1927). What is critical here in relation to inhibiting the nocebo effect is that the greater learning that develops to the more salient cue means that there is weaker learning to the less salient cues. That is, Pavlov found much less conditioning occurred to the weaker tone than otherwise would have because it was paired with a more intense tone that accrued substantial learning. In other words and as per its name, learning about a salient cue *overshadows* learning about the other less salient cues.

There are many examples of overshadowing in both the animal and human learning literature, including in the overshadowing of aversive outcomes, which provide support for the idea that this strategy could be used to inhibit the nocebo effect. For example, in rodent studies, a standard conditioning procedure involves pairing a context, e.g., a specific cage, with nausea via lithium chloride injections. When this is done, the rodents develop associations between that context and nausea, such that they demonstrate conditioned nausea when placed back in that context. However, if rodents have access to and consume a novel flavored drink while they are placed in the context and receive the lithium chloride injections, then they exhibit much weaker conditioned nausea when placed back in the context, because the flavored drink overshadows the context-nausea learning (Symonds & Hall, 1999).

To date there have been only three studies attempting to use overshadowing to reduce a nocebo effect in humans, all three of which involved nausea. These studies have focused specifically on reducing anticipatory nausea. Anticipatory nausea refers to the phenomenon whereby patients develop such strong learning between their chemotherapy context and the nausea they experience that they begin to feel nauseous purely when reentering this context (Bovbjerg, 2006), and is potentially the most well-studied conditioned nocebo effect in the literature. This focus on the anticipatory period is likely to be in part because in the period immediately before treatment the contribution of conditioned nausea is clearer, as there is no nauseogenic stimulus to attribute the experience of nausea to. However, learning strategies such as overshadowing are just as likely to be effective in reducing posttreatment symptoms.

In a small pilot study ($N=16$, with 8 in each group) of patients undergoing chemotherapy, Stockhorst, Wiener, Klosterhalfen, et al. (1998) randomized patients to receive novel tasting drinks during their chemotherapy infusions or no novel tasting drinks. The aim was to prevent the chemotherapy context forming associations with nausea by having these novel and hopefully salient flavors overshadow any context-nausea learning. The authors found that whereas the control group experienced anticipatory nausea at a rate comparable to population estimates, none of the patients in the experimental group did, with some suggestion that their postchemotherapy nausea was also lower. However, while promising, the small sample size of this pilot study makes it difficult to determine whether the results reflected a genuine overshadowing effect or was simply a spurious difference resulting from a patient population with a naturally large variance in side-effect experience.

The second study used a similar method to attempt to reduce anticipatory nausea in the laboratory setting with a slightly larger group of 24 healthy participants ($n = 12$ in each group). This study used a "rotation" paradigm, where rotation in a chair or a drum is used to make participants nauseous. Anticipatory nausea is assessed by placing the participants in the rotation context on test and observing whether they experience nausea before being rotated, and postrotation nausea is assessed after they have been rotated. In this overshadowing paradigm, novel flavored drinks were administered prior to rotation as potential overshadowing cues (Stockhorst, Hall, Enck, & Klosterhalfen, 2014). This procedure effectively reduced anticipatory nausea in terms of the number of nausea symptoms endorsed via self-report and in TNF-α levels for women when later exposed to the rotation context without the flavored drinks, with trends toward a significant difference from conditioning controls. These trends were supported by intragroup comparisons, where it appeared that participants in the overshadowing group showed a reduction in scores on the nausea profile over the days, whereas those in the control group did not.

The third study by the same group contained an identical overshadowing manipulation using the same rotation paradigm and overshadowing procedure with eight participants in each group. However, in this case, the authors observed evidence that overshadowing with novel flavors prior to rotation reduced postrotation nausea, but failed to replicate the effect of overshadowing on anticipatory nausea (Hall et al., 2015).

Together, the handful of existing studies on overshadowing provide some preliminary evidence that it could be used to inhibit nocebo effects. To this end, it is important to highlight that there is no deception inherent in overshadowing interventions, meaning that such interventions could be ethically applied in clinical practice. However, before recommending the use of overshadowing in clinical practice, it is clear that more empirical studies are required, particularly studies involving larger samples of participants compared with the relatively small pilot studies run to date. Further, given that the existing studies on overshadowing have been conducted exclusively on nausea, it would be very useful for future studies to test whether overshadowing can be used to inhibit nocebo effects in other areas, such as pain, itch, and cough, where it is clear that condition contributes to the nocebo effect.

3.3 Latent Inhibition

Latent inhibition is the phenomenon whereby simple pre-exposure to a neutral cue impairs learning about that same cue when it is later paired with a biologically salient event. In standard conditioning, rats exposed to pairings of a light with a shock will develop conditioned fear to the light. However, in latent inhibition, pre-exposure to the light without the shock inoculates the rats from developing as much fear to the light when it is later paired with the shock (Lubow & Moore, 1959). Latent inhibition is a well-documented and robust effect in nonhuman animals, having been observed for a wide range of stimuli and outcomes (for an early review, see Lubow, 1973). While the exact mechanisms of latent inhibition continue to be debated, it has been suggested that the initial pre-exposure of a cue reduces its novelty or salience, leading to the individual directing less attention toward it, which then reduces the extent of conditioning when it is subsequently followed by an outcome (Holmes & Harris, 2010). As such, it seems quite possible that pre-exposing individuals to features of the treatment context that could become associated with adverse outcomes could prevent or reduce nocebo learning from occurring and hence inhibit the nocebo effect.

To date, there have been four studies that have attempted to use a latent inhibition manipulation to reduce the nocebo effect, all four in conditioned nausea. The first two of these were conducted using the rotation paradigm described earlier in the overshadowing studies. In these studies, the pre-exposure intervention involved asking participants to spend the first session(s) sitting in the rotation chair in the absence of rotation so that no nausea was experienced. Participants were then rotated to create context-nausea learning in subsequent sessions, and both anticipatory and postrotation nausea were assessed. This strategy aims to reduce the associability of cues associated with sitting in the rotation chair so that they do not then come to be conditioned with nausea.

In the first of these studies, Klosterhalfen, Kellermann, Stockhorst, et al. (2005) found that pre-exposing healthy participants to the rotation context reduced reporting of anticipatory nausea when placed back in the rotation context. Conversely, the second study that aimed to replicate this effect and examine its interaction with overshadowing found no evidence of latent inhibition in anticipatory nausea. In fact, pre-exposure nonsignificantly *increased* reporting of anticipatory nausea on test (Hall et al., 2015). Further, neither study found that pre-exposure reduced postrotation nausea. Given that these studies had very similar designs the difference is perplexing and suggests that if latent inhibition is possible in this context, it is not a robust effect. However, both of these studies used rotation to induce nausea, and both aimed to latently inhibit environmental cues in the rotation environment. Using rotation chairs makes it difficult to distinguish between purely contextual features of the environment and those associated only with the sensations of rotation, as the chairs dominate the context in which they are placed. This then makes it hard to discern which cues have become associated with nausea, and hence which should be pre-exposed to latently inhibit this relationship and successfully reduce nocebo nausea.

In an attempt to address some of these issues, we recently used our GVS model of nausea described above to explore latent inhibition of nocebo nausea. As we have argued previously (Quinn et al., 2015), GVS is arguably a closer approximation of the types of treatments that have been typically found to induce nocebo nausea, e.g., chemotherapy. This is because the GVS involves application of a "treatment" without the use of rotation to evoke nausea. In a preliminary study, we first aimed to test whether nausea could be readily conditioned to a context using GVS (Quinn et al., 2015). Participants received active GVS in a specific context over 2 days and then were tested with placebo GVS either in the same or in a new context. If the nocebo nausea was conditioned, at least partly, to the context, then changing contexts on test should reduce the amount of nocebo nausea experienced. However, we found equally strong nocebo nausea irrespective of whether the context was the same or different at test, suggesting that nocebo nausea generalized strongly across context. To explain this, we proposed that cues associated with the salient and novel experience of receiving GVS were far more likely to become conditioned with nausea than purely exteroceptive environmental cues. If this is the case, then it stands to reason that to latently inhibition nocebo nausea, one would need to pre-expose these GVS or treatment-related cues, rather than the context itself, because it is the former that enters most strongly into an association with nausea.

We therefore conducted two experiments to test this, pre-exposing healthy participants to the sensations associated with receiving GVS before exposing them to nauseating GVS (Quinn et al., 2017). As depicted in Fig. 3A, we randomized participants either to receive pre-exposure to placebo GVS under the guise it was active GVS, or to instead participate in

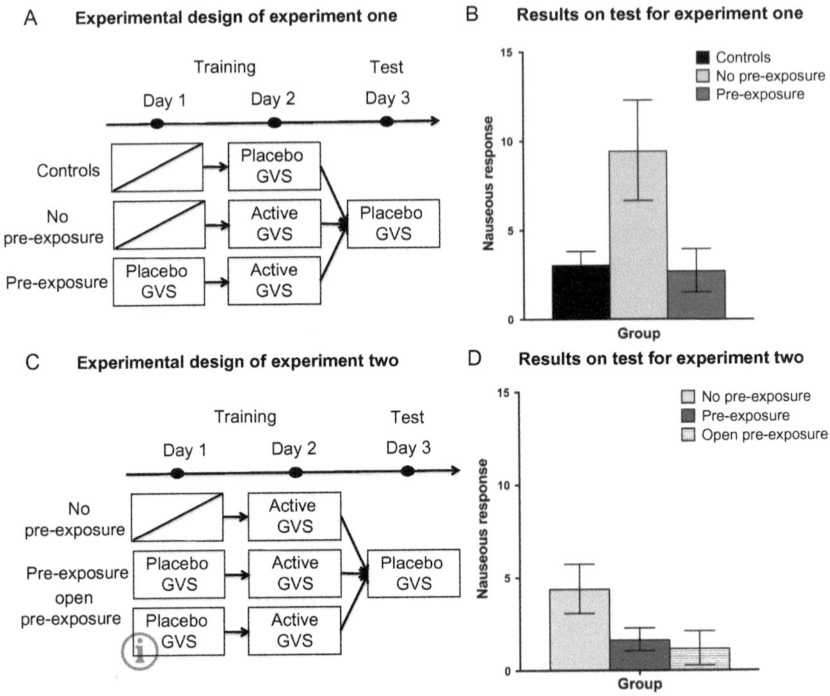

Fig. 3 The experimental design and results of two experiments published in Quinn et al. (2017).

what they were told was a baseline session where they were not introduced to the GVS machine at all. Participants in the two conditioning groups then experienced active GVS on day 2, with controls now receiving the placebo GVS to allow us to assess the contribution of conditioning. We found evidence that this pre-exposure inhibited, and in fact may have entirely eradicated, the nocebo nausea that otherwise developed in participants who did not experience pre-exposure (Fig. 3B). As such, this provides strong evidence that nocebo nausea can be inhibited via latent inhibition, providing that the pre-exposure occurs to the salient treatment cues.

However, although these findings provided promising evidence that latent inhibition could be used to reduce nausea in the clinic, our experiments involved deceiving participants into thinking the pre-exposure they received during training was "real" GVS. This limits the clinical applicability of these findings because it would not be ethically permissible in the clinic, where patients would have to be surreptitiously given placebos when expecting real treatment. To address this, as depicted in Fig. 3C we replicated the pre-exposure experiment but included a group who were told that they were being pre-exposed to sham GVS in order to reduce the unpleasantness of the experience of real GVS, i.e., open pre-exposure. Critically, we found that this "open" pre-exposure was just as effective as deceptive pre-exposure (Fig. 3D), suggesting that deception is not necessary to elicit latent inhibition.

Overall, the existing studies on latent inhibition suggest that it could be an effective way of blocking nocebo effects from forming, at least for nausea. However, they do highlight that the pre-exposure must be to the salient cues that are closely tied with nausea. Perhaps most interestingly, the evidence also suggests that open pre-exposure is just as effective as deceptive pre-exposure for inducing latent inhibition, which means that such techniques could be applied ethically in clinical settings. A potentially effective clinical intervention would therefore be to pre-expose participants to the sensations of infusion apparatus or olfactory and gustatory cues associated with oral treatments in the absence of nausea prior to receiving a nauseating treatment. But, it also remains necessary to test whether latent inhibition is effective for other symptom modalities beyond nausea.

3.4 Contingency Degradation

One learning strategy that has received less attention in the human nocebo literature is contingency degradation. A basic finding that underpins all learning models is that more exposure to a contingent cue–outcome pairing leads to stronger learning (e.g., Rescorla & Wagner, 1972). That is, the more the outcome occurs in the presence of the cue and the less the outcome occurs in the absence of the cue, the stronger the learning. The same has been shown within nocebo learning, where more opportunities to learn that a cue was followed by pain led to stronger pain responses to that cue (Colloca et al., 2010). The contingency of the cue and the outcome is critical; if the cue is presented without the outcome, learning is weaker. Similarly, if the outcome is presented without the cue, learning is weaker. Intentionally reducing the contingency between cues and outcomes in this way is referred to as "contingency degradation."

Interspersing sham treatments within a treatment schedule so that the adverse outcome is not always experienced with the treatment is one way to degrade treatment-outcome contingencies. This has been used in healthy participants to reduce nocebo hyperalgesia, and is referred to as "partial reinforcement" (Colagiuri, Quinn, et al., 2015). Partial reinforcement in the current context was first explored in placebo learning, where it led to weaker learning, but this learning was also more persistent (as would be predicted by learning theories) (Au Yeung, Colagiuri, Lovibond, & Colloca, 2014). We recently examined the effect of partial reinforcement on nocebo hyperalgesia (Colagiuri, Quinn, et al., 2015). Using a similar design for conditioning nocebo hyperalgesia shown in Fig. 2, we randomized healthy participants to either receive training where a sham treatment was always followed by a high shock and no treatment was always followed by a low shock (continuous reinforcement) or to receive training where a sham treatment led to a high shock roughly 60% and a low shock 40% of the time, with no sham always leading to a low shock (partial reinforcement). At test, when all participants received medium shocks, we found that participants who were given partial reinforcement exhibited significantly less nocebo hyperalgesia than those who received the standard continuous reinforcement. This suggests that contingency degradation in the form of partial reinforcement could reduce nocebo effects. However, one caveat to suggesting using partial reinforcement to reduce the size of nocebo hyperalgesia, is that it also may be slower to extinguish. Although we observed no differences in the rate of extinction of nocebo hyperalgesia established via continuous and partial reinforcement—with both failing to extinguish—it is possible that with enough extinction trials, partial reinforced nocebo effects could be found to outlast continuously reinforced nocebo effects. If so, then caution would be required in employing such a technique to attempt to reduce the nocebo effect.

4. LIMITATIONS OF THE CURRENT RESEARCH

The learning strategies described above offer appealing potentially effective methods of intervening to reduce nocebo effects, particularly latent inhibition and overshadowing. As noted earlier, a general limitation of research in this area is the fact that there are very few existing empirical studies in the context of the nocebo effect and even fewer with patients. However, even beyond this general limitation, there are a few specific limitations and unanswered questions in the literature that should be addressed in future research.

4.1 Limitations of Animal Models

Much of the theory for the learning strategies described above comes from the animal literature. While animal models are heavily relied upon for drug discovery, the translation to humans can be discrepant. In fact, one review found that only a third of highly cited animal papers testing medical therapies in animal in vivo models translated successfully in humans (Hackam & Redelmeier, 2006). Apart from the obvious differences in physiology, animal studies typically use otherwise healthy animals except for the presence of the specific disease (rather than having comorbidities), have much more homogeneous samples (e.g. strain of animal, gender of animals), and afford substantially more invasive study designs, including higher and more frequent doses (van der Worp, Howells, Sena, et al., 2010). Thus, one cannot take for granted that learning strategies observed in animals will translate to humans. Instead, the animal research plays a critical role for identifying potential strategies for intervention in humans that need to be tested empirically.

4.2 The Role of Cognition

Related to the reliance on animal research, there has been very little attempt to understand the role of cognition in most of the studies reviewed earlier. As discussed in Section 1, human learning and conditioning is an active rather than passive process that does not preclude conscious deliberation. Whereas it is harder for animals to require reliable information about the causality of events in their environment, most patients enter treatment with explicit knowledge about their treatment and its possible side effects. It is currently unclear how this explicit knowledge interacts with or is influenced by the conditioning procedures described earlier. The evidence indicating that open pre-exposure effectively induces latent inhibition described earlier (Quinn et al., 2017) certainly suggests that conditioning manipulations can be effective without deception, which is encouraging. However, given the general lack of data available regarding the role of cognition in the above learning strategies, we suggest that a research priority should be to investigate cognitive factors such as expectancy when studying these manipulations. It could be the case that we could enhance their efficacy by combining learning strategies with (nondeceptive) explicit information in order to produce the strongest effects.

4.3 Generalizability to Other Symptoms and Treatment Modalities

The way in which animals and humans learn about different adverse outcomes is mediated by the different neurological and behavioral responses elicited by the experience of those outcomes. The majority of the studies reviewed above have been in either pain or nausea. The well-accepted role that conditioned nausea plays in the chemotherapy context (Bovbjerg, 2006; Quinn & Colagiuri, 2015) is likely to have driven this research focus, as well as the availability of well-controlled paradigms for studying nocebo hyperalgesia. While these studies have provided promising results, it is important to test how well they extend to other symptoms and treatment modalities beyond pain and nausea.

4.4 Some Learning Mechanisms May Vary Across Sexes

There is also some evidence that sex may be an important moderator of the placebo and nocebo effects. Four studies in nocebo nausea found that females showed stronger responses to conditioning manipulations than male participants (Klosterhalfen et al., 2009; Klosterhalfen, Kellermann, Pan, et al., 2005; Quinn & Colagiuri, 2016; Stockhorst et al., 2014) with one study specifically only recruiting female participants for this reason (Hall et al., 2015). Given that all experimenters in those studies were females, these sex effects may be examples of participant–experimenter interactions that varied across sexes rather than biological differences in nocebo responsivity. However, if the sex of an individual, or whether their sex matches the experimenter's, changes the efficacy of an intervention then there are important clinical implications in terms of which approaches will be most beneficial to participants under different circumstances. Notably, however, many other conditioning-based nocebo studies examined but did not observe sex differences (Colagiuri, Quinn, et al., 2015; Quinn et al., 2017; Au Yeung et al., 2014). As such, the area would benefit from more systematic assessment of the role of sex, including comparing any possible biological effects of sex vs experimenter effects that interact with sex.

5. CONCLUSION AND FUTURE RESEARCH DIRECTIONS

This review describes a number of ways in which the learning literature can be leveraged to inform interventions to reduce nocebo effects. There is promising preliminary evidence that latent inhibition and overshadowing can be used to prevent or diminish nocebo effects. These preventive strategies are particularly important given that nocebo effects appear resistant to extinction. There are, however, a number of important research steps required before we can effectively capitalize on learning mechanisms to reduce the nocebo effect. First, the contribution that learning makes to nocebo effects other than in pain and nausea should be assessed in more detail, so that the applicability of learning strategies to these other outcomes can be explored. Overshadowing and latent inhibition also require replication and exploration in large-scale clinical studies. The fact that these interventions do not involve deception means that such studies can be applied and translated ethically to practice if they are found to be successful. As such, these strategies present exciting opportunities to reduce patient burden in the future.

REFERENCES

Au Yeung, S. T., Colagiuri, B., Lovibond, P., & Colloca, L. (2014). Partial reinforcement, extinction, and placebo analgesia. *Pain, 155*, 1110–1117.

Barsky, A. J., Saintfort, R., Rogers, M. P., & Borus, J. F. (2002). Nonspecific medication side effects and the nocebo phenomenon. *Journal of the American Medical Association, 287*, 622–627.

Bartels, D. J., van Laarhoven, A. I., Haverkamp, E. A., et al. (2014). Role of conditioning and verbal suggestion in placebo and nocebo effects on itch. *PLoS One, 9*, e91727.

Benedetti, F., Pollo, A., Lopiano, L., et al. (2003). Conscious expectation and unconscious conditioning in analgesic, motor, and hormonal placebo/nocebo responses. *The Journal of Neuroscience, 23*, 4315–4323.

Bouton, M. E. (2002). Context, ambiguity, and unlearning: Sources of relapse after behavioral extinction. *Biological Psychiatry, 52*, 976–986.

Bouton, M. E., & Bolles, R. C. (1979). Contextual control of the extinction of conditioned fear. *Learning and Motivation, 10*, 445–466.

Bovbjerg, D. H. (2006). The continuing problem of post chemotherapy nausea and vomiting: Contributions of classical conditioning. *Autonomic Neuroscience: Basic and Clinical, 129*, 92–98.

Colagiuri, B., & Livesey, E. (2016). Contextual cuing as a form of nonconscious learning: Theoretical and empirical analysis in large and very large samples. *Psychonomic Bulletin and Review, 23*, 1996–2009.

Colagiuri, B., & Quinn, V. F. (2018). Autonomic arousal as a mechanism of the persistence of nocebo hyperalgesia. *Journal of Pain*, [in press].

Colagiuri, B., Quinn, V. F., & Colloca, L. (2015). Nocebo hyperalgesia, partial reinforcement, and extinction. *Journal of Pain, 16*, 995–1004.

Colagiuri, B., Schenk, L. A., Kessler, M. D., Dorsey, S. G., & Colloca, L. (2015). The placebo effect: From concepts to genes. *Neuroscience, 307*, 171–190.

Colloca, L., & Miller, F. G. (2011). The nocebo effect and its relevance for clinical practice. *Psychosomatic Medicine, 73*, 598–603.

Colloca, L., Petrovic, P., Wager, T. D., Ingvar, M., & Benedetti, F. (2010). How the number of learning trials affects placebo and nocebo responses. *Pain, 151*, 430–439.

Colloca, L., Siguado, M., & Benedetti, F. (2008). The role of learning in nocebo and placebo effects. *Pain, 126*, 211–218.

Dunsmoor, J. E., Niv, Y., Daw, N., & Phelps, E. A. (2015). Rethinking extinction. *Neuron, 88*, 47–63.

Hackam, D. G., & Redelmeier, D. A. (2006). Translation of research evidence from animals to humans. *Journal of the American Medical Association, 296*, 1731–1732.

Hall, G., Stockhorst, U., Enck, P., & Klosterhalfen, S. (2015). Overshadowing and latent inhibition in nausea-based context conditioning in humans: Theoretical and practical implications. *The Quarterly Journal of Experimental Psychology, 69*, 1227–1238.

Hart, G., Harris, J. A., & Westbrook, R. F. (2009). Systemic or intra-amygdala injection of a benzodiazepine (midazolam) impairs extinction but spares re-extinction of conditioned fear responses. *Learning and Memory, 16*, 53–61.

Holmes, N. M., & Harris, J. A. (2010). Latent inhibition. In C. J. Mitchell & M. E. Le Pelley (Eds.), *Attention and associative learning: From brain to behaviour* (pp. 99–130). Oxford: Oxford University Press.

Jensen, K., Kirsch, I., Odmalm, S., Kaptchuk, T. J., & Ingvar, M. (2015). Classical conditioning of analgesic and hyperalgesic pain responses without conscious awareness. *Proceedings of the National Academy of Sciences, 112*, 7863–7867.

Kirsch, I. (1997). Response expectancy theory and application: A decennial review. *Applied and Preventive Psychology, 6*, 69–79.

Kirsch, I. (2004). Conditioning, expectancy, and the placebo effect: Comment on Stewart-Williams and Podd (2004). *Psychological Bulletin, 130*, 341–343.

Klosterhalfen, S., Kellermann, S., Braun, S., et al. (2009). Gender and the nocebo response following conditioning and expectancy. *Journal of Psychosomatic Research, 66*, 323–328.

Klosterhalfen, S., Kellermann, S., Pan, F., et al. (2005). Effects of ethnicity and gender on motion sickness susceptibility. *Aviation, Space, and Environmental Medicine, 76*, 1051–1057.

Klosterhalfen, S., Kellermann, S., Stockhorst, U., et al. (2005). Latent inhibition of rotation chair-induced nausea in healthy male and female volunteers. *Psychosomatic Medicine, 67*, 335–340.

Klosterhalfen, S., Ruttgers, A., Krumrey, E., et al. (2000). Pavlovian conditioning of taste aversion using a motion sickness paradigm. *Psychosomatic Medicine, 62*, 671–677.

Lubow, R. E. (1973). Latent inhibition. *Psychological Bulletin, 79*, 398–407.

Lubow, R. E., & Moore, A. U. (1959). Latent inhibition: The effect of nonreinforced pre-exposure to the conditional stimulus. *Journal of Comparative and Physiological Psychology, 54*, 415–419.

Mackintosh, N. J. (1974). *The psychology of animal learning*. Oxford, England: Academic Press.

Mitchell, C. J., De Houwer, J., & Lovibond, P. F. (2009). The propositional nature of human associative learning. *Behavioral and Brain Sciences, 32*, 183–246.

Pavlov, I. P. (1927). *Conditional reflexes: An investigation of the physiological activity of the cerebral cortex*. Oxford, England: Oxford University Press.

Perruchet, P., Cleeremans, A., & Destrebecqz, A. (2006). Dissociating the effects of automatic activation and explicit expectancy on reaction times in a simple associative learning task. *Journal of Experimental Psychology: Learning, Memory, and Cognition, 32*, 955–965.

Quinn, V. F., & Colagiuri, B. (2015). Placebo interventions for nausea: A systematic review. *Annals of Behavioral Medicine, 49*, 449–462.

Quinn, V. F., & Colagiuri, B. (2016). Sources of placebo-induced relief from nausea: The role of instruction and conditioning. *Psychosomatic Medicine, 78*, 365–372.

Quinn, V. F., Livesey, E. J., & Colagiuri, B. (2017). Latent inhibition reduces nocebo nausea, even without deception. *Annals of Behavioral Medicine, 51*, 432–441.

Quinn, V. F., MacDougall, H. G., & Colagiuri, B. (2015). Galvanic vestibular stimulation: A new model of placebo-induced nausea. *Journal of Psychosomatic Research, 78*, 484–488.

Rescorla, R. A. (1988). Pavlovian conditioning: It's not what you think it is. *American Psychologist, 43*, 151–160.

Rescorla, R., & Wagner, A. (1972). A theory of Pavlovian conditioning: Variations in the effectiveness of reinforcement and nonreinforcement. In A. Black & W. Prokasy (Eds.), *Classical conditioning II: Current research and theory* (pp. 64–99). New York: Appleton Century Crofts.

Rodriguez-Romaguera, J., Sotres-Bayon, F., Mueller, D., & Quirk, G. J. (2009). Systemic propranolol acts centrally to reduce conditioned fear in rats without impairing extinction. *Biological Psychiatry, 65*, 887–892.

Stewart-Williams, S., & Podd, J. (2004). The placebo effect: Dissolving the expectancy versus conditioning debate. *Psychological Bulletin, 130*, 324–340.

Stockhorst, U., Hall, G., Enck, P., & Klosterhalfen, S. (2014). Effects of overshadowing on conditioned and unconditioned nausea in a rotation paradigm with humans. *Experimental Brain Research*, *232*, 2651–2664.
Stockhorst, U., Wiener, J. A., Klosterhalfen, S., et al. (1998). Effects of overshadowing on conditioned nausea in cancer patients: An experimental study. *Physiology and Behavior*, *64*, 743–753.
Symonds, M., & Hall, G. (1999). Overshadowing not potentiation of illness-based contextual conditioning by a novel taste. *Animal Learning and Behavior*, *27*, 379–390.
van der Worp, H. B., Howells, D. W., Sena, E. S., et al. (2010). Can animal models of disease reliably inform human studies? *PLoS Medicine*, *7*, e1000245.
Voudouris, N. J., Peck, C. L., & Coleman, G. (1990). The role of conditioning and verbal expectancy in the placebo response. *Pain*, *43*, 121–128.

CHAPTER EIGHTEEN

Can knowledge of Placebo and Nocebo Mechanisms Help Improve Randomized Clinical Trials?

Elisa Carlino*, Lene Vase†,1
*University of Turin Medical School, Turin, Italy
†School of Business and Social Sciences, Aarhus University, Aarhus, Denmark
1Corresponding author: e-mail address: lenevase@psy.au.dk

Contents

Abstract

Over the last decade, there has been a substantial increase in negative results from randomized controlled trials (RCTs), which may be due to an increasing placebo response among other factors. Currently, identification and exclusion of placebo responders from trials are attempted to overcome this problem, but so far the success of these approaches has been limited. At the same time, the placebo-mechanism literature has highlighted how contextual factors, such as patients' expectations, interfere with the effect of drug administration, leading to a certain degree of uncertainty in RCTs. In this chapter, we review the current challenges of RCTs including the uncertainties of the active arm, the placebo arm, the additivity assumption, and the double-blind procedure. We use the placebo-mechanism literature to debate the strengths and weaknesses of attempts to identify and exclude placebo responders from trials. Finally, we illustrate how insights from the placebo-mechanism literature may point to new ways of improving RCTs.

International Review of Neurobiology, Volume 138
ISSN 0074-7742
https://doi.org/10.1016/bs.irn.2018.01.012

1. CURRENT CHALLENGES OF THE RCTs

Over the last 30 years, negative randomized controlled trials (RCTs) have become the rule rather than the exception (Fava, Evins, Dorer, & Schoenfeld, 2003) and several attempts are currently made to overcome this problem (Dworkin et al., 2010, 2012). At the same time, the placebo-mechanism literature has highlighted how contextual factors, such as patients' expectations of therapeutic benefit, interfere with the effect of the drug administration, leading to a certain degree of uncertainty in RCTs (Benedetti, Carlino, & Piedimonte, 2016). In this chapter, we outline the challenges of the RCT, discuss strengths and weaknesses of the current approaches toward the problem, and point to ways in which knowledge from placebo-mechanism studies may inform and possibly improve the way in which the challenges of the RCT are currently addressed.

1.1 The Uncertainty of the Treatment Arm

Treatments are typically tested by comparing the efficacy in an active treatment arm with the efficacy in a placebo arm (Cartwright, 2007; Kaptchuk, 1998). The failure to detect significant differences between active treatments and placebos is one of the main sources of uncertainty in the RCTs, especially as negative RCTs are increasing over time across diseases and therapeutic areas. The breakdown by therapeutic area in phase II and phase III shows that the highest percentages of failure were found in oncology (32%), central nervous system diseases (17%), and musculoskeletal diseases (13%). Crucially, in the majority of the trials (52%), the main reason for failure was inefficacy (Harrison, 2016); thus, the effect of the new drug was not significantly different from the effect of a placebo.

One vividly discussed reason for the failure of RCTs is that the placebo response has increased over time whereas the drug response has remained stable, thereby leading to a smaller difference between the effect in the drug and placebo arms (Fig. 1A). This trend is reported in clinical trials of antidepressants and neuropathic pain treatments (Andrews, 2001; Khan & Brown, 2015; Tuttle et al., 2015; Walsh, Seidman, Sysko, & Gould, 2002). In a meta-analysis of all relevant phase II–IV placebo-controlled trials of venlafaxine and desvenlafaxine for the treatment of depression, the authors calculated the placebo response and the difference between the drug and placebo response confirming the phenomenon of an increasing placebo response over time (Dunlop et al., 2012). Also, in a recent meta-analysis of

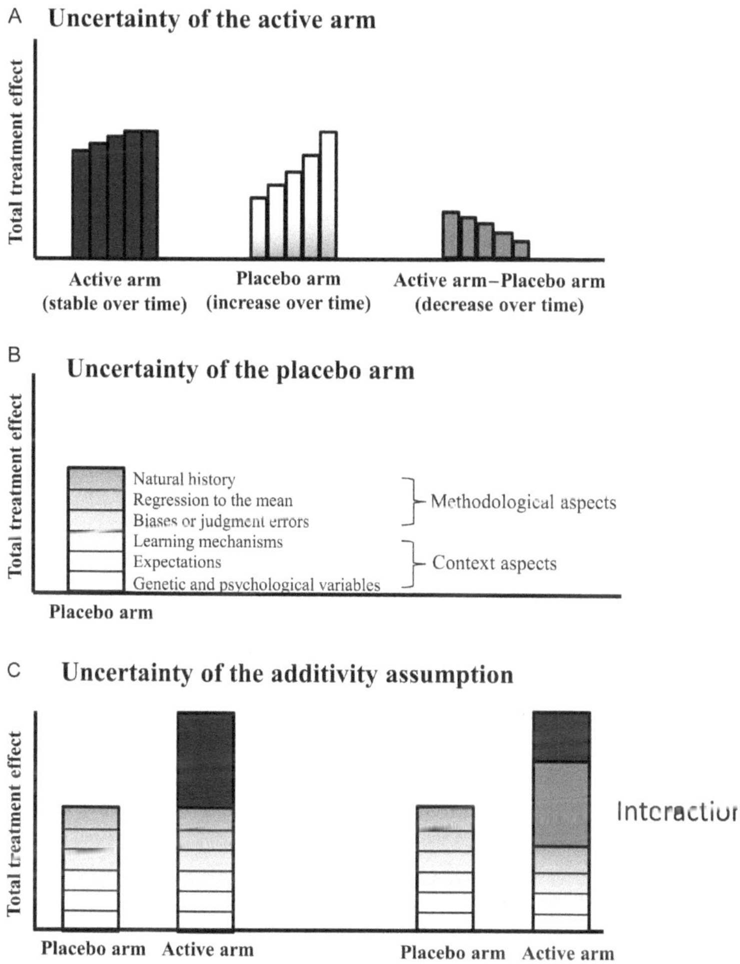

Fig. 1 Challenges of the RCTs. (A) Uncertainty of the active arm. Different clinical trials have reported that the drug response has remained stable over time (*black bars*), whereas the placebo response has increased (*light gray bars*). This leads to a smaller difference between drug and placebo arms (*dark gray bars*). (B) Uncertainty of the placebo arm. The improvement that may take place in the placebo arm can be attributed to different "methodological" or "context" aspects. (C) Uncertainty of the additivity assumption. The additivity model states that the total treatment effect is the sum of the effect in the placebo arm plus the effect in the active arm (*left side*). This model has been questioned since a placebo/drug interaction can occur (*right side*).

chronic neuropathic pain, the authors extracted data of 80 RCTs conducted over the last 23 years. In 1996, patients reported that active drugs reduced pain 27% more than a placebo, whereas this gap was reduced to 9% in 2013. As the drug responses did not increase over time, the authors concluded that this difference was due to a considerable increase of the placebo response (Tuttle et al., 2015). The reduced beneficial effects of antidepressants and pain medication over placebo echo the findings of reduced treatment efficacy in antipsychotic drugs and various treatments of Parkinson's disease (PD) (Alphs, Benedetti, Fleischhacker, & Kane, 2012; Rutherford et al., 2014). In a review of 28 atypical antipsychotic trials where the placebo response was analyzed, the mean change from baseline in total Positive and Negative Syndrome Scale (PANSS) between 1991 and 2006 increased over time. Moreover, the increased symptom improvement was correlated with the year in which the studies were conducted (Kemp et al., 2010). Also, substantial improvements in patients who received placebo were present in studies with pharmacological (Goetz, Laska, et al., 2008; Goetz, Leurgans, Raman, & Stebbins, 2000; Goetz, Wuu, et al., 2008; Shetty, Friedman, Kieburtz, Marshall, & Oakes, 1999), surgical (Olanow et al., 2003; Watts et al., 2001), or gene (Bartus, Weinberg, & Samulski, 2014; Christine et al., 2009; LeWitt et al., 2011) treatments for PD. For example, over the past decade, nine gene therapy clinical trials for PD have been initiated but without a robust clinical efficacy (Bartus et al., 2014).

From the placebo-mechanism literature, it is well known that the magnitude of the placebo effect is large and highly variable both within trials (Forsberg, Martinussen, & Flaten, 2017; Vase, Petersen, Riley, & Price, 2009; Vase, Riley, & Price, 2002) and across diseases (Howick et al., 2013). Thus, in order to reduce the uncertainty of the RCT, it is imperative to quantify the magnitude of the placebo effect (and response) across diseases and to understand the factors contributing to its variability.

1.2 The Uncertainty of the Placebo Arm

A main source of confusion and uncertainty in the RCT concerns the conceptualization of the placebo response and the meaning of the placebo arm (Fig. 1B). The improvement that may take place in the placebo arm can be attributed to different factors that, for simplicity, can be dichotomized and labeled as "context" and "methodological" aspects. The context aspects represent the actual placebo response, that is, the effect due to the psychosocial context surrounding the administration of a treatment, be it a real treatment or an inert treatment that the patient believes to be effective (Carlino & Benedetti, 2016). The methodological aspects include several statistical and procedural factors, such as spontaneous remission, regression to the mean, response biases, detection ambiguity, and cointervention factors (Benedetti, Carlino, & Pollo, 2011).

Thus, in order to provide a valid interpretation of RCTs, it is important to differentiate between contextual and methodological factors. In the placebo-mechanism literature, the contextual factors relating to placebo are differentiated from methodological confounding factors by the inclusion of a natural history condition (Benedetti, 2014a; Fields & Price, 1997). This has proved a rewarding way of determining the magnitude of placebo effects across diseases and of investigating factors that contribute to placebo effects (Benedetti, 2014a; Benedetti, Enck, Frisaldi, & Schedlowski, 2014; Colloca, Flaten, & Meissner, 2013). In RCTs, a natural history condition could be included by having an arm where patients did not receive a treatment (no treatment or wait-list). This would allow for an estimation of the effect of spontaneous fluctuation whereby it could be subtracted from the placebo arm and the active arm. The first comparison (placebo—natural history) yields the placebo effect, whereas the second (active treatment—natural history) shows the effect of administering a treatment, be it a placebo or an active treatment.

Three-armed trials, including a natural history arm, have shown that a high proportion of clinical improvement is attributable to spontaneous remission. In antidepressant trials, the natural history of the disease accounted for 24% of the overall treatment effect (Kirsch & Sapirstein, 1998). Similarly, in a more recent meta-analysis of patients with major depression, the improvement of patients in the wait-list group was compared with that observed in the placebo and active treatment groups; thus, patients in wait list control conditions experience approximately 33% of the improvement occurring with medication treatment and 40% of the improvement seen with placebo administration (Rutherford, Mori, Sneed, Pimontel, & Roose, 2012; Rutherford & Roose, 2013). In a Cochrane review of 37 RCTs with three-armed trials, Krogsbøll, Hróbjartsson, and Gøtzsche (2009) investigated to which extent spontaneous remission contributes to the observed symptom change in different clinical conditions. The relative contributions of spontaneous remission and of placebo to the change from baseline in the active intervention groups were 24% and 20%, respectively. The clinical conditions with the most pronounced spontaneous improvement were nausea (45%), smoking (40%), depression (35%), phobia (34%), and acute pain (25%) (ibidem). For ethical and/or design reasons, many clinical trials do not include a natural history group, thus making a precise quantification of the placebo response impossible. Especially in these trials, in which the inclusion of a natural history group is not feasible, the need to develop and use alternative designs is crucial.

1.3 The Uncertainty of the Additivity Assumption

One of the main assumption underlying RCTs is that the total treatment effect is the sum of the placebo effect plus the active drug effect: following this assumption, it is possible to deduce the active drug effect by subtracting the placebo effect from the total treatment effect (Fig. 1C). This is the so-called additivity model. Four implicit assumptions of this model are (1) the drug effect and the placebo effect do not interact, (2) the placebo component is identical for both the placebo and the drug arms, (3) the total treatment effect is the sum of the drug effect plus the placebo effect, and (4) the active drug effect is not influenced by whether or not patients believe having received the drug.

Placebo-mechanism studies and meta-analyses have started to investigate the additivity assumption. In an fMRI study, Atlas et al. (2012) tested and confirmed the assumption. Two experiments were performed, and in both experiments a painkiller (remifentanil) or a placebo along with the expectation of receiving the real painkiller was administered to healthy subjects during experimental thermal pain. Both remifentanil and placebo produced an analgesic effect without interacting with each other. Indeed, drug effects on pain reports and fMRI activity did not interfere with the expectation of receiving the drug. Moreover, remifentanil effects on pain-related brain responses were the same whether or not participants believed they were receiving the drug (Atlas et al., 2012), and verbal information of drug delivery was accompanied by changes in brain regions outside the traditional pain matrix.

Despite this confirmation, recent studies and meta-analyses have explicitly or implicitly questioned the additivity model assumptions (Boehm, Berger, Weger, & Heusser, 2017; Kube & Rief, 2017).

In an fMRI study, Schenk, Sprenger, Geuter, and Büchel (2014) investigated the effect of a possible interaction between topical analgesic treatment and treatment expectation using a 2 × 2 within-subject design (open active treatment, hidden active treatment, placebo, and control). The authors showed that the expectation effect was significantly larger in the active treatment conditions compared with the no-treatment conditions (placebo and control) and was associated with signal changes in the anterior insular cortex, the anterior cingulate cortex, and the ventral striatum. Thus, they concluded that expectation and treatment seem not necessarily additive.

In a different study, Lund, Vase, Petersen, Jensen, and Finnerup (2014) directly tested the assumption that the total treatment effect is the sum of the drug effect plus the placebo effect. In a within-subject balanced placebo design, experimental pain was induced by injections of hypertonic saline into the masseter. Subjects received four injections with hypertonic saline along with lidocaine or placebo and were told that they would receive a painkiller or a placebo. In this study, the sum of the drug effect and the placebo effect was larger than the total treatment effect. In particular, this underestimation of the treatment effect was higher for placebo responders. Moreover, experimental placebo-mechanism studies using the "open–hidden" approach have proved that a treatment delivered unbeknownst to the patient, i.e., hidden is less effective than the same treatment delivered in full view, i.e., open (Benedetti et al., 2011). Thus, the biochemical action of an active treatment is remarkably influenced by whether or not patients are aware of receiving an effective drug. Also, the experience of side effects due to the administration of active treatments can lead to the perception that a real treatment has been received, which in turn can boost the effect of the drug (Berna et al., 2017). Hence, it is essential to gain a better understanding of when the drug and the placebo treatment interact and how this may be influenced by patients' expectations in order to correctly interpret the results of RCTs.

1.4 The Uncertainty of the Double-Blind Procedure

Blinding is a procedure that prevents healthcare providers, participants, caregivers, or outcome assessors from knowing which intervention (active treatment or placebo) was given or received. It is a critical methodological feature of RCTs aimed at reducing bias. In particular, the blinding procedure should ensure control of the effects of expectations: since neither participants nor investigators know the group allocation (drug or placebo arm), expectancies should be balanced across groups (Feys, Bekkering, Singh, & Devroey, 2014). Even if blinding is highly recommended (Moher et al., 2012; Schulz et al., 2011), a first major issue is that in many RCTs the reporting of blinding is inadequate. A cohort study of 200 blinded randomized clinical trials published in 2001 documented that the blinding status of key trial persons was incompletely reported in most randomized clinical trials. Seventy eight percent of the articles described the trials as "double-blind" trials, but only 2% properly described the blinding status of patients, health care providers, and data collectors (Haahr & Hróbjartsson, 2006). The impact of blinding on the results was also investigated in a randomized, placebo-controlled multiple sclerosis clinical trial where all patients were examined by both a blinded and an unblinded neurologist (Noseworthy et al., 1994).

Comparing the blinded and the unblinded neurologist's judgment of treatment response, the unblinded neurologist's scores showed an apparent treatment benefit at 6, 12, and 24 months, whereas the blinded neurologist did not. The authors concluded that physician blinding prevented type-1 error, i.e., considering the active treatment more effective than the placebo. Noteworthy, the importance of blinding appears to be even more crucial in trials with subjective outcomes (Wood et al., 2008). In general, unblinding is a problem in pharmacological trials and to a high extent in non-pharmacological trials as most methods for studying, for example, acupuncture and surgery leave the healthcare provider unblinded (George et al., 2016; Kaptchuk, 2002).

A second major issue is the unintentional unblinding. It is well known that patients and healthcare provider often know if the active treatment or the placebo treatment has been administered (Kaptchuk, 2002; Probst et al., 2016; Turner, Jensen, Warms, & Cardenas, 2002). RCTs and placebo-mechanism studies have shown that patients' perception of treatment allocation can markedly influence the results of a study. In a clinical trial of human fetal mesencephalic transplantation for PD, the effect of fetal tissue implant was compared with placebo (sham surgery) for 12 months. The authors reported no differences between the real and the sham interventions for several outcome measures, such as physical scores and quality-of-life scores. However, when patients' perceived assignment to either active treatment or placebo was used to predict the outcomes, patients who believed they received transplanted tissue had significant improvements in their quality-of-life and motor outcomes, regardless of the received treatment. Thus, patients' perceived treatment assignment had a beneficial effect on the overall outcome, and this difference was still present 12 months after surgery (McRae et al., 2004). Similar results have been observed in alcohol and smoking dependencies (Colagiuri, Morley, Boakes, & Haber, 2009). Also, Correa et al. (2014) have determined the temporal relationship between treatment expectancies and smoking-related variables, assessing the impact of treatment guesses during a placebo-controlled laboratory study of varenicline in 88 participants. Generalized linear models confirmed that, irrespective of the actual treatment assignment, perceived medication allocation improved model fit for both subjective reports and behavioral smoking-related measures. Furthermore, placebo-mechanism studies have shown that patients' expectations of treatment effect increase overtime, so that patients who believe they have received the real analgesic treatment come to expect—and experience—larger amounts of pain relief over time

(Vase et al., 2013). The perception of being in a specific treatment group also affects neural responses to visceral pain. In an fMRI study, Kotsis and colleagues showed that visceral pain was significantly lower when subjects believed they had received a potent analgesic drug. In fact, this pain reduction was associated with significantly reduced activation of the insular, the posterior, and anterior cingulate cortices during pain anticipation, and of the anterior cingulate cortex during pain (Kotsis et al., 2012).

This implicit awareness of treatment allocation is often based on patients' experience of adverse events: indeed, the perception of side effects can lead to expectations of being assigned to the real treatment group. To overcome this problem the use of active placebos have been recommended, i.e., substances producing noticeable side effects so participants think they receive a legitimate treatment (Enck, Bingel, Schedlowski, & Rief, 2013). However, when directly testing blinding in relation to active placebo, this procedure does not seem to ensure blinding (Rief & Glombiewski, 2012; Turner et al., 2002). In a randomized experimental study on pain, even if active placebos show substantial differences to passive placebos when patients had a 50% probability of receiving the active treatment, this procedure did not rule out the unblinding issue (Rief & Glombiewski, 2012).

Thus, in future RCTs it is important to directly assess patients' perceived treatment allocation and to take this factor into consideration in the analyses of the data in order to fully and correctly interpret the results of the RCTs.

2. HOW ARE THE CHALLENGES CURRENTLY MET?

Given the increasing number of negative clinical trials and the apparent increase in the placebo response, a major focus of RCTs has been on placebo responsiveness, especially on reducing the placebo response and improving the drug–placebo comparison, which may also be termed assay sensitivity (Dworkin et al., 2012).

2.1 Stable vs Situational Factors as Predictors of the Placebo Response: Strengths and Weaknesses

One major problem in the identification of placebo predictors is the uncertainty as to whether placebo responsiveness constitutes a stable or a situational trait. In other words, placebo responsiveness could be a stable uni- or multidimensional trait consistently predicting the individual response to a placebo treatment independently of situational factors, or a situational state that affects the individual response in different ways according to the specific condition (Horing, Weimer, Muth, & Enck, 2014).

Different attempts have been made to identify stable factors to predict placebo responsiveness. These factors include demographic variables, such as age, sex, and race (Aslaksen, Bystad, Vambheim, & Flaten, 2011; Häuser, Bartram-Wunn, Bartram, Reinecke, & Tölle, 2011; Wrobel et al., 2015); illness characteristics (Bialik, Ravindran, Bakish, & Lapierre, 1995; Brown, Johnson, & Chen, 1992; Cepeda, Berlin, Gao, Wiegand, & Wada, 2012; Niklson, Edrich, & Verdru, 2006; Waxmonsky, Waschbusch, Glatt, & Faraone, 2011; Weimer, Colloca, & Enck, 2015); personality traits (Darragh, Booth, & Consedine, 2015, 2016; Geers, Helfer, Kosbab, Weiland, & Landry, 2005; Jakšić, Aukst-Margetić, & Jakovljević, 2013); culture (Bhugra & Ventriglio, 2015; Moerman, 2000); or genetic profiles (Hall et al., 2012; Hall, Loscalzo, & Kaptchuk, 2015; Peciña et al., 2014; Peciña & Zubieta, 2015; Wang et al., 2017). Using a psychological approach, Feltner and coworkers have developed a scale, the Placebo Response Screening Scale (PRSS), to identify specific domains associated with placebo response (Feltner, Hill, Lenderking, Williams, & Morlock, 2009). The scale was administered to more than 200 patients with generalized anxiety disorder, during the screening visit of a 4-week, placebo-controlled study of alprazolam. Twenty items reliably predicted patient global improvement rating (assessed by means of the Patient Global Impression of Change Scale), clinician global improvement rating (assessed by means of the Clinician Global Impression of Improvement Scale), or improvement on anxiety scores (assessed by means of Hamilton Rating Scale for Anxiety) in patients in the placebo arms of the clinical trial. Even if this tool requires further use and validation, it can represent a starting point for a psychological investigation of placebo responsiveness that could be easily applicable in the RCTs setting. Different genetic studies have tried to identify genome-related mediators that affect the individual placebo effect (Colagiuri, Schenk, Kessler, Dorsey, & Colloca, 2015; Hall et al., 2015). For example, the *catechol-O-methyltransferase* (*COMT*) gene was reported to be a placebo biomarker in patients with irritable bowel syndrome (Hall et al., 2012). Recently, Wang et al. (2017) conducted a network analysis of a group of genome-related mediators that affects an individual's response to a placebo (placebome) in a large cohort of the Woman's Genome Health Study trial for the prevention of cardiovascular disease and cancer. The authors identified a placebome module and assessed its relationships (defined "network proximity") with various diseases or drug categories, finding that the placebome module is significantly proximate to the network modules of specific diseases (and the targets of specific drug categories), and distant from others. Thus, they concluded that the network proximity measure can be used to predict the strength of placebo responses in different patients with different diseases (Wang et al., 2017).

Also, neuroimaging studies have been conducted to identify stable biomarkers related to placebo responsiveness. For example, Tétreault et al. (2016) examined the predictability of future placebo response in the clinical setting in patients with chronic osteoarthritis pain. In three different studies, the authors showed that clinical placebo response can be predicted from resting-state fMRI brain connectivity. In particular, the connectivity of the right midfrontal gyrus with the rest of the brain best identified placebo responders, with an accuracy of 95% and an accurate prediction of the magnitude of placebo effectiveness. This approach seems promising as it aims at identifying stable predictive brain markers of placebo responsiveness, but, again, further studies are necessary to extend this result to different controlled conditions. Despite the research interest in identifying stable factors that reliably predict placebo responsiveness, there are no consistent findings across studies yet. This may be related to methodological issues, such as the lack of a large body of comparable studies, or to the high number of factors that contribute to the placebo responsiveness (Vase, Amanzio, & Price, 2015). However, it may also relate to the circumstance that placebo responsiveness is highly influenced by situational factors in the treatment context.

Placebo-mechanism studies question whether placebo responding is a stable trait (Colloca & Grillon, 2014). For example, Whalley and colleagues reported that placebo analgesia was not correlated to personality traits, like acquiescence and absorption, but instead to response expectancies (Whalley, Hyland, & Kirsch, 2008). In their study on healthy volunteers exposed to experimental pain stimuli, they found that placebo effects across trials were highly correlated when the treatments were exactly similar. Yet, by simply changing the name of the treatment there was no longer a correlation between the placebo effect across trials (Whalley et al., 2008). Moreover, in patients suffering from irritable bowel syndrome, placebo responsiveness to a first treatment period does not reliably predict responsiveness to a second different treatment (Tack et al., 2005); thus, the same person can respond in different ways to different placebo treatments (Lieberman, 1967; Whalley et al., 2008). Not only can a placebo responder become a placebo nonresponder after the administration of a second treatment, but the opposite phenomenon has also been observed in different experimental settings

(Benedetti, Frisaldi, et al., 2016; Carlino, Guerra, & Piedimonte, 2016). Future larger scale studies are needed for addressing this point.

Another element of variability is represented by prior exposures to active treatments and learning effects (André-Obadia, Magnin, & Garcia-Larrea, 2011; Kessner, Wiech, Forkmann, Ploner, & Bingel, 2013). For example, if parkinsonian patients are exposed to repeated administrations of the anti-Parkinson's agent apomorphine, their placebo response is high. In fact, as the number of apomorphine administrations increase from one to four, both the clinical response and the neuronal activity in the ventral anterior and anterior ventrolateral thalamus increased during subsequent placebo administration (Benedetti, Frisaldi, et al., 2016). This relationship has been proved true for pain (Colloca, Petrovic, Wager, Ingvar, & Benedetti, 2010) and immune responses (Hadamitzky, Engler, & Schedlowski, 2013; Lückemann, Unteroberdörster, Kirchhof, Schedlowski, & Hadamitzky, 2017).

Taking these considerations into account, different attempts have been made to identify situational factors capable of predicting placebo responsiveness. Accumulating evidence from placebo-mechanism studies suggests that patients' expectations of treatment efficacy might help predict the responsiveness to a treatment, be it active or placebo (Laferton, Kube, Salzmann, Auer, & Shedden-Mora, 2017; Peerdeman et al., 2016). In a study of patients suffering from irritable bowel syndrome, it has been shown that patients' expected pain levels and patients' desire for pain relief accounted for 77% and 81% of the variance in pain intensity levels following placebo and lidocaine administration, respectively (Vase, Robinson, Verne, & Price, 2003). Also in patients with chronic neuropathic pain, expected pain levels significantly predicted the variance in spontaneous pain (53%) and evoked pain (24%–35%) (Petersen et al., 2012). Similarly, in a pooled analysis of four RCTs of acupuncture in patients with migraine, patients with higher expectations of acupuncture efficacy had larger clinical benefits than those with lower expectations, regardless of their allocation to real or sham groups (Linde et al., 2007). Expectations in RCTs are not fixed as they can change during the trial and be triggered by different factors (Stone, Kerr, Jacobson, Conboy, & Kaptchuk, 2005; Vase, Nørskov, Petersen, & Price, 2011). In RCTs, patients' expectations are typically not measured. Yet, it is possible to make approximations of expectations based on design characteristics such as the type of active medication (opioid vs nonopioid) and the number of planned face-to-face visits with healthcare professionals. Patients are likely to expect a better of effects of opioid vs nonopioid and a higher number of interactions with the healthcare provider is likely to boost the therapeutic relationship (Vase, Vollert, et al., 2015). In a meta-analysis of nine industrially sponsored, randomized, double-blind, placebo-controlled trials in patients suffering from chronic painful osteoarthritis or low back pain, approximations of expectations significantly predicted the placebo response whereas stable factors generally did not (Vase, Vollert, et al., 2015). Moreover, since interaction among patients from the same trial is quite common, patients may be influenced by the experience of others belonging to the same trial (Benedetti & Frisaldi, 2014; Frisaldi, Shaibani, & Benedetti, 2017; Stone et al., 2005).

Based on these studies, the magnitude of the placebo response and effect seems highly variable not only across diseases and trials but also in the same individual. In addition, interactions between stable and situational factors seem to exist and it will be essential to further understand how they influence not only the placebo response but also the drug response and the interaction between the two. Finally, it will be important to specify the factors that influence expectation, in order to fully understand how, for example, adverse events can influence the efficacy of a treatment.

2.2 Complex Designs to Reduce the Placebo Response: Strength and Weaknesses

Several new designs and strategies have been developed to minimize the placebo response in the context of clinical trials (Enck et al., 2013, Enck, Klosterhalfen, Weimer, Horing, & Zipfel, 2011). Based on the evidence that within-subject variability of responses is lower than between-subject variability, crossover designs have traditionally been used (Wellek & Blettner, 2012). In crossover trials, patients receive both drug and placebo in separate phases in a double-blinded, randomized, and balanced order (Fig. 2A). Since each patient serves as his/her own control, these designs can effectively minimize errors associated with patient's characteristics. However, this increased precision is susceptible to carryover effects. Indeed, the major pitfall of crossover designs is the need of standardized washout periods to effectively control possible carryover effects. Another method is to use a placebo run-in phase prior to the traditional clinical trial in order to identify and exclude possible placebo responders, and thus randomize only placebo nonresponders to either placebo or active treatment (Pablos-Méndez, Barr, & Shea, 1998) (Fig. 2B). More recently, the sequential parallel comparison design (SPCD) (Fava et al., 2003) and the two-way enriched design (TED) (Ivanova & Tamura, 2015) have been proposed (Fig. 2C). The SPCD is

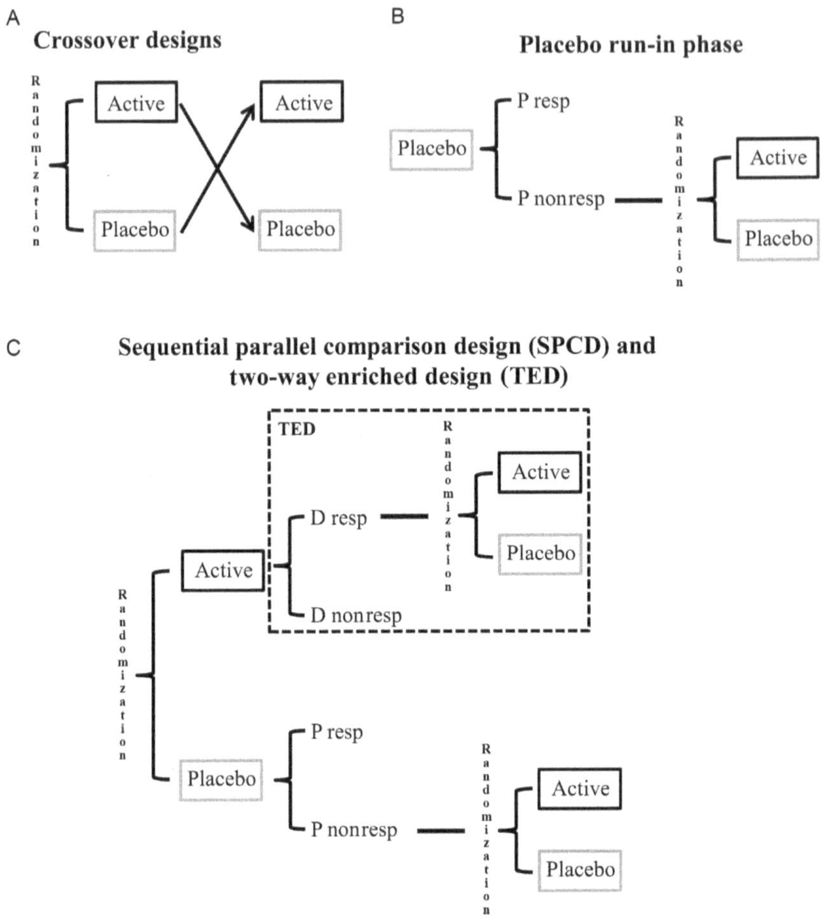

Fig. 2 Examples of clinical trials designs. (A) Crossover designs. In crossover designs, a group of subjects first receive placebo and then are switched to active treatment. Another group first receives active treatment and then is switched to placebo. (B) Placebo run-in designs. In the placebo run-in design, after a first phase in which placebo is administered, only placebo nonresponders are randomized to either placebo or active treatment. (C) SPCD and TED. In the SPCD, two different phases are planned: the first phase involves an unbalanced randomization between placebo and active treatment with more patients randomized to placebo, whereas in the second phase placebo nonresponders treated with placebo are randomized to either an active or a placebo treatment (Fava et al., 2003). In the TED (*broken line*) also drug responders are rerandomized between drug and placebo arms (Ivanova & Tamura, 2015).

aimed at reducing both the placebo response rate and the sample size requirement. Two different phases of treatment are planned: the first phase involves an unbalanced randomization between placebo and active treatment with more patients randomized to placebo, whereas in the second phase placebo nonresponders treated with placebo are randomized to either an active or a placebo treatment (Fava et al., 2003). The TED is similar to the SPCD with the addition that also drug responders are rerandomized between drug and placebo arms (Ivanova & Tamura, 2015). SPCD and TED may offer more efficient alternatives to crossover designs and potential applications in device trials in chronic diseases with a documented high placebo response, such as psychiatry (e.g., depression, schizophrenia, bipolar disorder, and attention deficit disorder), neurology (e.g., PD, behavioral symptoms of various dementias), pain (e.g., chronic arthritis, neuropathic pain, and migraine), asthma, hypertension, and diabetes (Ivanova et al., 2016).

Yet, as demonstrated by placebo-mechanism studies, these approaches present several methodological problems. The risk of unblinding can occur when the active treatment produces side effects that are not perceived in the placebo arm or, especially in the run-in phase, when the investigator is aware of the procedure (Enck et al., 2011). Furthermore, the risk of learning interference and carryon effects is typical of within-subject studies and is not easy to work out (Benedetti, 2014b; Suchman & Ader, 1992). For example, the increase of placebo responsiveness after a pharmacological preconditioning with an active drug has been extensively documented in different conditions, such as pain, immune system, and PD (Albring et al., 2012; Benedetti, Frisaldi, et al., 2016; Colloca et al., 2010; Frisaldi, Carlino, et al., 2017; Goebel et al., 2002; Piedimonte, Guerra, Vighetti, & Carlino, 2017). Thus, enriched designs in which previous exposures to active treatments are induced should be avoided because they can produce substantial placebo responses due to learning and reinforced expectation mechanisms. Finally, the risk of withdrawal effects has been documented, particularly when exposure to the active treatment can induce dependency and when a short washout is used (Rothwell, 2006). In general, although these designs are intended to improve assay sensitivity, they do not always do so (Enck et al., 2013), and they have been questioned for practical and theoretical reasons (Lee, Walker, Jakul, & Sexton, 2004). For example, pain studies have shown that they may not reduce the placebo response (Simpson et al., 2014) and meta-analyses of antidepressant trials comparing trials with or without a run-in period suggest that this method does not improve drug–placebo comparisons (Posternak, Zimmerman, Keitner, & Miller, 2002; Trivedi & Rush, 1994). From a theoretical point of view, these approaches present several problems. First, they are based on the assumption that placebo responsiveness is a stable trait unrelated to situational fluctuations, yet as shown above stable factors are not consistent predictors of placebo responsiveness across diseases and trials. Second, some studies suggest that participants with the highest drug response also tend to have the highest placebo response (Lund et al., 2014); thus, the reduction of placebo responsiveness can lead to a reduction of the drug responsiveness (Kam-Hansen et al., 2014). Hence, eliminating the placebo response may not be the optimum solution for improving RCTs, (Kam-Hansen et al., 2014; Kaptchuk, 1998). Third, the exclusion of placebo responders can compromise the generalizability (external validity) of clinical trial results (Pablos-Méndez et al., 1998); indeed, placebo responders cannot be excluded in clinical practice (Dworkin et al., 2010; Vase, Amanzio, et al., 2015) and placebo responders are likely good drug responders. Thus, the effectiveness of a drug artificially measured in a clinical trial setting would not represent the "real-world" effectiveness of a medical treatment measured in a routine clinical practice.

3. CAN KNOWLEDGE OF PLACEBO/NOCEBO MECHANISMS POINT TO A NEW APPROACH?

In recent years, various attempts have been made to minimize the placebo response in RCTs with the aim of increasing the sensitivity of the clinical trial and thereby discern specific effects due to the pharmacological agent in the drug. As mentioned earlier, these attempts failed, so new strategies to reduce the uncertainties of RCTs are needed. A promising approach is to increase our knowledge of placebo and nocebo mechanisms and to use this to improve the prediction of both the placebo and the active treatment responses.

3.1 Increase the Knowledge and Answer Questions

Our knowledge of the mechanisms underlying placebo effects has substantially increased over the last 20 years, and the effects have been documented in several medical conditions as well as in different experimental and clinical contexts (Benedetti, 2014a). Different approaches have been used to estimate the magnitude of placebo effects (Ashar, Chang, & Wager, 2017). Comparing the placebo treatment with a no-treatment group, studies have documented that patients in the placebo group experience statistically superior symptom improvements (Bausell, Lao, Bergman, Lee, & Berman, 2005; Kaptchuk et al., 2008; Verne, Robinson, Vase, & Price, 2003). Also, comparing the effectiveness of open drug treatment with hidden drug treatment, substantial placebo effects have been demonstrated in experimental and postoperative pain (Amanzio, Pollo, Maggi, & Benedetti, 2001;

Benedetti, Maggi, Lopiano, Lanotte, & Rainero, 2003; Bingel et al., 2011) and PD (Mercado et al., 2006; Pollo et al., 2002). Similarly, comparison of placebo effect with active drug effect has shown that the placebo response can be as large as the active treatment response across conditions, such as chronic pain (Tuttle et al., 2015), depression (Fournier et al., 2010; Kirsch et al., 2008), and PD (Olanow et al., 2003; Watts et al., 2001). Moreover, different types of placebo treatments have been compared, showing larger placebo responses for sham surgery compared with sham injections and pills (Bannuru et al., 2015; Meissner et al., 2013). As documented in this chapter, placebo responses in RCT and effects in studies including a no-intervention arm are highly variable, and it is therefore crucial to carefully quantify the magnitude of these effects via meta-analyses and studies using the same methodological approach across diseases.

First, it is desirable to understand in which medical conditions placebo and nocebo responses and effects are large and in which they are small. Moreover, a clear estimation of the variation between settings (RCTs, research, and clinical settings) is essential. In experimental studies, the aim is to investigate the placebo mechanisms, whereas in clinical trial studies the aim is to estimate the effectiveness of a treatment. Thus, in placebo-mechanism studies the magnitude of placebo effects is typically larger than placebo responses in clinical RCTs (Forsberg et al., 2017; Vase et al., 2009, 2002). However, in RCTs, patients usually know that they have a 50% chance of receiving the active treatment and here the magnitude of the placebo effect is typically small (Hróbjartsson & Gøtzsche, 2001; Vase et al., 2002). Moreover, in the routine medical practice patients know they receive the active treatment and the placebo response is therefore likely to be large, but thorough estimation of the effect in this setting is still needed. Indeed, it seems likely that, due to the intrinsic differences among settings, RCTs can provide only a rough quantification of placebo effects (Linde, Fassler, & Meissner, 2011), different from the placebo effect that is typically boosted in placebo-mechanism studies or a clinical routine practice.

Several questions need to be answered in order to better understand the placebo phenomenon and develop strategies to help improve for the methodology of RCTs. First, since the prediction of placebo responsiveness is crucial, the role of stable and situational factors and their possible interaction (Oken, 2008) needs further investigation. As for different psychological conditions, such as anxiety, it is likely that a combination of dispositional and situational factors leads individuals to respond to specific placebo treatment: in other words, it is likely that specific contexts could trigger the individual disposition to respond to a specific placebo treatment. For example, the optimism–pessimism scale can predict individual response to placebo only when participants are certain of receiving the active drug, and not when they have a 50% chance (Geers et al., 2005). Thus, future studies should investigate if these interactions better account for the variability of placebo and nocebo responsiveness as studies of traits in themselves often yields contrasting results (Darragh, Booth, & Consedine, 2014; Watson, Power, Brown, El-Deredy, & Jones, 2012).

Second, findings from recent studies have questioned the additivity assumption and, in order to improve the assessment of pharmacological treatments, further studies are necessary which directly test the drug/placebo interaction and test an interactive model to explain the complex interplay between drug and placebo effect (Enck et al., 2011). In contrast to the additive model, the interactive model assumes that an interaction effect may exist in addition to the single genuine drug and placebo effects (Schenk et al., 2014). The interactive model explains the results of different behavioral, neuroimaging, and clinical studies better (Kube & Rief, 2017), but it makes the interpretation of RCTs highly complex. For this reason, it is crucial to use new predictors of placebo and drug responsiveness to explain this interaction and to quantify the drug effect properly.

Third, considering that placebo effects are basically a learning phenomenon involving patient expectations (Colloca & Miller, 2011; Colloca et al., 2010), it is essential to investigate the role of treatment history in RCTs. RCTs usually involve patients with a personal treatment history; in particular, RCTs testing the effectiveness of treatment for chronic conditions involve patients with a personal positive or negative treatment history, i.e., a series of treatment successes or failures. In healthy volunteers, a negative treatment history reduced subsequent analgesic responses to a new treatment (Kessner et al., 2014). Also, although a change of route of administration induced expectations of positive treatment effect, it did not counteract the negative carryover effects on treatment efficacy in healthy volunteers (Zunhammer et al., 2017). The only study in chronic pain patients showed that treatment history modulated the magnitude of the placebo response in the opposite way: patients with a more negative pain-related treatment history reported significantly larger placebo responses to their own chronic pain (Müller et al., 2016). These data highlight the importance of future studies investigating the role not only of the natural history of the disease and expectation but also of the treatment history in clinical trials, as treatment history may explain part of the treatment response.

3.2 Use the Knowledge

To date, placebo-mechanism studies have provided useful insight for the design of future clinical trials. Traditionally, RCTs do not explicitly consider patient expectations. However, as demonstrated by placebo-mechanism studies, expectations can substantially influence the outcome of a treatment, and assessment of expectations in RCTs is highly recommended in order to lower the uncertainty of the outcomes. Different strategies are needed to investigate patients' expectations (Laferton et al., 2017). Younger, Gandhi, Hubbard, and Mackey (2012) created a six-item scale. After the validation process, the authors found that the scale predicted a significant amount of outcome variance (between 12% and 18%) in patients receiving surgical and pain interventions. An easier way consists is directly asking patients their expectations about the treatment and about the expected pain levels (Petersen et al., 2014; Vase, Baram, et al., 2015; Vase et al., 2003). Moreover, since patients' expectations can change during the course of a trial, as a result of self-monitoring, it could be useful to assess them at different stages: before entering the trial, after randomization, during the course of the trial, and at the end of the trial (Stone et al., 2005). Also, patients' perception of treatment allocation should be tested since patients' perception of the treatment can compromise the blinding of RCTs and markedly influence the results of a trial. Again, a practical and easy way consists in directly asking patients "Which treatment do you believe that you have received?" (Vase, Baram, et al., 2015; Vase et al., 2013). Hence, by measuring patients perceived treatment allocation and expectation it is possible to calculate to which extent placebo-related factors influence the outcome of the trial, thereby leading to a better estimation of the true drug effect. Also, after a proper assessment of patient expectations it could be possible to balance the weight of expectations across groups in order to compare the effective ness placebos and active treatments in homogeneous groups or to use expectations and perception of treatment allocation as covariate factors to predict placebo and treatment responses (Frisaldi, Shaibani, et al., 2017).

Thus, placebo-mechanism studies have yielded sufficient knowledge to identify some of the current problems in RCTs and to point to pitfalls in the attempts to overcome these problems. Also, placebo-mechanism studies have shown areas of importance for future studies. In future studies, it is central to investigate (1) the magnitude of the placebo responses and effects across diseases and settings; (2) the interaction between stable and situational factors in placebo responses, drug responses as well as the interaction between the two; and (3) the importance of learning history, perception of treatment allocation, and expectation of treatment efficacy. Yet, in order to gain a full understanding of these topics, it will be crucial to undertake a large investigation of these aspects across various diseases and settings using similar designs. Only then can we make truly evidence-based recommendations for each of the specific treatments and diseases of interest. In addition, it will be pivotal to collaborate with regulatory agencies and medical companies to actually use this evidence-based knowledge to change the way in which clinical trials are conducted. Such changes may ultimately lead to development of more efficient treatments and to a better understanding of how to optimize the efficacy of treatment in clinical practice, which could greatly improve the treatment of patients in clinical practice.

REFERENCES

Albring, A., Wendt, L., Benson, S., Witzke, O., Kribben, A., Engler, H., et al. (2012). Placebo effects on the immune response in humans: The role of learning and expectation. *PLoS One*, 7(11), e49477. https://doi.org/10.1371/journal.pone.0049477.

Alphs, L., Benedetti, F., Fleischhacker, W. W., & Kane, J. M. (2012). Placebo-related effects in clinical trials in schizophrenia: What is driving this phenomenon and what can be done to minimize it? *The International Journal of Neuropsychopharmacology*, *15*(7), 1003–1014. https://doi.org/10.1017/S1461145711001738.

Amanzio, M., Pollo, A., Maggi, G., & Benedetti, F. (2001). Response variability to analgesics: A role for non-specific activation of endogenous opioids. *Pain*, *90*(3), 205–215.

André-Obadia, N., Magnin, M., & Garcia-Larrea, L. (2011). On the importance of placebo timing in rTMS studies for pain relief. *Pain*, *152*(6), 1233–1237. https://doi.org/10.1016/j.pain.2010.12.027.

Andrews, G. (2001). Placebo response in depression: Bane of research, boon to therapy. *The British Journal of Psychiatry: The Journal of Mental Science*, *178*, 192–194.

Ashar, Y. K., Chang, L. J., & Wager, T. D. (2017). Brain mechanisms of the placebo effect: An affective appraisal account. *Annual Review of Clinical Psychology*, *13*, 73–98. https://doi.org/10.1146/annurev-clinpsy-021815-093015.

Aslaksen, P. M., Bystad, M., Vambheim, S. M., & Flaten, M. A. (2011). Gender differences in placebo analgesia: Event-related potentials and emotional modulation. *Psychosomatic Medicine*, *73*(2), 193–199. https://doi.org/10.1097/PSY.0b013e3182080d73.

Atlas, L. Y., Whittington, R. A., Lindquist, M. A., Wielgosz, J., Sonty, N., & Wager, T. D. (2012). Dissociable influences of opiates and expectations on pain. *The Journal of Neuroscience: The Official Journal of the Society for Neuroscience*, *32*(23), 8053–8064. https://doi.org/10.1523/JNEUROSCI.0383-12.2012.

Bannuru, R. R., McAlindon, T. E., Sullivan, M. C., Wong, J. B., Kent, D. M., & Schmid, C. H. (2015). Effectiveness and implications of alternative placebo treatments: A systematic review and network meta-analysis of osteoarthritis trials. *Annals of Internal Medicine*, *163*(5), 365–372. https://doi.org/10.7326/M15-0623.

Bartus, R. T., Weinberg, M. S., & Samulski, R. J. (2014). Parkinson's disease gene therapy: Success by design meets failure by efficacy. *Molecular Therapy: The Journal of the American Society of Gene Therapy*, *22*(3), 487–497. https://doi.org/10.1038/mt.2013.281.

Bausell, R. B., Lao, L., Bergman, S., Lee, W.-L., & Berman, B. M. (2005). Is acupuncture analgesia an expectancy effect? Preliminary evidence based on participants' perceived assignments in two placebo-controlled trials. *Evaluation & the Health Professions*, *28*(1), 9–26. https://doi.org/10.1177/0163278704273081.

Benedetti, F. (2014a). *Placebo effects* (2nd ed.). Oxford: OUP Oxford.

Benedetti, F. (2014b). Placebo effects: From the neurobiological paradigm to translational implications. *Neuron*, *84*(3), 623–637. https://doi.org/10.1016/j.neuron.2014.10.023.

Benedetti, F., Carlino, E., & Piedimonte, A. (2016). Increasing uncertainty in CNS clinical trials: The role of placebo, nocebo, and Hawthorne effects. *The Lancet. Neurology*, *15*(7), 736–747. https://doi.org/10.1016/S1474-4422(16)00066-1.

Benedetti, F., Carlino, E., & Pollo, A. (2011). Hidden administration of drugs. *Clinical Pharmacology and Therapeutics*, *90*(5), 651–661. https://doi.org/10.1038/clpt.2011.206.

Benedetti, F., Enck, P., Frisaldi, E., & Schedlowski, M. (2014). *Placebo. Handbook of experimental pharmacology*. Germany, Berlin: Springer.

Benedetti, F., & Frisaldi, E. (2014). Creating placebo responders and nonresponders in the laboratory: Boons and banes. *Pain Management*, *4*(3), 165–167. https://doi.org/10.2217/pmt.14.11.

Benedetti, F., Frisaldi, E., Carlino, E., Giudetti, L., Pampallona, A., Zibetti, M., et al. (2016). Teaching neurons to respond to placebos. *The Journal of Physiology*, *594*(19), 5647–5660. https://doi.org/10.1113/JP271322.

Benedetti, F., Maggi, G., Lopiano, L., Lanotte, M., & Rainero, I. (2003). *Open versus hidden medical treatments: The Patient's knowledge about a therapy affects the therapy outcome.* Recuperato da. https://www.scienceopen.com/document?vid=57741f4e-e734-479c-af27-0f1573dbfed6.

Berna, C., Kirsch, I., Zion, S. R., Lee, Y. C., Jensen, K. B., Sadler, P., et al. (2017). Side effects can enhance treatment response through expectancy effects: An experimental analgesic randomized controlled trial. *Pain*, *158*(6), 1014–1020. https://doi.org/10.1097/j.pain.0000000000000870.

Bhugra, D., & Ventriglio, A. (2015). Do cultures influence placebo response? *Acta Psychiatrica Scandinavica*, *132*(4), 227–230. https://doi.org/10.1111/acps.12422.

Bialik, R. J., Ravindran, A. V., Bakish, D., & Lapierre, Y. D. (1995). A comparison of placebo responders and nonresponders in subgroups of depressive disorder. *Journal of Psychiatry and Neuroscience*, *20*(4), 265–270.

Bingel, U., Wanigasekera, V., Wiech, K., Ni Mhuircheartaigh, R., Lee, M. C., Ploner, M., et al. (2011). The effect of treatment expectation on drug efficacy: Imaging the analgesic benefit of the opioid remifentanil. *Science Translational Medicine*, *3*(70), 70ra14. https://doi.org/10.1126/scitranslmed.3001244.

Boehm, K., Berger, B., Weger, U., & Heusser, P. (2017). Does the model of additive effect in placebo research still hold true? A narrative review. *JRSM Open*, *8*(3), 2054270416681434. https://doi.org/10.1177/2054270416681434.

Brown, W. A., Johnson, M. F., & Chen, M. G. (1992). Clinical features of depressed patients who do and do not improve with placebo. *Psychiatry Research*, *41*(3), 203–214.

Carlino, E., & Benedetti, F. (2016). Different contexts, different pains, different experiences. *Neuroscience*, *338*, 19–26. https://doi.org/10.1016/j.neuroscience.2016.01.053.

Carlino, E., Guerra, G., & Piedimonte, A. (2016). Placebo effects: From pain to motor performance. *Neuroscience Letters*, *632*, 224–230. https://doi.org/10.1016/j.neulet.2016.08.046.

Cartwright, N. (2007). Are RCTs the gold standard? *BioSocieties*, *2*(1), 11–20. https://doi.org/10.1017/S1745855207005029.

Cepeda, M. S., Berlin, J. A., Gao, C. Y., Wiegand, F., & Wada, D. R. (2012). Placebo response changes depending on the neuropathic pain syndrome: Results of a systematic review and meta-analysis. *Pain Medicine (Malden, Mass.)*, *13*(4), 575–595. https://doi.org/10.1111/j.1526-4637.2012.01340.x.

Christine, C. W., Starr, P. A., Larson, P. S., Eberling, J. L., Jagust, W. J., Hawkins, R. A., et al. (2009). Safety and tolerability of putaminal AADC gene therapy for Parkinson disease. *Neurology*, *73*(20), 1662–1669. https://doi.org/10.1212/WNL.0b013e3181c29356.

Colagiuri, B., Morley, K., Boakes, R., & Haber, P. (2009). Expectancy in double-blind placebo-controlled trials: An example from alcohol dependence. *Psychotherapy and Psychosomatics*, *78*(3), 167–171. https://doi.org/10.1159/000206871.

Colagiuri, B., Schenk, L. A., Kessler, M. D., Dorsey, S. G., & Colloca, L. (2015). The placebo effect: From concepts to genes. *Neuroscience*, *307*, 171–190. https://doi.org/10.1016/j.neuroscience.2015.08.017.

Colloca, L., Flaten, M. A., & Meissner, K. (2013). *Placebo and pain: From bench to bedside* (1st ed.). . Amsterdam; Boston: Academic Press.

Colloca, L., & Grillon, C. (2014). Understanding placebo and nocebo responses for pain management. *Current Pain and Headache Reports*, *18*(6), 419. https://doi.org/10.1007/s11916-014-0419-2.

Colloca, L., & Miller, F. G. (2011). How placebo responses are formed: A learning perspective. *Philosophical Transactions of the Royal Society of London. Series B, Biological Sciences*, *366*(1572), 1859–1869. https://doi.org/10.1098/rstb.2010.0398.

Colloca, L., Petrovic, P., Wager, T. D., Ingvar, M., & Benedetti, F. (2010). How the number of learning trials affects placebo and nocebo responses. *Pain*, *151*(2), 430–439. https://doi.org/10.1016/j.pain.2010.08.007.

Correa, J. B., Heckman, B. W., Marquinez, N. S., Drobes, D. J., Unrod, M., Roetzheim, R. G., et al. (2014). Perceived medication assignment during a placebo-controlled laboratory study of varenicline: Temporal associations of treatment expectancies with smoking-related outcomes. *Psychopharmacology*, *231*(13), 2559–2566. https://doi.org/10.1007/s00213-013-3420-2.

Darragh, M., Booth, R. J., & Consedine, N. S. (2014). Investigating the "placebo personality" outside the pain paradigm. *Journal of Psychosomatic Research*, *76*(5), 414–421. https://doi.org/10.1016/j.jpsychores.2014.02.011.

Darragh, M., Booth, R. J., & Consedine, N. S. (2015). Who responds to placebos? Considering the "placebo personality" via a transactional model. *Psychology, Health & Medicine*, *20*(3), 287–295. https://doi.org/10.1080/13548506.2014.936885.

Darragh, M., Booth, R. J., & Consedine, N. S. (2016). Trait predictors of placebo responses in itch. *Psychology, Health & Medicine*, *21*(6), 769–775. https://doi.org/10.1080/13548506.2015.1115106.

Dunlop, B. W., Thase, M. E., Wun, C.-C., Fayyad, R., Guico-Pabia, C. J., Musgnung, J., et al. (2012). A meta-analysis of factors impacting detection of antidepressant efficacy in clinical trials: The importance of academic sites. *Neuropsychopharmacology*, *37*(13), 2830–2836. https://doi.org/10.1038/npp.2012.153.

Dworkin, R. H., Turk, D. C., Peirce-Sandner, S., Baron, R., Bellamy, N., Burke, L. B., et al. (2010). Research design considerations for confirmatory chronic pain clinical trials: IMMPACT recommendations. *Pain*, *149*(2), 177–193. https://doi.org/10.1016/j.pain.2010.02.018.

Dworkin, R. H., Turk, D. C., Peirce-Sandner, S., Burke, L. B., Farrar, J. T., Gilron, I., et al. (2012). Considerations for improving assay sensitivity in chronic pain clinical trials: IMMPACT recommendations. *Pain*, *153*(6), 1148–1158. https://doi.org/10.1016/j.pain.2012.03.003.

Enck, P., Bingel, U., Schedlowski, M., & Rief, W. (2013). The placebo response in medicine: Minimize, maximize or personalize? *Nature Reviews. Drug Discovery*, *12*(3), 191–204. https://doi.org/10.1038/nrd3923.

Enck, P., Klosterhalfen, S., Weimer, K., Horing, B., & Zipfel, S. (2011). The placebo response in clinical trials: More questions than answers. *Philosophical Transactions of the*

Royal Society of London. Series B, Biological Sciences, 366(1572), 1889–1895. https://doi.org/10.1098/rstb.2010.0384.

Fava, M., Evins, A. E., Dorer, D. J., & Schoenfeld, D. A. (2003). The problem of the placebo response in clinical trials for psychiatric disorders: Culprits, possible remedies, and a novel study design approach. *Psychotherapy and Psychosomatics, 72*(3), 115–127. https://doi.org/69738.

Feltner, D., Hill, C., Lenderking, W., Williams, V., & Morlock, R. (2009). Development of a patient-reported assessment to identify placebo responders in a generalized anxiety disorder trial. *Journal of Psychiatric Research, 43*(15), 1224–1230. https://doi.org/10.1016/j.jpsychires.2009.04.001.

Feys, F., Bekkering, G. E., Singh, K., & Devroey, D. (2014). Do randomized clinical trials with inadequate blinding report enhanced placebo effects for intervention groups and nocebo effects for placebo groups? *Systematic Reviews, 3*, 14. https://doi.org/10.1186/2046-4053-3-14.

Fields, H. L., & Price, D. D. (1997). Toward a neurobiology of placebo analgesia. In A. Harrington (Ed.), *The placebo effect* (pp. 93–116). Boston, MA: Harvard University Press.

Forsberg, J. T., Martinussen, M., & Flaten, M. A. (2017). The placebo analgesic effect in healthy individuals and patients: A meta-analysis. *Psychosomatic Medicine, 79*(4), 388–394. https://doi.org/10.1097/PSY.0000000000000432.

Fournier, J. C., DeRubeis, R. J., Hollon, S. D., Dimidjian, S., Amsterdam, J. D., Shelton, R. C., et al. (2010). Antidepressant drug effects and depression severity: A patient-level meta-analysis. *JAMA, 303*(1), 47–53. https://doi.org/10.1001/jama.2009.1943.

Frisaldi, E., Carlino, E., Zibetti, M., Barbiani, D., Dematteis, F., Lanotte, M., et al. (2017). The placebo effect on bradykinesia in Parkinson's disease with and without prior drug conditioning. *Movement Disorders, 32*, 1474–1478.

Frisaldi, E., Shaibani, A., & Benedetti, F. (2017). Why we should assess patients' expectations in clinical trials. *Pain and Therapy, 6*(1), 107–110. https://doi.org/10.1007/s40122-017-0071-8.

Geers, A. L., Helfer, S. G., Kosbab, K., Weiland, P. E., & Landry, S. J. (2005). Reconsidering the role of personality in placebo effects: Dispositional optimism, situational expectations, and the placebo response. *Journal of Psychosomatic Research, 58*(2), 121–127. https://doi.org/10.1016/j.jpsychores.2004.08.011.

George, A. J. T., Collett, C., Carr, A. J., Holm, S., Bale, C., Burton, S., et al. (2016). When should placebo surgery as a control in clinical trials be carried out? *The Bulletin of the Royal College of Surgeons of England, 98*(2), 75–79.

Goebel, M. U., Trebst, A. E., Steiner, J., Xie, Y. F., Exton, M. S., Frede, S., et al. (2002). Behavioral conditioning of immunosuppression is possible in humans. *FASEB Journal: Official Publication of the Federation of American Societies for Experimental Biology, 16*(14), 1869–1873. https://doi.org/10.1096/fj.02-0389com.

Goetz, C. G., Laska, E., Hicking, C., Damier, P., Müller, T., Nutt, J., et al. (2008). Placebo influences on dyskinesia in Parkinson's disease. *Movement Disorders: Official Journal of the Movement Disorder Society, 23*(5), 700–707. https://doi.org/10.1002/mds.21897.

Goetz, C. G., Leurgans, S., Raman, R., & Stebbins, G. T. (2000). Objective changes in motor function during placebo treatment in PD. *Neurology, 54*(3), 710–714.

Goetz, C. G., Wuu, J., McDermott, M. P., Adler, C. H., Fahn, S., Freed, C. R., et al. (2008). Placebo response in Parkinson's disease: Comparisons among 11 trials covering medical and surgical interventions. *Movement Disorders: Official Journal of the Movement Disorder Society, 23*(5), 690–699. https://doi.org/10.1002/mds.21894.

Haahr, M. T., & Hróbjartsson, A. (2006). Who is blinded in randomized clinical trials? A study of 200 trials and a survey of authors. *Clinical Trials (London, England), 3*(4), 360–365. https://doi.org/10.1177/1740774506069153.

Hadamitzky, M., Engler, H., & Schedlowski, M. (2013). Learned immunosuppression: Extinction, renewal, and the challenge of reconsolidation. *Journal of Neuroimmune Pharmacology: The Official Journal of the Society on NeuroImmune Pharmacology*, *8*(1), 180–188. https://doi.org/10.1007/s11481-012-9388-6.

Hall, K. T., Lembo, A. J., Kirsch, I., Ziogas, D. C., Douaiher, J., Jensen, K. B., et al. (2012). Catechol-O-methyltransferase val158met polymorphism predicts placebo effect in irritable bowel syndrome. *PLoS One*, 7(10), e48135, https://doi.org/10.1371/journal.pone.0048135.

Hall, K. T., Loscalzo, J., & Kaptchuk, T. J. (2015). Genetics and the placebo effect: The placebome. *Trends in Molecular Medicine*, *21*(5), 285–294. https://doi.org/10.1016/j.molmed.2015.02.009.

Harrison, R. K. (2016). Phase II and phase III failures: 2013–2015. *Nature Reviews Drug Discovery*, *15*(12), 817–818. https://doi.org/10.1038/nrd.2016.184.

Häuser, W., Bartram-Wunn, E., Bartram, C., Reinecke, H., & Tölle, T. (2011). Systematic review: Placebo response in drug trials of fibromyalgia syndrome and painful peripheral diabetic neuropathy—Magnitude and patient-related predictors. *Pain*, *152*(8), 1709–1717. https://doi.org/10.1016/j.pain.2011.01.050.

Horing, B., Weimer, K., Muth, E. R., & Enck, P. (2014). Prediction of placebo responses: A systematic review of the literature. *Frontiers in Psychology*, *5*, 1079. https://doi.org/10.3389/fpsyg.2014.01079.

Howick, J., Friedemann, C., Tsakok, M., Watson, R., Tsakok, T., Thomas, J., et al. (2013). Are treatments more effective than placebos? A systematic review and meta-analysis. *PLoS One*, *8*(5), e62599. https://doi.org/10.1371/journal.pone.0062599.

Hróbjartsson, A., & Gøtzsche, P. C. (2001). Is the placebo powerless? An analysis of clinical trials comparing placebo with no treatment. *The New England Journal of Medicine*, *344*(21), 1594–1602. https://doi.org/10.1056/NEJM200105243442106.

Ivanova, A., & Tamura, R. N. (2015). A two-way enriched clinical trial design: Combining advantages of placebo lead-in and randomized withdrawal. *Statistical Methods in Medical Research*, *24*(6), 871–890. https://doi.org/10.1177/0962280211431023.

Ivanova, A., Zhang, Z., Thompson, L., Yang, Y., Kotz, R. M., & Fang, X. (2016). Can sequential parallel comparison design and two-way enriched design be useful in medical device clinical trials? *Journal of Biopharmaceutical Statistics*, *26*(1), 167–177. https://doi.org/10.1080/10543406.2015.1092028.

Jakšić, N., Aukst-Margetić, B., & Jakovljević, M. (2013). Does personality play a relevant role in the placebo effect? *Psychiatria Danubina*, *25*(1), 17–23.

Kam-Hansen, S., Jakubowski, M., Kelley, J. M., Kirsch, I., Hoaglin, D. C., Kaptchuk, T. J., et al. (2014). Altered placebo and drug labeling changes the outcome of episodic migraine attacks. *Science Translational Medicine*, *6*(218), 218ra5. https://doi.org/10.1126/scitranslmed.3006175.

Kaptchuk, T. J. (1998). Powerful placebo: The dark side of the randomised controlled trial. *Lancet (London, England)*, *351*(9117), 1722–1725. https://doi.org/10.1016/S0140-6736(97)10111-8.

Kaptchuk, T. J. (2002). Acupuncture: Theory, efficacy, and practice. *Annals of Internal Medicine*, *136*(5), 374–383.

Kaptchuk, T. J., Kelley, J. M., Conboy, L. A., Davis, R. B., Kerr, C. E., Jacobson, E. E., et al. (2008). Components of placebo effect: Randomised controlled trial in patients with irritable bowel syndrome. *BMJ (Clinical Research Ed.)*, *336*(7651), 999–1003. https://doi.org/10.1136/bmj.39524.439618.25.

Kemp, A. S., Schooler, N. R., Kalali, A. H., Alphs, L., Anand, R., Awad, G., et al. (2010). What is causing the reduced drug-placebo difference in recent schizophrenia clinical trials and what can be done about it? *Schizophrenia Bulletin*, *36*(3), 504–509. https://doi.org/10.1093/schbul/sbn110.

Kessner, S., Forkmann, K., Ritter, C., Wiech, K., Ploner, M., & Bingel, U. (2014). The effect of treatment history on therapeutic outcome: Psychological and neurobiological underpinnings. *PLoS One*, *9*(10), e109014. https://doi.org/10.1371/journal.pone.0109014.

Kessner, S., Wiech, K., Forkmann, K., Ploner, M., & Bingel, U. (2013). The effect of treatment history on therapeutic outcome: An experimental approach. *JAMA Internal Medicine*, *173*(15), 1468–1469. https://doi.org/10.1001/jamainternmed.2013.6705.

Khan, A., & Brown, W. A. (2015). Antidepressants versus placebo in major depression: An overview. *World Psychiatry*, *14*(3), 294–300. https://doi.org/10.1002/wps.20241.

Kirsch, I., Deacon, B. J., Huedo-Medina, T. B., Scoboria, A., Moore, T. J., & Johnson, B. T. (2008). Initial severity and antidepressant benefits: A meta-analysis of data submitted to the Food and Drug Administration. *PLoS Medicine*, *5*(2), e45. https://doi.org/10.1371/journal.pmed.0050045.

Kirsch, I., & Sapirstein, G. (1998). Listening to Prozac but hearing placebo: A meta-analysis of antidepressant medication. *Prevention and Treatment*, *1*(2), 0002a. https://doi:10.1037/1522-3736.1.1.12a.

Kotsis, V., Benson, S., Bingel, U., Forsting, M., Schedlowski, M., Gizewski, E. R., et al. (2012). Perceived treatment group affects behavioral and neural responses to visceral pain in a deceptive placebo study. *Neurogastroenterology and Motility: The Official Journal of the European Gastrointestinal Motility Society*, *24*(10), 935-e462. https://doi.org/10.1111/j.1365-2982.2012.01968.x.

Krogsbøll, L. T., Hróbjartsson, A., & Gøtzsche, P. C. (2009). Spontaneous improvement in randomised clinical trials: Meta-analysis of three-armed trials comparing no treatment, placebo and active intervention. *BMC Medical Research Methodology*, *9*, 1. https://doi.org/10.1186/1471-2288-9-1.

Kube, T., & Rief, W. (2017). Are placebo and drug-specific effects additive? Questioning basic assumptions of double-blinded randomized clinical trials and presenting novel study designs. *Drug Discovery Today*, *22*(4), 729–735. https://doi.org/10.1016/j.drudis.2016.11.022.

Laferton, J. A. C., Kube, T., Salzmann, S., Auer, C. J., & Shedden-Mora, M. C. (2017). Patients' expectations regarding medical treatment: A critical review of concepts and their assessment. *Frontiers in Psychology*, *8*, 233. https://doi.org/10.3389/fpsyg.2017.00233.

Lee, S., Walker, J. R., Jakul, L., & Sexton, K. (2004). Does elimination of placebo responders in a placebo run-in increase the treatment effect in randomized clinical trials? A meta-analytic evaluation. *Depression and Anxiety*, *19*(1), 10–19. https://doi.org/10.1002/da.10134.

LeWitt, P. A., Rezai, A. R., Leehey, M. A., Ojemann, S. G., Flaherty, A. W., Eskandar, E. N., et al. (2011). AAV2-GAD gene therapy for advanced Parkinson's disease: A double-blind, sham-surgery controlled, randomised trial. *The Lancet. Neurology*, *10*(4), 309–319. https://doi.org/10.1016/S1474-4422(11)70039-4.

Lieberman, R. P. (1967). The elusive placebo reactor. *NeuroPsychopharmacology*, *5*, 557–566.

Linde, K., Fässler, M., & Meissner, K. (2011). Placebo interventions, placebo effects and clinical practice. *Philosophical Transactions of the Royal Society of London. Series B, Biological Sciences*, *366*(1572), 1905–1912. https://doi.org/10.1098/rstb.2010.0383

Linde, K., Witt, C. M., Streng, A., Weidenhammer, W., Wagenpfeil, S., Brinkhaus, B., et al. (2007). The impact of patient expectations on outcomes in four randomized controlled trials of acupuncture in patients with chronic pain. *Pain*, *128*(3), 264–271. https://doi.org/10.1016/j.pain.2006.12.006.

Lückemann, L., Unteroberdörster, M., Kirchhof, J., Schedlowski, M., & Hadamitzky, M. (2017). Applications and limitations of behaviorally conditioned immunopharmacological responses. *Neurobiology of Learning and Memory*, *142*(Pt. A), 91–98. https://doi.org/10.1016/j.nlm.2017.02.012.

Lund, K., Vase, L., Petersen, G. L., Jensen, T. S., & Finnerup, N. B. (2014). Randomised controlled trials may underestimate drug effects: Balanced placebo trial design. *PLoS One*, *9*(1), e84104. https://doi.org/10.1371/journal.pone.0084104.
McRae, C., Cherin, E., Yamazaki, T. G., Diem, G., Vo, A. H., Russell, D., et al. (2004). Effects of perceived treatment on quality of life and medical outcomes in a double-blind placebo surgery trial. *Archives of General Psychiatry*, *61*(4), 412–420. https://doi.org/10.1001/archpsyc.61.4.412.
Meissner, K., Fässler, M., Rücker, G., Kleijnen, J., Hróbjartsson, A., Schneider, A., et al. (2013). Differential effectiveness of placebo treatments: A systematic review of migraine prophylaxis. *JAMA Internal Medicine*, *173*(21), 1941–1951. https://doi.org/10.1001/jamainternmed.2013.10391.
Mercado, R., Constantoyannis, C., Mandat, T., Kumar, A., Schulzer, M., Stoessl, A. J., et al. (2006). Expectation and the placebo effect in Parkinson's disease patients with subthalamic nucleus deep brain stimulation. *Movement Disorders: Official Journal of the Movement Disorder Society*, *21*(9), 1457–1461. https://doi.org/10.1002/mds.20935.
Moerman, D. E. (2000). Cultural variations in the placebo effect: Ulcers, anxiety, and blood pressure. *Medical Anthropology Quarterly*, *14*(1), 51–72.
Moher, D., Hopewell, S., Schulz, K. F., Montori, V., Gøtzsche, P. C., Devereaux, P. J., et al. (2012). CONSORT 2010 explanation and elaboration: Updated guidelines for reporting parallel group randomised trials. *International Journal of Surgery (London, England)*, *10*(1), 28–55. https://doi.org/10.1016/j.ijsu.2011.10.001.
Müller, M., Kamping, S., Benrath, J., Skowronek, H., Schmitz, J., Klinger, R., et al. (2016). Treatment history and placebo responses to experimental and clinical pain in chronic pain patients. *European Journal of Pain (London, England)*, *20*(9), 1530–1541. https://doi.org/10.1002/ejp.877.
Niklson, I., Edrich, P., & Verdru, P. (2006). Identifying baseline characteristics of placebo responders versus nonresponders in randomized double-blind trials of refractory partial-onset seizures. *Epileptic Disorders: International Epilepsy Journal With Videotape*, *8*(1), 37–44.
Noseworthy, J. H., Ebers, G. C., Vandervoort, M. K., Farquhar, R. E., Yetisir, E., & Roberts, R. (1994). The impact of blinding on the results of a randomized, placebo-controlled multiple sclerosis clinical trial. *Neurology*, *44*(1), 16–20.
Oken, B. S. (2008). Placebo effects: Clinical aspects and neurobiology. *Brain*, *131*(11), 2812–2823. https://doi.org/10.1093/brain/awn116.
Olanow, C. W., Goetz, C. G., Kordower, J. H., Stoessl, A. J., Sossi, V., Brin, M. F., et al. (2003). A double-blind controlled trial of bilateral fetal nigral transplantation in Parkinson's disease. *Annals of Neurology*, *54*(3), 403–414. https://doi.org/10.1002/ana.10720.
Pablos-Méndez, A., Barr, R. G., & Shea, S. (1998). Run-in periods in randomized trials: Implications for the application of results in clinical practice. *JAMA*, *279*(3), 222–225.
Peciña, M., Martínez-Jauand, M., Hodgkinson, C., Stohler, C. S., Goldman, D., & Zubieta, J. K. (2014). FAAH selectively influences placebo effects. *Molecular Psychiatry*, *19*(3), 385–391. https://doi.org/10.1038/mp.2013.124.
Peciña, M., & Zubieta, J.-K. (2015). Molecular mechanisms of placebo responses in humans. *Molecular Psychiatry*, *20*(4), 416–423. https://doi.org/10.1038/mp.2014.164.
Peerdeman, K. J., van Laarhoven, A. I. M., Keij, S. M., Vase, L., Rovers, M. M., Peters, M. L., et al. (2016). Relieving patients' pain with expectation interventions: A meta-analysis. *Pain*, *157*(6), 1179–1191. https://doi.org/10.1097/j.pain.0000000000000540.
Petersen, G. L., Finnerup, N. B., Grosen, K., Pilegaard, H. K., Tracey, I., Benedetti, F., et al. (2014). Expectations and positive emotional feelings accompany reductions in ongoing and evoked neuropathic pain following placebo interventions. *Pain*, *155*(12), 2687–2698. https://doi.org/10.1016/j.pain.2014.09.036.

Petersen, G. L., Finnerup, N. B., Nørskov, K. N., Grosen, K., Pilegaard, H. K., Benedetti, F., et al. (2012). Placebo manipulations reduce hyperalgesia in neuropathic pain. *Pain, 153*(6), 1292–1300. https://doi.org/10.1016/j.pain.2012.03.011.

Piedimonte, A., Guerra, G., Vighetti, S., & Carlino, E. (2017). Measuring expectation of pain: Contingent negative variation in placebo and nocebo effects. *European Journal of Pain (London, England), 21*(5), 874–885. https://doi.org/10.1002/ejp.990.

Pollo, A., Torre, E., Lopiano, L., Rizzone, M., Lanotte, M., Cavanna, A., et al. (2002). Expectation modulates the response to subthalamic nucleus stimulation in parkinsonian patients. *Neuroreport, 13*(11), 1383–1386.

Posternak, M. A., Zimmerman, M., Keitner, G. I., & Miller, I. W. (2002). A reevaluation of the exclusion criteria used in antidepressant efficacy trials. *The American Journal of Psychiatry, 159*(2), 191–200. https://doi.org/10.1176/appi.ajp.159.2.191.

Probst, P., Grummich, K., Heger, P., Zaschke, S., Knebel, P., Ulrich, A., et al. (2016). Blinding in randomized controlled trials in general and abdominal surgery: Protocol for a systematic review and empirical study. *Systematic Reviews, 5*, 48. https://doi.org/10.1186/s13643-016-0226-4.

Rief, W., & Glombiewski, J. A. (2012). The hidden effects of blinded, placebo-controlled randomized trials: An experimental investigation. *Pain, 153*(12), 2473–2477. https://doi.org/10.1016/j.pain.2012.09.007.

Rothwell, P. M. (2006). Factors that can affect the external validity of randomised controlled trials. *PLoS Clinical Trials, 1*(1), e9. https://doi.org/10.1371/journal.pctr.0010009.

Rutherford, B. R., Mori, S., Sneed, J. R., Pimontel, M. A., & Roose, S. P. (2012). Contribution of spontaneous improvement to placebo response in depression: A meta-analytic review. *Journal of Psychiatric Research, 46*(6), 697–702. https://doi.org/10.1016/j.jpsychires.2012.02.008.

Rutherford, B. R., Pott, E., Tandler, J. M., Wall, M. M., Roose, S. P., & Lieberman, J. A. (2014). Placebo response in antipsychotic clinical trials: A meta-analysis. *JAMA Psychiatry, 71*(12), 1409–1421. https://doi.org/10.1001/jamapsychiatry.2014.1319.

Rutherford, B. R., & Roose, S. P. (2013). A model of placebo response in antidepressant clinical trials. *The American Journal of Psychiatry, 170*(7), 723–733. https://doi.org/10.1176/appi.ajp.2012.12040474.

Schenk, L. A., Sprenger, C., Geuter, S., & Büchel, C. (2014). Expectation requires treatment to boost pain relief: An fMRI study. *Pain, 155*(1), 150–157. https://doi.org/10.1016/j.pain.2013.09.024.

Schulz, K. F., Altman, D. G., Moher, D., & CONSORT Group (2011). CONSORT 2010 statement: Updated guidelines for reporting parallel group randomised trials. *International Journal of Surgery (London, England), 9*(8), 672–677. https://doi.org/10.1016/j.ijsu.2011.09.004.

Shetty, N., Friedman, J. H., Kieburtz, K., Marshall, F. J., & Oakes, D. (1999). The placebo response in Parkinson's disease. Parkinson study group. *Clinical Neuropharmacology, 22*(4), 207–212.

Simpson, D. M., Rice, A. S. C., Emir, B., Landen, J., Semel, D., Chew, M. L., et al. (2014). A randomized, double-blind, placebo-controlled trial and open-label extension study to evaluate the efficacy and safety of pregabalin in the treatment of neuropathic pain associated with human immunodeficiency virus neuropathy. *Pain, 155*(10), 1943–1954. https://doi.org/10.1016/j.pain.2014.05.027.

Stone, D. A., Kerr, C. E., Jacobson, E., Conboy, L. A., & Kaptchuk, T. J. (2005). Patient expectations in placebo-controlled randomized clinical trials. *Journal of Evaluation in Clinical Practice, 11*(1), 77–84. https://doi.org/10.1111/j.1365-2753.2004.00512.x.

Suchman, A. L., & Ader, R. (1992). Classic conditioning and placebo effects in crossover studies. *Clinical Pharmacology and Therapeutics, 52*(4), 372–377.

Tack, J., Müller-Lissner, S., Bytzer, P., Corinaldesi, R., Chang, L., Viegas, A., et al. (2005). A randomised controlled trial assessing the efficacy and safety of repeated tegaserod therapy in women with irritable bowel syndrome with constipation. *Gut*, *54*(12), 1707–1713. https://doi.org/10.1136/gut.2005.070789.

Tétreault, P., Mansour, A., Vachon-Presseau, E., Schnitzer, T. J., Apkarian, A. V., & Baliki, M. N. (2016). Brain connectivity predicts placebo response across chronic pain clinical trials. *PLoS Biology*, *14*(10), e1002570. https://doi.org/10.1371/journal.pbio.1002570.

Trivedi, M. H., & Rush, H. (1994). Does a placebo run-in or a placebo treatment cell affect the efficacy of antidepressant medications? *Neuropsychopharmacology: Official Publication of the American College of Neuropsychopharmacology*, *11*(1), 33–43. https://doi.org/10.1038/npp.1994.63.

Turner, J. A., Jensen, M. P., Warms, C. A., & Cardenas, D. D. (2002). Blinding effectiveness and association of pretreatment expectations with pain improvement in a double-blind randomized controlled trial. *Pain*, *99*(1–2), 91–99.

Tuttle, A. H., Tohyama, S., Ramsay, T., Kimmelman, J., Schweinhardt, P., Bennett, G. J., et al. (2015). Increasing placebo responses over time in U.S. clinical trials of neuropathic pain. *Pain*, *156*(12), 2616–2626. https://doi.org/10.1097/j.pain.0000000000000333.

Vase, L., Amanzio, M., & Price, D. D. (2015). Nocebo vs. placebo: The challenges of trial design in analgesia research. *Clinical Pharmacology and Therapeutics*, *97*(2), 143–150. https://doi.org/10.1002/cpt.31.

Vase, L., Baram, S., Takakura, N., Takayama, M., Yajima, H., Kawase, A., et al. (2015). Can acupuncture treatment be double-blinded? An evaluation of double-blind acupuncture treatment of postoperative pain. *PLoS One*, *10*(3), e0119612. https://doi.org/10.1371/journal.pone.0119612.

Vase, L., Baram, S., Takakura, N., Yajima, H., Takayama, M., Kaptchuk, T. J., et al. (2013). Specifying the nonspecific components of acupuncture analgesia. *Pain*, *154*(9), 1659–1667. https://doi.org/10.1016/j.pain.2013.05.008.

Vase, L., Nørskov, K. N., Petersen, G. L., & Price, D. D. (2011). Patients' direct experiences as central elements of placebo analgesia. *Philosophical Transactions of the Royal Society of London. Series B, Biological Sciences*, *366*(1572), 1913–1921. https://doi.org/10.1098/rstb.2010.0402.

Vase, L., Petersen, G. L., Riley, J. L., & Price, D. D. (2009). Factors contributing to large analgesic effects in placebo mechanism studies conducted between 2002 and 2007. *Pain*, *145*(1–2), 36–44. https://doi.org/10.1016/j.pain.2009.04.008.

Vase, L., Riley, J. L., & Price, D. D. (2002). A comparison of placebo effects in clinical analgesic trials versus studies of placebo analgesia. *Pain*, *99*(3), 443–452.

Vase, L., Robinson, M. E., Verne, G. N., & Price, D. D. (2003). The contributions of suggestion, desire, and expectation to placebo effects in irritable bowel syndrome patients. An empirical investigation. *Pain*, *105*(1–2), 17–25.

Vase, L., Vollert, J., Finnerup, N. B., Miao, X., Atkinson, G., Marshall, S., et al. (2015). Predictors of the placebo analgesia response in randomized controlled trials of chronic pain: A meta-analysis of the individual data from nine industrially sponsored trials. *Pain*, *156*(9), 1795–1802. https://doi.org/10.1097/j.pain.0000000000000217.

Verne, G. N., Robinson, M. E., Vase, L., & Price, D. D. (2003). Reversal of visceral and cutaneous hyperalgesia by local rectal anesthesia in irritable bowel syndrome (IBS) patients. *Pain*, *105*(1–2), 223–230.

Walsh, B. T., Seidman, S. N., Sysko, R., & Gould, M. (2002). Placebo response in studies of major depression: Variable, substantial, and growing. *JAMA*, *287*(14), 1840–1847.

Wang, R.-S., Hall, K. T., Giulianini, F., Passow, D., Kaptchuk, T. J., & Loscalzo, J. (2017). Network analysis of the genomic basis of the placebo effect. *JCI Insight*, *2*(11), e93911. https://doi.org/10.1172/jci.insight.93911.

Watson, A., Power, A., Brown, C., El-Deredy, W., & Jones, A. (2012). Placebo analgesia: Cognitive influences on therapeutic outcome. *Arthritis Research & Therapy*, *14*(2), 206. https://doi.org/10.1186/ar3783.

Watts, R. L., Freeman, T. B., Hauser, R. A., Bakay, R. A. E., Ellias, S. A., Stoessl, A. J., et al. (2001). A double-blind, randomized, controlled, multicenter clinical trial of the safety and efficacy of stereotaxic intrastriatal implantation of fetal porcine ventral mesencephalic tissue (Neurocell-PD) vs imitation surgery in patients with Parkinson disease (PD). *Parkinsonism and Related Disorders*, 7, S87.

Waxmonsky, J. G., Waschbusch, D. A., Glatt, S. J., & Faraone, S. V. (2011). Prediction of placebo response in 2 clinical trials of lisdexamfetamine dimesylate for the treatment of ADHD. *The Journal of Clinical Psychiatry*, *72*(10), 1366–1375. https://doi.org/10.4088/JCP.10m05979pur.

Weimer, K., Colloca, L., & Enck, P. (2015). Placebo effects in psychiatry: Mediators and moderators. *The Lancet. Psychiatry*, *2*(3), 246–257. https://doi.org/10.1016/S2215-0366(14)00092-3.

Wellek, S., & Blettner, M. (2012). On the proper use of the crossover design in clinical trials. *Deutsches Ärzteblatt International*, *109*(15), 276–281. https://doi.org/10.3238/arztebl.2012.0276.

Whalley, B., Hyland, M. E., & Kirsch, I. (2008). Consistency of the placebo effect. *Journal of Psychosomatic Research*, *64*(5), 537–541. https://doi.org/10.1016/j.jpsychores.2007.11.007.

Wood, L., Egger, M., Gluud, L. L., Schulz, K. F., Jüni, P., Altman, D. G., et al. (2008). Empirical evidence of bias in treatment effect estimates in controlled trials with different interventions and outcomes: Meta-epidemiological study. *BMJ*, *336*(7644), 601–605. https://doi.org/10.1136/bmj.39465.451748.AD.

Wrobel, N., Fadai, T., Sprenger, C., Hebebrand, J., Wiech, K., & Bingel, U. (2015). Are children the better placebo analgesia responders? An experimental approach. *The Journal of Pain*, *16*(10), 1005–1011. https://doi.org/10.1016/j.jpain.2015.06.013.

Younger, J., Gandhi, V., Hubbard, E., & Mackey, S. (2012). Development of the Stanford expectations of treatment scale (SETS): A tool for measuring patient outcome expectancy in clinical trials. *Clinical Trials*, *9*(6), 767–776. https://doi.org/10.1177/1740774512465064.

Zunhammer, M., Ploner, M., Engelbrecht, C., Bock, J., Kessner, S. S., & Bingel, U. (2017). The effects of treatment failure generalize across different routes of drug administration. *Science Translational Medicine*, *9*(393). https://doi.org/10.1126/scitranslmed.aal2999.

FSC
www.fsc.org
MIX
Papier | Fördert gute Waldnutzung
FSC® C083411

Kolibri 360 GmbH
Ferdinand-Jühlke-Straße 7
99095 Erfurt, Deutschland
produktsicherheit@kolibri360.de